D1482981

Analytic Geometry

SIXTH EDITION

Analytic Geometry

Gordon Fuller

Professor Emeritus of Mathematics
Texas Tech University

Dalton Tarwater

Professor of Mathematics
Texas Tech University

Addison-Wesley Publishing Company

Reading, Massachusetts • Menlo Park, California • Don Mills, Ontario
Wokingham, England • Amsterdam • Sydney • Singapore • Tokyo • Mexico City
Bogotá • Santiago • San Juan

Sponsoring Editor   Jeffrey Pepper

Production Supervisor   Susanah H. Michener
Production Service   Bookwrights, Inc.
Text Designer   Patricia O. Williams
Manufacturing Supervisor   Ann DeLacey
Cover Designer   Marshall Henrichs
Cover Photography   Slide Graphics

Library of Congress Cataloging in Publication Data

Fuller, Gordon, 1894–1985
  Analytic geometry.

  Includes index.
  1. Geometry, Analytic.   I. Tarwater, Dalton.
II. Title.
QA551.F87   1986   516.3   84-28218
ISBN 0-201-10861-5

Reprinted with corrections, July 1986

BCDEFGHIJ-DO-89876

Dedication

Gordon Fuller was born in Joshua, Texas, on January 17, 1894. He received degrees in mathematics from West Texas State (B.A., 1926), and the University of Michigan (M.A., Ph.D., 1933). After teaching at Auburn for 12 years, he joined the faculty at Texas Tech in 1950, where he remained until he retired in 1968. He authored or co-authored texts in College Algebra, Plane Trigonometry, Analytic Geometry, and Calculus.

He is remembered as a tough-but-fair professor, a lucid expositor, a fine gentleman, and a warm colleague.

Gordon Fuller died in Dallas, Texas, on March 17, 1985. This text is dedicated to his memory.

Preface

This sixth edition of *Analytic Geometry*, like the earlier editions, has been tailored for a first course in the study of the subject. It emphasizes the essential elements of analytic geometry and stresses those concepts that are needed in calculus, whether the traditional calculus or the calculus taken by business majors in college.

Users of previous editions of the book will find many improvements in this edition, while the best features of the earlier editions have been left intact. All changes have been made with an eye toward improving the readability of the text. Some of the more important changes in this book are as follows:

1. Modern applications to business, social sciences, and physical science are included to show off the breadth of usefulness of analytic geometry.
2. Many exercises that require a hand calculator are included. They are marked with a ⓒ .
3. The chapter on transcendental functions has been completely rewritten.
4. Many of the answers are accompanied with a graph or a sketch.
5. The chapter on algebraic curves occurs earlier than before.
6. The section on addition of ordinates occurs later, at the end of the chapter on transcendental functions, so that functions like $x + \sin x$ may be sketched.

The text continues to serve as a semester course for senior high students or for freshman students in colleges or universities. The student should have prior knowledge of algebra, geometry, and trigonometry. While important formulas are included in the Appendix, the student should know how to solve quadratics (using the formula *and* completing the square) and linear systems. Determinants are relegated to exercises until Chapter 10. The student should also be acquainted with the trigonometric functions, sine, cosine, and tangent, and with the double-angle formulas.

We have not included computer programs or software, since the availability of hardware varies so widely. Nonetheless, teachers are urged to exploit computer graphics in their classes if they are able to do so.

We gratefully acknowledge the helpful guidance provided by Patricia Clark (Indiana State University) and Richard J. Palmaccio (Pine Crest School, Ft. Lauderdale, Florida), whose suggestions have been particularly valuable. Derald Walling (Texas Tech University) has been very helpful in suggesting several applications to social and behavioral sciences. Joyce Martin has done a superb typing job and is owed our sincere thanks.

Lubbock, Texas   D.T.

To the Student

Welcome to the study of analytic geometry. You are in good company. Throughout the past two thousand years, millions have studied some aspect of this subject. Among them are many of the greatest intellects of historical and of modern times. Many of these students learned analytic geometry for its intrinsic values. Nonetheless, it is fair to say that the subject is studied today primarily as a preparatory course before calculus.

We have tried to show you, in examples and in exercises, that the ideas you encounter here are applicable in many fields of study. Unfortunately, one may need to take calculus, or even courses beyond calculus, to see the full depth of the applications. It is hoped that the several applications that are presented will be sufficient to indicate the widespread utility of these concepts.

You should have completed courses in algebra, geometry, and trigonometry. You will be expected to solve quadratics by formula and/or completing the square, to solve systems of equations, and to use the trigonometric functions sine, cosine, and tangent, and some identities involving them. Knowledge of determinants will be useful for several exercises.

You are urged to read the text *before* attempting the exercises. While reading the text, use scratch paper to fill in missing steps in examples, and to copy theorems and formulas until you know them.

In recent years the popularity of hand calculators and of computers has increased. If you have a calculator, you will be able to work problems which involve more tedious calculations than the person who does not. In this case, you are urged to work the problems marked **C** . They incorporate the same concepts as their neighbors, but they require more computational capability. A hand calculator with the trigonometric functions, log, ln, and e^x or y^x should serve you well.

It is important to note that possession of a calculator or of a computer with a lot of software does not free the mathematician, scientist, or engineer from knowing the principles of analytic geometry. It simply enhances his or her ability to deal with more difficult applications. The student with access to computer graphics will profit from using that capability to graph many of the functions in this text. While the text does not *require* the student to use a computer, the authors certainly urge any student who has access to one to apply the knowledge gained herein to program a computer to plot graphs whenever possible.

Contents

1

FUNDAMENTAL CONCEPTS 1

1.1 Basic Concepts 1
1.2 Inclination and Slope of a Line 12
1.3 Division of a Line Segment 23
1.4 Analytic Proofs of Geometric Theorems 29
1.5 Relations and Functions 35
1.6 The Equation of a Graph 42
   Review Exercises 47

2

THE STRAIGHT LINE AND THE CIRCLE 49

2.1 Lines and First-Degree Equations 49
2.2 Other Forms of First-Degree Equations 56
2.3 Intersection of Lines 60
2.4 Directed Distance from a Point to a Line 63
2.5 Families of Lines 71
2.6 Circles 77
2.7 Families of Circles 85
2.8 Translation of Axes 90
   Review Exercises 93

3

CONICS 95

3.1 The Parabola 96
3.2 Parabola with Vertex at (h, k) 103

3.3 The Ellipse 112
3.4 The Hyperbola 125
     Review Exercises 134

4
———

SIMPLIFICATION OF EQUATIONS 136

4.1 Simplification by Translation 136
4.2 Rotation of Axes 140
4.3 Simplification by Rotations and Translations 143
4.4 Identification of a Conic 149
     Review Exercises 153

5
———

ALGEBRAIC CURVES 154

5.1 Vertical and Horizontal Asymptotes 154
5.2 Irrational Equations 161
5.3 Slant Asymptotes 163
     Review Exercises 170

6
———

TRANSCENDENTAL FUNCTIONS 172

6.1 Trigonometric Functions 172
6.2 The Exponential Function 181
6.3 Logarithms 185
6.4 Addition of Ordinates 190
     Review Exercises 195

7
———

POLAR COORDINATES 196

7.1 The Polar Coordinate System 196
7.2 Relations between Rectangular and Polar Coordinates 202
7.3 Graphs of Polar-Coordinate Equations 206

7.4 Aids in Graphing Polar-Coordinate Equations 210
7.5 Polar Equations of Lines and Circles 219
7.6 Polar Equations of Conics 224
7.7 Trigonometric Equations 229
7.8 Intersections of Polar-Coordinate Graphs 232
      Review Exercises 237

8

PARAMETRIC EQUATIONS 240

8.1 Parametric Equations of the Conics 240
8.2 Applications of Parametric Equations 248
      Review Exercises 254

9

SPACE COORDINATES AND SURFACES 255

9.1 Space Coordinates 255
9.2 Surface of Revolution and Quadric Surfaces 265
      Review Exercises 283

10

VECTORS, PLANES, AND LINES 285

10.1 Operations on Vectors 285
10.2 Vectors in Space 297
10.3 The Scalar Product of Two Vectors 301
10.4 The Equation of a Plane 307
10.5 Vector Equation of a Line 312
10.6 The Vector Product 319
      Review Exercises 328

Appendix A 329
Tables 332
Answers to Selected Exercises 338
Index 385

1

Fundamental Concepts

For several centuries geometry and algebra developed slowly, bit by bit, as distinct mathematical disciplines. In 1637, however, a French mathematician and philosopher, René Descartes, published his *La Géométrie*, which introduced a device for unifying these two branches of mathematics. The basic feature of this new process, now called **analytic geometry**, is the use of a coordinate system. By means of coordinate systems algebraic methods can be applied powerfully in the study of geometry, and perhaps of still greater importance is the advantage accruing to algebra by the graphical representation of algebraic equations. Indeed, Descartes' remarkable contribution paved the way for rapid and far-reaching developments in mathematics, because it provided the very framework for the creation of calculus.

Many concepts discussed in this book are of ancient origin. Do not be misled into thinking that they are studied only for historical value. On the contrary, these ideas have withstood the test of time and are studied today because of their usefulness in dealing with today's (and probably tomorrow's) problems. The topics that are of historical interest only, those that are not fruitful today, have mostly disappeared from active study.

The topics that follow in this book have meaningful applications in various mathematical investigations and in such diverse disciplines as astronomy, the natural sciences, engineering, business, medicine, the social sciences, psychology, statistics, and economics.

1.1

BASIC CONCEPTS

A line on which one direction is chosen as positive and the opposite direction as negative is called a **directed line**. A segment of the line, consisting of any two points and the part between, is called a **directed line segment**. In Fig. 1.1, the positive direction is indicated by the arrowhead. The points A and B determine a

1

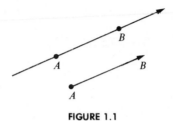

FIGURE 1.1

segment, which we denote by AB or BA. We specify that the distance from A to B, measured in the positive direction, is positive; and the distance from B to A, measured in the negative direction, is negative. These two distances, which we denote by $\overrightarrow{AB}$ and $\overrightarrow{BA}$, are called **directed distances**. If the length of the line segment is 3, then $\overrightarrow{AB} = 3$, and $\overrightarrow{BA} = -3$. Distances, therefore, on a directed line segment satisfy the equation

$$\overrightarrow{AB} = -\overrightarrow{BA}.$$

Another concept with respect to distance on the segment AB is that of the **undirected distances** between A and B. The undirected distance is the length of the segment, which we take as positive. We will use the notation $|AB|$ or $|BA|$ to indicate the positive measurement of distance between A and B, or the length of the line segment AB.

In view of the preceding discussion we may write

$$\overrightarrow{AB} = |AB| = |BA| = 3,$$
$$\overrightarrow{BA} = -|AB| = -|BA| = -3.$$

Frequently the concept of the absolute value of a number is of particular significance. Relative to this concept, we have the following definition.

DEFINITION 1.1 ● *The **absolute value** of a real number a, denoted by $|a|$, is the real number such that*

$$|a| = a \text{ when } a \text{ is positive or zero,}$$
$$|a| = -a \text{ when } a \text{ is negative.}$$

According to this definition, the absolute value of every nonzero number is positive and the absolute value of zero is zero. Thus,

$$|5| = 5, \qquad |-5| = -(-5) = 5, \qquad |0| = 0.$$

We see then, that

$$|a| = \sqrt{a^2}$$

for any real number a, since the square root of any nonnegative number is nonnegative.

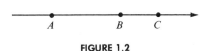

FIGURE 1.2

THEOREM 1.1 ● *If A, B, and C are three points of a directed line, then the directed distances determined by these points satisfy the equations*

$$\overrightarrow{AB} + \overrightarrow{BC} = \overrightarrow{AC}, \qquad \overrightarrow{AC} + \overrightarrow{CB} = \overrightarrow{AB}, \qquad \overrightarrow{BA} + \overrightarrow{AC} = \overrightarrow{BC}.$$

PROOF. If B is between A and C, the distance $\overrightarrow{AB}$, $\overrightarrow{BC}$, and $\overrightarrow{AC}$ all have the same sign, and $\overrightarrow{AC}$ is obviously equal to the sum of the other two (Fig. 1.2). The second and third equations follow readily from the first. To establish the second equation, we add $-\overrightarrow{BC}$ to both sides of the first equation and then use the condition that $-\overrightarrow{BC} = \overrightarrow{CB}$. Thus,

$$\overrightarrow{AB} = \overrightarrow{AC} - \overrightarrow{BC} = \overrightarrow{AC} + \overrightarrow{CB}.$$

The Real Number Line

A basic concept of analytic geometry is the representation of all real numbers by points on a directed line. The real numbers, we note, consist of the positive numbers, the negative numbers, and zero.

To establish the desired representation, we first choose a direction on a line as positive (to the right in Fig. 1.3) and select a point O of the line, which we call the **origin**, to represent the number zero. Next we mark points at distances 1, 2, 3, and so on, units to the right of the origin. We let the points thus located represent the numbers 1, 2, 3, and so on. In the same way we locate points to the left of the origin to represent the numbers -1, -2, -3, and so on. We now have points assigned to the positive integers, the negative integers, and the integer zero. Numbers whose values are between two consecutive integers have their corresponding points between the points associated with those integers. Thus the number $2\frac{1}{4}$ corresponds to the point $2\frac{1}{4}$ units to the right of the origin. And, in general, any positive number p is represented by the point p units to the right of the origin, and a negative number $-n$ is represented by the point n units to the left of the origin. Further, we assume that every real number corresponds to one

FIGURE 1.3

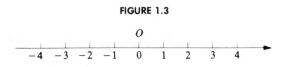

point on the line and, conversely, every point on the line corresponds to one real number. This relation of the set of real numbers and the set of points on a directed line is called a **one-to-one correspondence.**

The directed line of Fig. 1.3, with its points corresponding to real numbers, is called a **real number line.** The number corresponding to a point on the line is called the **coordinate** of the point. Since the positive numbers correspond to points in the chosen positive direction from the origin and the negative numbers correspond to points in the opposite or negative direction from the origin, we shall consider the coordinates of points on a number line to be **directed distances** from the origin. For convenience, we shall sometimes speak of a point as being a number, and vice versa. For example, we may say "the point 5" when we mean "the number 5," and "the number 5" when we mean "the point 5."

Rectangular Coordinates

Having obtained a one-to-one correspondence between the points on a line and the system of real numbers, we next develop a scheme for putting the points of a plane into a one-to-one correspondence with a set of ordered pairs of real numbers.

DEFINITION 1.2 ● *A pair of numbers* (x, y) *in which the order of occurrence of the numbers is distinguished is an* **ordered pair** *of numbers. Two ordered pairs,* (x, y) *and* (x', y'), *are equal if and only if* $x = x'$ *and* $y = y'$.

Note that $(3, 2) \neq (2, 3)$, and $(1, 1) = (x, y)$ if and only if $x = 1$ and $y = 1$.

We draw a horizontal line and a vertical line meeting at the origin O (Fig. 1.4). The horizontal line OX is called the **x axis** and the vertical line OY, the **y axis.** The x axis and the y axis, taken together, are called the **coordinate axes,** and the plane determined by the coordinate axes is called the **coordinate plane.** The x axis, usually drawn horizontally, is called the **horizontal** axis and the y axis the **vertical** axis. With a convenient unit of length, we make a real number scale on each coordinate axis, letting the origin be the zero point. The positive direction is chosen to the right on the x axis and upward on the y axis, as indicated by the arrowheads in the figure.

It is extremely important that coordinate axes be labeled. The student should form this habit immediately. A simple arrow in the positive direction of each coordinate axis will suffice, but the name of each coordinate, x or y in our present situation, must also be indicated, just as they are in Fig. 1.4 and throughout the remainder of the text.

If P is a point on the coordinate plane, we define the distances of the point from the coordinate axes to be **directed distances.** That is, the distance from the y axis is positive if P is to the right of the y axis and negative if P is to the left, and the distance from the x axis is positive if P is above the x axis and negative if P is below the x axis. Each point P of the plane has associated with it a pair of

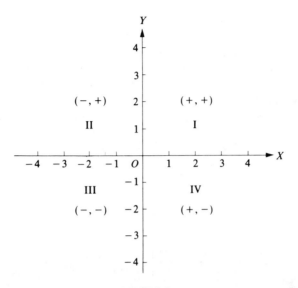

FIGURE 1.4

numbers called **coordinates**. The coordinates are defined in terms of the perpendicular distances from the axes to the point.

DEFINITION 1.3 ● *The x coordinate, or **abscissa**, of a point P is the directed distance from the y axis to the point. The y coordinate, or **ordinate**, of a point P is the directed distance from the x axis to the point.*

A point whose abscissa is x and whose ordinate is y is designated by (x, y), in that order, the abscissa always coming first. Hence the coordinates of a point are an ordered pair of numbers. Although a pair of coordinates determines a point, the coordinates themselves are often referred to as a point.

We assume that to any pair of real numbers (coordinates) there corresponds one definite point. Conversely, we assume that to each point of the plane there corresponds one definite pair of coordinates. This relation of points on a plane and pairs of real numbers is called a one-to-one correspondence. The device we have described for obtaining this correspondence is called a **rectangular coordinate system**.

A point of given coordinates is **plotted** by measuring the proper distances from the axes and marking the point thus located. For example, if the coordinates of a point are $(-4, 3)$, the abscissa -4 means the point is 4 units to the left of the y axis and the ordinate 3 (plus sign understood) means the point is 3 units above the x axis. Consequently, we locate the point by going from the origin 4 units to the left along the x axis and then 3 units upward parallel to the y axis (Fig. 1.5).

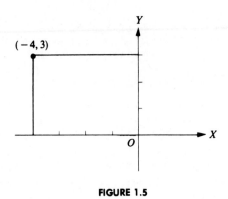

FIGURE 1.5

Similarly, if we wish to plot the points $(5, -3)$, we move 5 units to the right of the origin along the x axis and then 3 units downward (since the ordinate is negative) parallel to the y axis. We have now located the desired point.

Some coordinates and their corresponding points are plotted in Fig. 1.6.

The coordinate axes divide the plane into four parts, called **quadrants**, which are numbered I to IV in Fig. 1.4. The coordinates of a point in the first quadrant

FIGURE 1.6

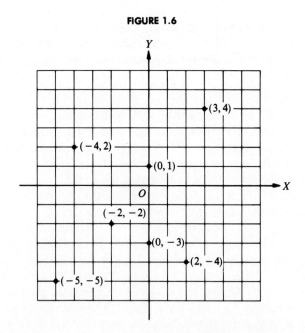

are both positive, which is indicated in the figure by $(+, +)$. The signs of the coordinates in each of the other quadrants are similarly indicated.

Distance Between Two Points

In many problems the distance between two points of the coordinate plane is required. The distance between any two points, or the length of the line segment connecting them, can be determined from the coordinates of the points. We shall classify a line segment (or line) as **horizontal, vertical,** or **slant,** depending on whether the segment is parallel to the x axis, to the y axis, or to neither axis. In deriving appropriate formulas for the lengths of these kinds of segments, we shall use the idea of directed segments.

Let $P_1(x_1, y)$ and $P_2(x_2, y)$ be two points on a horizontal line, and let A be the point where the line cuts the y axis (Fig 1.7). We have, by Theorem 1.1,

$$\overrightarrow{AP_1} + \overrightarrow{P_1P_2} = \overrightarrow{AP_2}$$
$$\overrightarrow{P_1P_2} = \overrightarrow{AP_2} - \overrightarrow{AP_1}$$
$$= x_2 - x_1.$$

Similarly, for the vertical distance, $\overrightarrow{Q_1Q_2}$, we have

$$\overrightarrow{Q_1Q_2} = \overrightarrow{Q_1B} + \overrightarrow{BQ_2}$$
$$= \overrightarrow{BQ_2} - \overrightarrow{BQ_1}$$
$$= y_2 - y_1.$$

Hence the directed distance from a first point to a second point on a horizontal line is equal to the abscissa of the second point minus the abscissa of the first point. The distance is positive or negative depending on whether the second point is to the right or left of the first point. A corresponding statement can be made relative to a vertical segment.

FIGURE 1.7

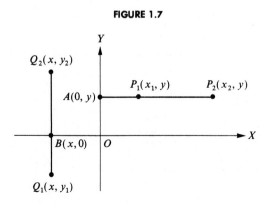

Inasmuch as the lengths of segments, without regard to direction, are often desired, we state a rule that gives results in positive quantities.

The length of a horizontal line segment joining two points is the abscissa of the point on the right minus the abscissa of the point on the left.

The length of a vertical line segment joining two points is the ordinate of the upper point minus the ordinate of the lower point.

If it is not known which point is to the right of the other, we may use the equivalent expression

$$|P_1P_2| = |x_1 - x_2| = \sqrt{(x_1 - x_2)^2} \qquad (1.1)$$

for the undirected distance between $P_1(x_1, y)$ and $P_2(x_2, y)$. Similarly,

$$|Q_1Q_2| = |y_1 - y_2| = \sqrt{(y_1 - y_2)^2}$$

is the distance between $Q_1(x, y_1)$ and $Q_2(x, y_2)$.

We apply these rules to find the lengths of the line segments in Fig. 1.8:

$$|AB| = 5 - 1 = 4, \qquad |CD| = 6 - (-2) = 6 + 2 = 8,$$
$$|EF| = 1 - (-4) = 1 + 4 = 5, \qquad |GH| = -2 - (-5) = -2 + 5 = 3.$$

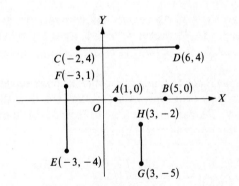

FIGURE 1.8

We next consider the points $P_1(x_1, y_1)$ and $P_2(x_2, y_2)$, which determine a slant line. Draw a line through P_1 parallel to the x axis and a line through P_2 parallel to the y axis (Fig. 1.9). These two lines intersect at the point R, whose abscissa is x_2 and whose ordinate is y_1. Hence

$$\overrightarrow{P_1R} = x_2 - x_1 \quad \text{and} \quad \overrightarrow{RP_2} = y_2 - y_1.$$

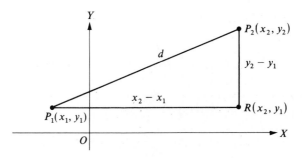

FIGURE 1.9

By the Pythagorean theorem,*

$$|P_1P_2|^2 = (x_2 - x_1)^2 + (y_2 - y_1)^2.$$

Denoting the length of the segment P_1P_2 by d, we have the formula

$$d = \sqrt{(x_2 - x_1)^2 + (y_2 - y_1)^2}.$$

(1.2)

To find the distance between two points, add the square of the difference of the abscissas to the square of the difference of the ordinates and take the square root of the sum.

In employing the distance formula, we may designate either point by (x_1, y_1) and the other by (x_2, y_2). This results from the fact that the two differences involved are squared. The square of the difference of the two numbers is unchanged when the order of subtraction is reversed.

EXAMPLE 1 ● Find the lengths of the sides of the triangle (Fig 1.10) with the vertices $A(-2, -3)$, $B(6, 1)$, and $C(-2, 5)$.

SOLUTION. The abscissas of A and C are the same, and therefore side AC is vertical. The length of the vertical side is the difference of the ordinates. The other sides are slant segments, and the general distance formula yields their lengths.

---

* The **Pythagorean theorem** states that the *sum of the squares on the perpendicular sides of a right triangle is equal to the square on the hypotenuse.* That is, if a and b are the lengths of the perpendicular sides and c is the length of the hypotenuse, then $a^2 + b^2 = c^2$.

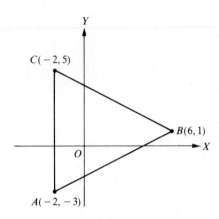

FIGURE 1.10

Hence we get

$$|AC| = 5 - (-3) = 5 + 3 = 8,$$

$$|AB| = \sqrt{(6 + 2)^2 + (1 + 3)^2} = \sqrt{80} = 4\sqrt{5},$$

$$|BC| = \sqrt{(6 + 2)^2 + (1 - 5)^2} = \sqrt{80} = 4\sqrt{5}.$$

The lengths of the sides show that the triangle is isosceles. ●

Exercises

1. Plot the points $A(-1,0)$, $B(2,0)$, and $C(5,0)$. Then find the directed distances $\overrightarrow{AB}$, $\overrightarrow{AC}$, $\overrightarrow{BC}$, $\overrightarrow{CB}$, $\overrightarrow{CA}$, and $\overrightarrow{BA}$.

2. Plot the points $A(-3,2)$, $B(0,2)$, and $C(4,2)$. Then find the directed distances $\overrightarrow{AB}$, $\overrightarrow{BA}$, $\overrightarrow{AC}$, $\overrightarrow{CA}$, $\overrightarrow{BC}$, and $\overrightarrow{CB}$.

3. Plot the points $A(-2,-3)$, $B(-2,0)$, and $C(-2,4)$, and verify the following equations by numerical substitutions:

$$\overrightarrow{AC} + \overrightarrow{CB} = \overrightarrow{AB}, \qquad \overrightarrow{BA} + \overrightarrow{AC} = \overrightarrow{BC}, \qquad \overrightarrow{AB} + \overrightarrow{BC} = \overrightarrow{AC}.$$

In each of Exercises 4 through 12, plot the pairs of points and find the distance between them.

4. $(3,1)$, $(7,4)$ ⓒ 5. $(4.137, -2.394)$, $(-8.419, 2.843)$

6. $(2,3)$, $(-1,0)$ 7. $(12, -5)$, $(0,0)$

8. $(0,4)$, $(-3,0)$ 9. $(-1,4)$, $(2, -1)$

10. $(6, 3), (-1, -1)$ 11. $(5, 5), (-5, 1)$

12. $(-3, -3), (2, 2)$

In each of Exercises 13 through 16, draw the triangle with the given vertices and find the lengths of the sides.

13. $A(-1, 1)$, $B(-1, 4)$, $C(3, 4)$ 14. $A(2, -1)$, $B(4, 2)$, $C(5, 0)$

15. $A(0, 0)$, $B(5, -2)$, $C(-3, 3)$ 16. $A(0, -3)$, $B(3, 0)$, $C(0, -4)$

In each of Exercises 17 through 20, draw the triangle having the given vertices and show that the triangle is isosceles.

17. $A(6, 2)$, $B(2, -3)$, $C(-2, 2)$ 18. $A(5, 4)$, $B(2, 0)$, $C(-2, 3)$

Ⓒ 19. $A(2.107, -1.549)$, $B(2.107, 6.743)$, $C(9.167, 2.597)$

20. $A(-2, -3)$, $B(4, 3)$, $C(-3, 4)$

In each of Exercises 21 through 24, draw the triangle with the given vertices and show that the triangle is a right triangle. That is, the square on the longest side is equal to the sum of the squares on the remaining sides.

21. $A(1, 3)$, $B(10, 5)$, $C(2, 1)$ 22. $A(-1, 1)$, $B(6, -2)$, $C(4, 3)$

23. $A(0, 3)$, $B(-3, -3)$, $C(2, 2)$ 24. $A(5, -2)$, $B(1, 1)$, $C(7, 9)$

25. Show that the points $A(-2, 0)$, $B(2, 0)$, and $C(0, 2\sqrt{3})$ are vertices of an equilateral triangle.

26. Show that the points $A(-\sqrt{3}, 1)$, $B(2\sqrt{3}, -2)$, and $C(2\sqrt{3}, 4)$ are vertices of an equilateral triangle.

27. Show that the points $A(1, -1)$, $B(5, 2)$, $C(2, 6)$, and $D(-2, 3)$ are equal sides of the quadrilateral $ABCD$.

28. Determine whether the points $(-5, 6)$, $(2, 5)$, and $(1, -2)$ are the same distances from $(-2, 2)$.

29. Prove that the points $A(-2, 7)$, $B(5, 4)$, $C(-1, -10)$, and $D(-8, -7)$ are vertices of the rectangle $ABCD$.

Determine, using the distance formula, whether the points in each of Exercises 30 through 33 lie on a straight line.

30. $(3, 3), (0, 1), (9, 7)$ Ⓒ 31. $(8.104, 0.478), (-2.502, 3.766), (2.801, 2.122)$

32. $(-3, 1), (1, 3), (10, 8)$ 33. $(-2, -2), (5, -2), (-11, 2)$

34. If $(x, 4)$ is equidistant from $(5, -2)$ and $(3, 4)$, find x.

35. If $(-3, y)$ is equidistant from $(2, 6)$ and $(7, -2)$, find y.

36. Find the point on the y axis that is equidistant from $(-4, -2)$ and $(3, 1)$.

37. Find the point on the x axis that is equidistant from $(-2, 5)$ and $(4, 1)$.

38. (For students who know determinants.) We may find the area of the triangle ABC by adding the areas of trapezoids $DECA$ and $EFBC$ and then subtracting the area of $DFBA$, as in Fig. 1.11. Recall that the area of a trapezoid is equal to one-half the sum of the parallel sides times the altitude. Show that the area S of the triangle ABC is:

$$S = \tfrac{1}{2}\big|[x_1(y_2 - y_3) - y_1(x_2 - x_3) + (x_2 y_3 - x_3 y_2)]\big|,$$

and that this equals one-half the absolute value of the determinant

$$\begin{vmatrix} x_1 & y_1 & 1 \\ x_2 & y_2 & 1 \\ x_3 & y_3 & 1 \end{vmatrix}.$$

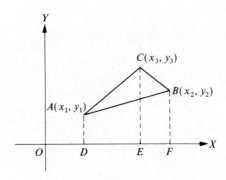

FIGURE 1.11

1.2

INCLINATION AND SLOPE OF A LINE

The inclination of a line is a concept used extensively in calculus and other areas of mathematics. Relative to this concept, we have the following definition.

DEFINITION 1.4 ● *The **inclination** of a line that intersects the x axis is the smallest angle, greater than or equal to 0°, that the line makes with the positive direction of the x axis. The inclination of a horizontal line is 0.*

According to this definition, the inclination θ of a line is such that

$$0° \le \theta < 180°, \quad \text{or, in radian measure,} \quad 0 \le \theta < \pi.$$

In Fig. 1.12 the inclination of the line L is indicated by the curved arrows. MX is the initial side and ML is the terminal side.

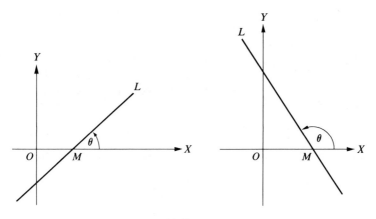

FIGURE 1.12

DEFINITION 1.5 ● *The* **slope** *of a line is the tangent of the inclination.*

A line that leans to the right has a positive slope because the inclination is an acute angle. The slope of a line that leans to the left is negative. Vertical lines do not have a slope, however, since 90° has no tangent.

If the inclination of a nonvertical line is known, the slope can be determined by the use of a table of trigonometric functions. Conversely, if the slope of a line is known, its inclination can be found. In most problems, however, it is more convenient to deal with the slope of a line rather than with its inclination.

EXAMPLE 1 ● Draw a line through $P(2, 2)$ with inclination 35°.

SOLUTION. We draw a line through P making an angle of 35° with the positive x direction, as shown in Fig. 1.13. The figure also shows a line through $(-4, 0)$ with inclination 135°. ●

FIGURE 1.13

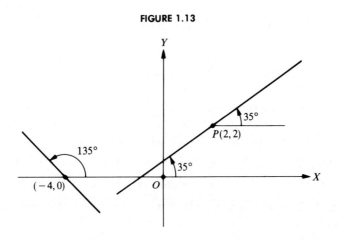

EXAMPLE 2 ● Draw a line through the point $P(-2, 2)$ with slope $-\frac{2}{3}$.

SOLUTION. We move 3 units to the left of P and then 2 units upward. The line through the point thus located and the given point P clearly has the required slope (Fig. 1.14). ●

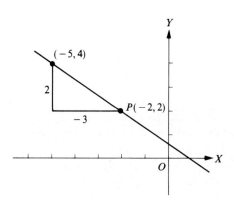

FIGURE 1.14

The definitions of inclination and slope lead immediately to a theorem concerning parallel lines. If two lines have the same slope, their inclinations are equal. Hence we know from geometry that they are parallel. Conversely, if two nonvertical lines are parallel, they have equal inclinations and thus equal slopes.

THEOREM 1.2 ● *Two nonvertical lines are parallel if, and only if, their slopes are equal.*

If the coordinates of two points on a line are known, we may find the slope of the line from the given coordinates. We now derive a formula for this purpose. Let $P_1(x_1, y_1)$ and $P_2(x_2, y_2)$ be the two given points, and indicate the slope by m. Then, referring to Fig. 1.15, we have

$$m = \tan \theta = \frac{\overrightarrow{RP_2}}{\overrightarrow{P_1R}} = \frac{y_2 - y_1}{x_2 - x_1}.$$

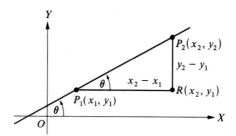

FIGURE 1.15

In Fig. 1.16 the line slants to the left. The quantities $y_1 - y_2$ and $x_2 - x_1$ are both positive and the angles θ and ϕ are supplementary. Consequently

$$\frac{y_1 - y_2}{x_2 - x_1} = \tan\phi = -\tan\theta.$$

FIGURE 1.16

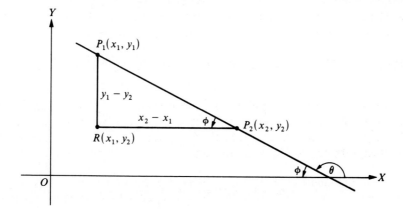

Therefore

$$m = \tan\theta = -\frac{y_1 - y_2}{x_2 - x_1} = \frac{y_2 - y_1}{x_2 - x_1}.$$

Hence the slope is determined in the same way for lines slanting either to the left or right.

THEOREM 1.3 ● *The slope m of a line passing through two given points $P_1(x_1, y_1)$ and $P_2(x_2, y_2)$ is equal to the difference of the ordinates divided by the difference of the abscissas taken in the same order; that is,*

$$\boxed{m = \frac{y_2 - y_1}{x_2 - x_1}.}$$

This formula yields the slope if the two points are on a slant or a horizontal line. If the line is vertical, the denominator of the formula becomes zero, a result in keeping with the fact that slope is not defined for a vertical line. We observe further that either of the points may be designated as $P_1(x_1, y_1)$ and the other as $P_2(x_2, y_2)$, since

$$\frac{y_2 - y_1}{x_2 - x_1} = \frac{y_1 - y_2}{x_1 - x_2}.$$

EXAMPLE 3 ● Given the points $A(-1, -1)$, $B(5,0)$, $C(4,3)$, and $D(-2,2)$, show that $ABCD$ is a parallelogram.

SOLUTION. We determine from the slopes of the sides if the figure is a parallelogram.

$$\text{Slope of } AB = \frac{0 - (-1)}{5 - (-1)} = \frac{1}{6}. \qquad \text{Slope of } BC = \frac{3 - 0}{4 - 5} = -3.$$

$$\text{Slope of } CD = \frac{2 - 3}{-2 - 4} = \frac{1}{6}. \qquad \text{Slope of } DA = \frac{2 - (-1)}{-2 - (-1)} = -3.$$

The opposite sides have equal slopes, and therefore $ABCD$ is a parallelogram. ●

Angle between Two Lines

Two intersecting lines form two pairs of equal angles, and an angle of one pair is the supplement of an angle of the other pair. We shall show how to find a measure of each angle in terms of the slopes of the lines. Noting Fig. 1.17 and recalling that an exterior angle of a triangle is equal to the sum of the remote

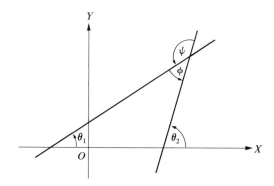

FIGURE 1.17

interior angles, we see that

$$\phi + \theta_1 = \theta_2 \quad \text{or} \quad \phi = \theta_2 - \theta_1.$$

Using the formula for the tangent of the difference of two angles, we find

$$\tan\phi = \tan(\theta_2 - \theta_1) = \frac{\tan\theta_2 - \tan\theta_1}{1 + \tan\theta_1 \tan\theta_2}.$$

If we let $m_2 = \tan\theta_2$ and $m_1 = \tan\theta_1$, then we have

$$\tan\phi = \frac{m_2 - m_1}{1 + m_1 m_2},$$

where m_2 is the slope of the terminal side and m_1 is the slope of the initial side, and ϕ is measured in a counterclockwise direction.

The angle ψ is the supplement of ϕ, and therefore

$$\tan\psi = -\tan\phi = \frac{m_1 - m_2}{1 + m_1 m_2}.$$

This formula for $\tan\psi$ is the same as the one for $\tan\phi$ except that the terms in the numerator are reversed. We observe from the diagram, however, that the terminal side of ψ is the initial side of ϕ and the initial side of ψ is the terminal side of ϕ, as indicated by the counterclockwise arrows. Hence the numerator for $\tan\psi$ is equal to the slope of the terminal side of ψ minus the slope of the initial side of ψ. The same wording holds for $\tan\phi$; that is, the numerator for $\tan\phi$ is equal to the slope of the terminal side of ϕ minus the slope of the initial side of ϕ. Accordingly, in terms of the slopes of the initial and terminal sides, the tangent of either angle may be found by the same rule. We state this conclusion as a theorem.

THEOREM 1.4 ● *If ϕ is an angle, measured counterclockwise, between two lines, then*

$$\tan\phi = \frac{m_2 - m_1}{1 + m_1 m_2},$$ (1.3)

where m_2 is the slope of the terminal side and m_1 is the slope of the initial side.

This formula will not apply if either of the lines is vertical, since a vertical line does not possess slope. For this case the problem would be that of finding the angle, or a trigonometric function of the angle, that a line of known slope makes with the vertical. Hence no new formula is needed.

For any two slant lines that are not perpendicular, Eq. (1.3) will yield a definite number as the value of $\tan\phi$. Conversely, if the slopes of the lines are such that the formula yields a definite value, the lines could not be perpendicular, because the tangent of a right angle does not exist. Since the formula fails to yield a value only when the denominator is equal to zero, it appears that the lines are perpendicular when and only when $1 + m_1 m_2 = 0$ or

$$m_2 = -\frac{1}{m_1}.$$

We note, additionally, that if α_2 and α_1 are the inclinations of slant lines that are perpendicular, then

$$\alpha_2 = \alpha_1 + 90° \quad \text{or} \quad \alpha_2 = \alpha_1 - 90°.$$

In either case, $\tan\alpha_2 = -\cot\alpha_1$ and $m_2 = -1/m_1$.

THEOREM 1.5 ● *Two slant lines are perpendicular if, and only if, the slope of one is the negative reciprocal of the slope of the other.*

Perpendicularity of two lines occurs, of course, if one line is parallel to the x axis and the other parallel to the y axis. The slope of the line parallel to the x axis is zero, but the line parallel to the y axis does not possess slope.

EXAMPLE 4 ● Find the tangents of the angles of the triangle whose vertices are $A(3, -2)$, $B(-5, 8)$, and $C(4, 5)$. Then refer to Table II in the Appendix or use a calculator to express each angle to the nearest degree.

SOLUTION. We first find the slope of each side. Thus, from Fig. 1.18,

$$\text{Slope of } AB = \frac{-2 - 8}{3 - (-5)} = -\frac{5}{4}.$$

$$\text{Slope of } BC = \frac{8 - 5}{-5 - 4} = -\frac{1}{3}.$$

$$\text{Slope of } AC = \frac{-2 - 5}{3 - 4} = 7.$$

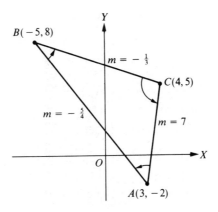

FIGURE 1.18

We now substitute in Eq. (1.3) and get

$$\tan A = \frac{-\frac{5}{4} - 7}{1 + \left(-\frac{5}{4}\right)(7)} = \frac{33}{31} = 1.06, \qquad A = 47°.$$

$$\tan B = \frac{-\frac{1}{3} - \left(-\frac{5}{4}\right)}{1 + \left(-\frac{1}{3}\right)\left(-\frac{5}{4}\right)} = \frac{11}{17} = 0.647, \qquad B = 33°.$$

$$\tan C = \frac{7 - \left(-\frac{1}{3}\right)}{1 + 7\left(-\frac{1}{3}\right)} = -\frac{22}{4} = -5.5, \qquad C = 100°. \ \bullet$$

EXAMPLE 5 ● The cross section of an A-frame cottage is an isosceles triangle. If the slope of one of the sides is 1.8 and it is 19 ft high at the peak, what is the width of the cottage?

SOLUTION. If the axes are set as in Fig. 1.19,

FIGURE 1.19

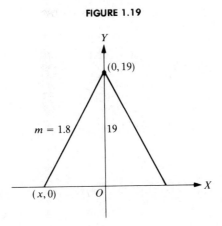

then

$$\frac{19 - 0}{0 - x} = 1.8$$

$$\frac{19}{-x} = 1.8$$

$$x = -\frac{19}{1.8} = -\frac{95}{9}.$$

Hence the width of the cottage is $2(\frac{95}{9}) = 21\frac{1}{9}$ ft. ●

EXAMPLE 6 ● A television camera is located along the 40-yd line at a football game. If the camera is 20 yds back from the sideline, through what angle should it be able to pan in order to cover the entire field of play, including end zones?

SOLUTION. Locate the camera at the origin so that it is able to cover all action from the line through (70, 20) to the line through (−50, 20). If ϕ is the angle in question, measured counterclockwise, then we see from Fig. 1.20 that

$$\tan\phi = \frac{-\frac{2}{5} - \frac{2}{7}}{1 - \frac{2}{5} \cdot \frac{2}{7}} = -\frac{24}{31},$$

$$\phi = 142°.$$

FIGURE 1.20

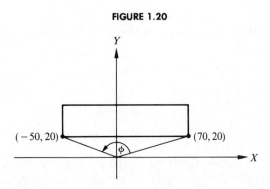

The camera should be able to sweep through an angle of 142°. ●

Exercises

Draw a line through the given point with the indicated inclination θ in each of Exercises 1 through 6.

1. $(2, 3)$, $\theta = 30°$ 2. $(-2, 1)$, $\theta = 45°$ 3. $(4, -3)$, $\theta = 150°$

4. $(-3, -1)$, $\theta = 60°$ 5. $(5, -4)$, $\theta = 0°$ 6. $(0, 0)$, $\theta = 75°$

Draw a line through the given point having the given slope m in each of Exercises 7 through 12.

7. $(2, 2)$, $m = 3$ 8. $(-1, 3)$, $m = 1$ 9. $(3, 1)$, $m = -1$

10. $(2, -2)$, $m = \frac{1}{2}$ 11. $(4, 0)$, $m = \frac{2}{3}$ 12. $(-3, 3)$, $m = -\frac{3}{4}$

13. A flat board leans against a wall. The upper edge is 6 ft above the floor and the lower edge is 2 ft out from the wall. What is the slope of the board?

14. A ladder 10 ft long leans against a wall, touching it 8 ft above the ground. What is the slope of the ladder? Can a person 6 ft tall pass under the ladder 1 ft away from the wall? Can the same person pass under the ladder 2 ft away from the wall?

15. A cross section of a cottage, 18 ft wide, is an isosceles triangle. If the slope of a side is 1.5, find the height of the cottage.

In Exercises 16 through 21, find the slope of the line passing through the two points. Find also the inclination to the nearest degree by using a hand calculator or by using Table II in the Appendix.

16. $(2, 3)$, $(3, 7)$ 17. $(-13, 6)$, $(4, 0)$ 18. $(3, -7)$, $(4, 8)$

19. $(-9, 0)$, $(3, 20)$ Ⓒ 20. $(11.7142, 4.0015)$, $(-3.8014, -2.8117)$

21. $(-2, 8)$, $(4, -3)$

Show that each of the four points in Exercises 22 through 25 are vertices of the parallelogram $ABCD$.

22. $A(3, 0)$, $B(7, 0)$, $C(5, 3)$, $D(1, 3)$

23. $A(-2, 3)$, $B(6, 1)$, $C(5, -2)$, $D(-3, 0)$

24. $A(-1, -2)$, $B(3, -6)$, $C(11, -1)$, $D(7, 3)$

25. $A(0, 0)$, $B(6, 3)$, $C(9, 9)$, $D(3, 6)$

Verify that each triangle with the given points as vertices in Exercises 26 through 31 is a right triangle since the slope of one side is the negative reciprocal of the slope of another side.

26. $(4, -4)$, $(4, 4)$, $(0, 0)$ 27. $(-1, 2)$, $(3, -6)$, $(3, 4)$

28. $(7, 1)$, $(0, -2)$, $(5, -4)$ 29. $(2, 5)$, $(-5, 7)$, $(-2, -9)$

30. $(1, 1)$, $(4, -1)$, $(3, 4)$ 31. $(-1, -1)$, $(16, -1)$, $(0, 3)$

Show that the four points in each of Exercises 32 through 35 are vertices of the rectangle $ABCD$.

32. $A(-4, 3)$, $B(0, -2)$, $C(5, 2)$, $D(1, 7)$

33. $A(2, 2)$, $B(7, -3)$, $C(10, 0)$, $D(5, 5)$

34. $A(5, -1)$, $B(7, 6)$, $C(0, 8)$, $D(-2, 1)$

35. $A(5, 7)$, $B(1, 1)$, $C(4, -1)$, $D(8, 5)$

Using slopes, determine which of the sets of three points in Exercises 36 through 39 lie on a straight line.

36. $(0, -2)$, $(3, 0)$, $(9, 4)$ 37. $(-1, 2)$, $(2, 1)$, $(5, 0)$

38. $(0, 1)$, $(9, 6)$, $(-4, -1)$ 39. $(-10, 2)$, $(1, -2)$, $(6, -5)$

In Exercises 40 through 43, find the tangents of the angles in each triangle ABC. Then use a hand calculator or Table II in the Appendix to find each angle to the nearest degree.

40. $A(1, 1)$, $B(5, 2)$, $C(3, 5)$ 41. $A(-1, 1)$, $B(2, -1)$, $C(3, 5)$

42. $A(2, 2)$, $B(-4, -1)$, $C(6, -5)$ 43. $A(3, 8)$, $B(-4, -3)$, $C(6, -1)$

44. The line through the points $(3, 4)$ and $(-5, 0)$ intersects the line through $(0, 0)$ and $(-5, 0)$. Find the angles of intersection.

45. Two lines passing through $(3, 2)$ make an angle of $45°$. If the slope of one of the lines is 1, find the slope of the other line (two solutions).

46. What acute angle does a line of slope $-\frac{3}{2}$ make with a vertical line?

In each of Exercises 47 through 52, find the slopes of the lines passing through the two pairs of points. Then decide whether the lines are parallel, perpendicular, or intersect obliquely.

47. $(1, -1)$, $(-5, -5)$; $(1, -2)$, $(7, 2)$ 48. $(1, -1)$, $(-4, -4)$; $(1, 1)$, $(4, -4)$

49. $(1, 8)$, $(-3, -4)$; $(-1, 8)$, $(0, 10)$ 50. $(2, -3)$, $(0, 2)$; $(1, 0)$, $(6, 2)$

51. $(6, 5)$, $(11, 9)$; $(2, 5)$, $(12, 9)$ 52. $(-6, -4)$, $(22, 8)$; $(-5, 7)$, $(7, -8)$

53. A cross section of a cottage, 18 ft wide, is an isosceles triangle. If the slope of a side is 1.75 and there is a second floor 8 ft above the ground floor, what is the width of the second floor?

54. A bridge is trussed as in Fig. 1.21. Find the slopes and inclinations of the sections AB and BC.

55. A television camera is 30 ft from the sideline of a basketball court 94 ft long. The camera is located 7 ft from midcourt. Through what angle must it sweep in order to cover all action on the court?

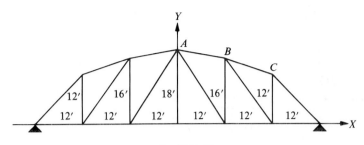

FIGURE 1.21

1.3

DIVISION OF A LINE SEGMENT

In this section we will show how to find the coordinates of a point which divides a line segment into two parts that have a specified relation. We first find formulas for the coordinates of a point that is midway between two points of given coordinates.

Let $A(x_1, y_1)$ and $B(x_2, y_2)$ be the extremities of a line segment, and let $P(x, y)$ be the midpoint of $\overrightarrow{AB}$. From similar triangles (Fig 1.22), we have

$$\frac{\overrightarrow{AP}}{\overrightarrow{AB}} = \frac{\overrightarrow{AM}}{\overrightarrow{AN}} = \frac{\overrightarrow{MP}}{\overrightarrow{NB}} = \frac{1}{2}.$$

Hence

$$\frac{\overrightarrow{AM}}{\overrightarrow{AN}} = \frac{x - x_1}{x_2 - x_1} = \frac{1}{2} \quad \text{and} \quad \frac{\overrightarrow{MP}}{\overrightarrow{NB}} = \frac{y - y_1}{y_2 - y_1} = \frac{1}{2}.$$

FIGURE 1.22

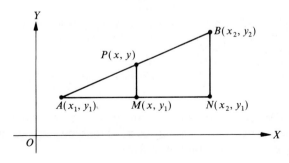

Solving for x and y gives

$$x = \frac{x_1 + x_2}{2} \quad \text{and} \quad y = \frac{y_1 + y_2}{2}. \tag{1.4}$$

THEOREM 1.6 ● *The abscissa of the midpoint of a line segment is half the sum of the abscissas of the endpoints; the ordinate is half the sum of the ordinates.*

This theorem may be generalized by letting $P(x, y)$ be any division point of the line through A and B. If the ratio of $\overrightarrow{AP}$ to $\overrightarrow{AB}$ is a number r instead of $\frac{1}{2}$, then

$$\frac{\overrightarrow{AP}}{\overrightarrow{AB}} = \frac{x - x_1}{x_2 - x_1} = r \quad \text{and} \quad \frac{\overrightarrow{AP}}{\overrightarrow{AB}} = \frac{y - y_1}{y_2 - y_1} = r.$$

These equations, when solved for x and y, give

$$x = x_1 + r(x_2 - x_1), \qquad y = y_1 + r(y_2 - y_1). \tag{1.5}$$

It should be clear that formulas (1.5) reduce to the midpoint formulas (1.4), if $r = \frac{1}{2}$.

It is probably better that the student *remember how to derive* formulas (1.5) by using similar triangles than that the student memorize them. The student may have many opportunities in this and in subsequent mathematics courses to use similar triangles to solve a problem. There are comparatively few occasions to use formulas (1.5).

EXAMPLE 1 ● Find the coordinates of the midpoint of the line segment joining $A(3, -4)$ and $B(7, 2)$.

SOLUTION. Applying the midpoint formulas (1.4), we have

$$x = \frac{x_1 + x_2}{2} = \frac{3 + 7}{2} = 5, \qquad y = \frac{y_1 + y_2}{2} = \frac{-4 + 2}{2} = -1.$$

Hence the coordinates of the midpoint are $(5, -1)$. ●

EXAMPLE 2 ● One endpoint of a line segment is $A(6, 4)$, and the midpoint of the segment is $P(-2, 9)$. Find the coordinates of the other endpoint.

SOLUTION. We let (x_2, y_2) stand for the unknown coordinates. Then in formulas (1.4), we replace x by -2, y by 9, x_1 by 6, and y_1 by 4, and have

$$-2 = \frac{6 + x_2}{2} \quad \text{and} \quad 9 = \frac{4 + y_2}{2}.$$

These equations yield $x_2 = -10$ and $y_2 = 14$. Hence the desired coordinates are $(-10, 14)$. ●

EXAMPLE 3 ● Find the two trisection points of the line segment joining $A(-3, -4)$ and $B(6, 11)$.

SOLUTION. We let $P_1(x, y)$ and $P_2(x, y)$ stand for trisection points. Then we use $r = \frac{1}{3}$ in formulas (1.5) to find P_1, and have

$$x = x_1 + r(x_2 - x_1) = -3 + \tfrac{1}{3}(6 + 3) = 0,$$
$$y = y_1 + r(y_2 - y_1) = -4 + \tfrac{1}{3}(11 + 4) = 1.$$

Next we use $r = \frac{2}{3}$ in formulas (1.5) to find P_2, and have

$$x = x_1 + r(x_2 - x_1) = -3 + \tfrac{2}{3}(6 + 3) = 3,$$
$$y = y_1 + r(y_2 - y_1) = -4 + \tfrac{2}{3}(11 + 4) = 6.$$

Hence the coordinates of the trisection points are $(0, 1)$ and $(3, 6)$, as shown in Fig. 1.23. ●

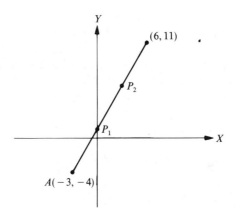

FIGURE 1.23

In formulas (1.5), the point P is between A and B if and only if $0 < r < 1$. However, if P is a point on the segment $\overrightarrow{AB}$ extended through B, then the length of the segment $\overrightarrow{AP}$ is greater than the length of $\overrightarrow{AB}$ and r is greater than 1. Conversely, if r is greater than 1, formulas (1.5) yield the coordinates of a point on the extension of the segment through B. In order to find a point on the segment extended in the other direction (through A), we may either use formulas (1.5) with r negative or we may use a similar triangles argument similar to that used in the derivation of (1.5).

EXAMPLE 4 ● A point $P(x, y)$ is on the line through $A(-4, 4)$ and $B(5, 2)$. Find (a) the coordinates of P given that the segment $\overrightarrow{AB}$ is extended through B to P

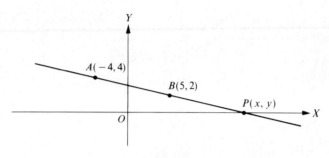

FIGURE 1.24

so that P is twice as far from A as from B, and (b) the coordinates of P given that $\overrightarrow{AB}$ is extended through A to P so that P is three times as far from B as from A.

SOLUTION. (a) Since $\overrightarrow{AP} = 2\overrightarrow{BP}$, it follows that $\overrightarrow{BP} = \overrightarrow{AB}$ (Fig 1.24). Hence the ratio of $\overrightarrow{AP}$ to $\overrightarrow{AB}$ is 2. Accordingly, we use $r = 2$ in formulas (1.5), and write

$$x = -4 + 2(5 + 4) = 14, \qquad y = 4 + 2(2 - 4) = 0.$$

The desired coordinates are $(14, 0)$.

SOLUTION. (b) First we sketch a graph (Fig 1.25) so that we may use similar triangles. From the figure, we see that

$$\frac{\overrightarrow{PA}}{\overrightarrow{PB}} = \frac{-4 - x}{5 - x} = \frac{1}{3}$$
$$3(4 + x) = x - 5$$
$$x = -8.5,$$

FIGURE 1.25

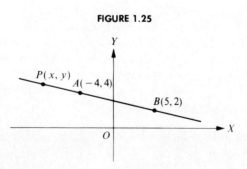

and

$$\frac{\overline{PA}}{\overline{PB}} = \frac{4-y}{2-y} = \frac{1}{3}$$
$$3(4-y) = 2 - y$$
$$y = 5.$$

On the other hand, we could use formulas (1.5) with $r = -\frac{1}{2}$, to get

$$x = -4 + \left(-\tfrac{1}{2}\right)(5+4) = -8.5$$

and

$$y = 4 + \left(-\tfrac{1}{2}\right)(2-4) = 5.$$

Either way, we see that the coordinates of P are $(-8.5, 5)$. ●

EXAMPLE 5 ● The line segment joining a vertex of a triangle and the midpoint of the opposite side is called a **median** of the triangle. Fig. 1.26 shows a triangle with vertices $A(4, -4)$, $B(10, 4)$, $C(2, 6)$, and the respective midpoints of the opposite sides $D(6, 5)$, $E(3, 1)$, $F(7, 0)$. Find the point on each median that is two-thirds of the distance from the vertex to the midpoint of the opposite side.

SOLUTION. Using $r = \frac{2}{3}$ in the point of division formulas (1.5), we get, for the medians AD, BE, and CF, respectively,

$$x = 4 + \tfrac{2}{3}(6-4) = \tfrac{16}{3}, \qquad y = -4 + \tfrac{2}{3}(5+4) = 2,$$
$$x = 10 + \tfrac{2}{3}(3-10) = \tfrac{16}{3}, \qquad y = 4 + \tfrac{2}{3}(1-4) = 2,$$
$$x = 2 + \tfrac{2}{3}(7-2) = \tfrac{16}{3}, \qquad y = 6 + \tfrac{2}{3}(0-6) = 2.$$

These results tell use that the medians are concurrent at the point $\left(\tfrac{16}{3}, 2\right)$. ●

FIGURE 1.26

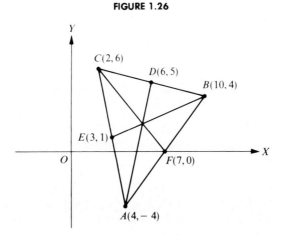

Exercises

Find the coordinates of the midpoint of each pair of points in Exercises 1 through 6.

1. $(4, 3)$, $(-4, -3)$ 2. $(3, 2)$, $(1, 6)$ 3. $(2, 3)$, $(3, 4)$

4. $(7, -4)$, $(-9, 6)$ 5. $(-7, -11)$, $(5, 15)$ 6. $(5, 7)$, $(-3, 3)$

Find the coordinates of the midpoints of the sides of each triangle whose vertices are given in Exercises 7 through 10.

7. $(1, 2)$, $(2, 5)$, $(6, 3)$ 8. $(4, 4)$, $(2, 3)$, $(5, 1)$

9. $(8, 3)$, $(2, -4)$, $(7, -6)$ 10. $(-1, -6)$, $(-3, -5)$, $(-2, -2)$

11. The line segment connecting $(x_1, 6)$ and $(9, y_2)$ is bisected by the point $(7, 3)$. Find the values of x_1 and y_2.

12. Find the coordinates of the midpoint of the hypotenuse of the right triangle whose vertices are $(2, 2)$, $(6, 3)$, and $(5, 7)$, and show that the midpoint is equidistant from the three vertices.

In Exercises 13 through 24, find the point $P(x, y)$ so that the ratio of $\overrightarrow{AP}$ to $\overrightarrow{AB}$ is equal to r.

13. $A(4, 3)$, $B(5, 1)$, $r = \frac{1}{3}$ 14. $A(2, -4)$, $B(-3, 3)$, $r = \frac{2}{3}$

15. $A(-1, 0)$, $B(3, 2)$, $r = \frac{4}{3}$ 16. $A(5, 6)$, $B(0, -5)$, $r = \frac{2}{5}$

17. $A(6, -2)$, $B(-1, 7)$, $r = 2$ 18. $A(-5, 1)$, $B(3, 3)$, $r = \frac{5}{2}$

19. $A(0, 0)$, $B(6, 2)$, $r = 3$ Ⓒ 20. $A(4.1001, 1.0952)$, $B(-2.8763, 0.0018)$, $r = 0.2412$

21. $A(-5, -5)$, $B(1, 1)$, $r = \frac{1}{5}$ 22. $A(1, 5)$, $B(6, 3)$, $r = \frac{4}{5}$

23. $A(2, 9)$, $B(-4, -3)$, $r = -\frac{1}{3}$ 24. $A(2, 5)$, $B(5, -2)$, $r = \frac{3}{4}$

25. Find the coordinates of the point which divides the line segment from $(-1, 4)$ to $(2, -3)$ in the ratio 3 to 4 (two solutions).

26. Find the coordinates of P if it divides the line segment joining $A(2, -5)$ and $B(6, 3)$ so that $\overrightarrow{AP}/\overrightarrow{PB} = 3/4$.

27. The line segment joining $A(2, 4)$ and $B(-3, -5)$ is extended through each end by a distance equal to twice its original length. Find the coordinates of the new endpoints.

28. A line passes through $A(2, 3)$ and $B(5, 7)$. Find (a) the coordinates of the point P on $\overrightarrow{AB}$ extended through B to P so that P is twice as far from A as from B; (b) the coordinates if P is on $\overrightarrow{AB}$ extended through A so that P is twice as far from B as from A.

29. A line passes through $A(-2, -1)$ and $B(3, 4)$. Find (a) the point P on $\overrightarrow{AB}$ extended through B so that P is three times as far from A as from B; (b) the point, if P is on $\overrightarrow{AB}$ extended through A so that P is three times as far from B as from A.

30. Find the point of the line passing through $A(-1, -1)$ and $B(4, 4)$ which is
    (a) twice as far from A as from B (two cases).
    (b) three times as far from B as from A (two cases).

31. The line segment joining $A(1, 3)$ and $B(-2, -1)$ is extended through each end by a distance equal to its original length. Find the coordinates of the new endpoints.

In each of Exercises 32 through 35, find the intersection point of the diagonals of the parallelogram $ABCD$.

32. $A(3, 0)$, $B(7, 0)$, $C(9, 3)$, $D(5, 3)$

33. $A(-2, 3)$, $B(6, 1)$, $C(5, -2)$, $D(-3, 0)$

34. $A(-1, -2)$, $B(3, -6)$, $C(11, -1)$, $D(7, 3)$

35. $A(0, 2)$, $B(-3, 1)$, $C(2, -1)$, $D(5, 0)$

36. A 30-lb child is sitting at $A(2, 3)$ and a 50-lb child is at $B(12, 7)$, where units are feet. Find the point P between A and B which could be used as the fulcrum of a teeterboard putting the two children in equilibrium. [*Hint.* $30AP = 50PB$ or $(AP/PB) = \frac{5}{3}$. Now use the point of division formulas.]

37. A 60-lb child is sitting on a teeterboard at $(1, 4)$ and the fulcrum is at $(6, 5)$, where units are feet. At what point should a 40-lb child sit to be in equilibrium? See hint in Exercise 36.

38. A person 6 ft tall is standing near a street light so that he is 4/10 of the distance from the pole to the tip of his shadow. How high above the ground is the lightbulb? If the person's head is exactly 5 ft from the lightbulb, how far is the person from the pole, and how long is the shadow?

1.4

ANALYTIC PROOFS OF GEOMETRIC THEOREMS

By the use of a coordinate system, many theorems of geometry can be proved with surprising simplicity and directness. We illustrate the procedure in the following examples.

EXAMPLE 1 ● Prove that the diagonals of a parallelogram bisect each other.

SOLUTION. We first draw a parallelogram and then introduce a coordinate system. A judicious location of the axes relative to the figure makes the writing of the coordinates of the vertices easier and also simplifies the algebraic operations involved in making the proof. Therefore we choose a vertex as the origin and a coordinate axis along a side of the parallelogram (Fig. 1.27). Then we write the coordinates of the vertices as $O(0,0)$, $P_1(a,0)$, $P_2(b,c)$, and $P_3(a+b,c)$. It is essential that the coordinates of P_2 and P_3 express the fact that P_2P_3 and OP_1 are parallel and have the same length. This is achieved by making the ordinates of P_2 and P_3 the same and making the abscissa of P_3 exceed the abscissa of P_2 by a.

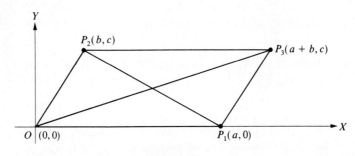

FIGURE 1.27

To show that OP_3 and P_1P_2 bisect each other, we find the coordinates of the midpoint of each diagonal.

$$\text{Midpoint of } OP_3: \quad x = \frac{a+b}{2}, \qquad y = \frac{c}{2}.$$

$$\text{Midpoint of } P_1P_2: \quad x = \frac{a+b}{2}, \qquad y = \frac{c}{2}.$$

Since the midpoint of each diagonal is $\left(\dfrac{a+b}{2}, \dfrac{c}{2}\right)$ the theorem is proved. ●

Note. In making a proof it is essential that a general figure be used. For example, neither a rectangle nor a rhombus (a parallelogram with all sides equal) should be used for a parallelogram. A proof of a theorem based on a special case would not constitute a general proof.

EXAMPLE 2 ● Prove that in any triangle the line segment joining the midpoints of two sides is parallel to, and one-half as long as, the third side.

SOLUTION. The triangle and midpoints of two sides are shown in Fig. 1.28. Note that the coordinate axes are positioned relative to the triangle so that it is easy to write the coordinates of the vertices. According to Theorem 1.3, the slope of DC is

$$\frac{(c/2)-(c/2)}{\frac{1}{2}(a+b)-\frac{1}{2}b} = 0.$$

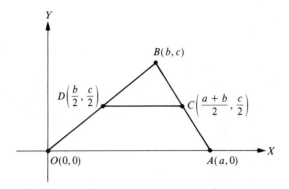

FIGURE 1.28

Hence the line segment DC and the third side are parallel since the slope of each is 0. To get the length of DC, we use the distance formula and find

$$\sqrt{\left(\frac{a+b}{2}-\frac{b}{2}\right)^2+\left(\frac{c}{2}-\frac{c}{2}\right)^2}=\frac{a}{2},$$

which is one-half the third side, as required. ●

EXAMPLE 3 ● Prove that a parallelogram whose diagonals are perpendicular is a rhombus.

SOLUTION. First, we recall that a parallelogram whose sides are all equal is called a **rhombus**. So we start the proof with the parallelogram $OACB$ and the perpendicular diagonals AB and OC (Fig. 1.29). If the sides of this parallelogram are all of

FIGURE 1.29

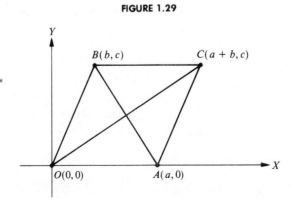

the same length, the figure satisfies the definition of a rhombus. We know that the opposite sides of a parallelogram are equal. Then if a side of one of the pairs of opposite sides has the same length as one of the sides of the other pair of opposite sides, all the sides are equal and $OACB$ is a rhombus. Let us now show that side OA is equal to side OB.

$$\text{Slope of } OC = \frac{c - 0}{a + b - 0} = \frac{c}{a + b}.$$

$$\text{Slope of } AB = \frac{c - 0}{b - a} = \frac{c}{b - a}.$$

Each of these slopes is the negative reciprocal of the slope of the other (Theorem 1.5). In other words, their product is -1. Hence

$$\frac{c}{b - a} \cdot \frac{c}{a + b} = -1 \quad \text{or} \quad c^2 = a^2 - b^2 \quad \text{and} \quad a = \sqrt{b^2 + c^2}.$$

The left-hand side of this last equation is the length of OA and the right-hand side is the length of OB. Hence $OACB$ is a rhombus. ●

EXAMPLE 4 ● The points $A(x_1, y_1)$, $B(x_2, y_2)$, and $C(x_3, y_3)$ are vertices of a triangle. Find the coordinates of the point on each median that is two-thirds of the way from the vertex to the midpoint of the opposite side.

SOLUTION. Figure 1.30 shows the triangle and the coordinates of the midpoints of the sides. We let (x, y) stand for the coordinates of the desired point on median

FIGURE 1.30

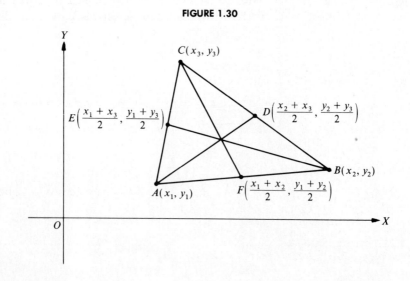

AD. Then using $r = \frac{2}{3}$ in the division formulas, Eq. (1.5), we obtain

$$x = x_1 + \frac{2}{3}\left(\frac{x_2 + x_3}{2} - x_1\right) = \frac{x_1 + x_2 + x_3}{3},$$

$$y = y_1 + \frac{2}{3}\left(\frac{y_2 + y_3}{2} - y_1\right) = \frac{y_1 + y_2 + y_3}{3}.$$

Similarly, we let (x, y) stand for the desired point on median BE and find

$$x = x_2 + \frac{2}{3}\left(\frac{x_1 + x_3}{2} - x_2\right) = \frac{x_2 + x_1 + x_3}{3},$$

$$y = y_2 + \frac{2}{3}\left(\frac{y_1 + y_3}{2} - y_2\right) = \frac{y_2 + y_1 + y_3}{3}.$$

From the above results, we see that two of the medians intersect at the point

$$\left(\frac{x_1 + x_2 + x_3}{3}, \frac{y_1 + y_2 + y_3}{3}\right).$$

We can now conclude that all three medians pass through this point. Could we have made this conclusion by considering only one median? ●

We have now established the following theorem:

THEOREM 1.7 ● *The three medians of a triangle intersect at the point whose abscissa is one-third the sum of the abscissas of the vertices of the triangle and whose ordinate is one-third the sum of the ordinates of the vertices.*

EXAMPLE 5 ● The vertices of a triangle are at $(-7, 3)$, $(4, -2)$, and $(6, 5)$. Find the point of intersection of the medians.

SOLUTION. The abscissa of the intersection point is $\frac{1}{3}(-7 + 4 + 6) = 1$, and the ordinate is $\frac{1}{3}(3 - 2 + 5) = 2$. Hence the medians intersect at $(1, 2)$. ●

Exercises

Give analytic proofs of the following theorems.

1. The diagonals of a rectangle have the same length and bisect each other.

2. If the diagonals of a parallelogram are of equal length, the figure is a rectangle.

3. The diagonals of a square are perpendicular to each other.

4. The segments that connect the midpoints of consecutive sides of a square form a square of one-half the area of the original figure.*

---

*This problem is mentioned in Plato's *Meno*.

5. If the diagonals of a rectangle are perpendicular to each other, the figure is a square.

6. The diagonals of a rhombus are perpendicular.

7. The segments that join the midpoints of the consecutive sides of a plane quadrilateral form a parallelogram.

8. The segments that join the midpoints of the consecutive sides of a rhombus form a rectangle.

9. The line segments joining the midpoints of the opposite sides of a quadrilateral bisect each other.

10. The sum of the squares of the diagonals of a rhombus is equal to four times the square of a side.

11. The midpoint of the hypotenuse of a right triangle is equidistant from the vertices.

12. If the midpoint of one side of a triangle is equidistant from the three vertices, the triangle is a right triangle.

13. If the sum of the squares of two sides of a triangle is equal to the square of the third side, the figure is a right triangle.

14. If two medians of a triangle are equal, it is isosceles.

15. The line segment joining the midpoints of two sides of a triangle bisects the median drawn to the third side.

16. The line through the vertex of an isosceles triangle parallel to the base bisects the exterior angle.

17. The vertex and the midpoints of the three sides of an isosceles triangle are the vertices of a rhombus.

18. The line segment joining the midpoints of the nonparallel sides of a trapezoid is parallel to the bases and its length is the average of the lengths of the bases.

19. The diagonals of an isosceles trapezoid are equal.

20. If the diagonals of a trapezoid are equal, the figure is an isosceles trapezoid.

21. The sum of the squares of the four sides of a parallelogram is equal to the sum of the squares of the two diagonals.

22. The lines drawn from a vertex of a parallelogram to the midpoints of the opposite sides trisect a diagonal.

23. The line joining each of the trisection points of a diagonal of a rectangle with the other vertices form a parallelogram.

24. If $P(a, b)$ is on a circle with center at the origin and radius r, then $a^2 + b^2 = r^2$.

25. If P is a point on the circumference of a circle, then the line segments joining P to the extremities of a diameter are perpendicular. [*Hint.* Choose the center at the origin and the diameter along one axis and use the result of Exercise 24.]

1.5

RELATIONS AND FUNCTIONS

The concepts of **relations** and **functions**, which we introduce in this section, pervade all of mathematics. They are perhaps the most fundamental ideas in the many branches of mathematics. Indeed, the reader has already encountered these notions in algebra and in trigonometry. Nonetheless, the concepts will be introduced now, because they are central to much of the remainder of this book.

DEFINITION 1.6 ● *A **relation** is a set of ordered pairs of numbers. The set of all first elements that occur in a relation is the **domain** of the relation, and the set of all second elements is the **range** of the relation.*

EXAMPLE 1 ● The set of ordered pairs

$$R = \{(-5, -5), (-4, 2), (-2, -2), (0, 1), (0, -3), (2, -4), (3, 4)\}$$

defines a relation with domain $\{-5, -4, -2, 0, 2, 3\}$ and range $\{-5, -4, -3, -2, 1, 2, 4\}$. This relation R does not exhibit an apparent connection between the elements of the ordered pairs; thus a listing of the pairs is the best way to present the relation. ●

It often happens that there is a specified "relationship" between the elements of the ordered pairs of a relation. For instance, the second element may always be twice the first element. When we have a rule or a recipe to show how the elements of the ordered pairs are related, we need not resort to a listing of the pairs as we did in Example 1. We may describe the relation by using the rule.

EXAMPLE 2 ● The relation S, whose domain is the set of real numbers and which has the property that each ordered pair is of the form $(x, 2x)$ for some real number x, has infinitely many ordered pairs. It can be denoted by the rule $y = 2x$. ●

DEFINITION 1.7 ● *A **function** is a relation in which no two ordered pairs have the same first element and distinct second elements.*

If a relation is a function, then for each member of the domain there corresponds one and only one member of the range. A function, then, is a particular type of relation. The relation R of Example 1 is not a function because $(0, 1)$ and $(0, -3)$ are in R, that is, the number 0 in the domain is related to distinct numbers 1 and -3 in the range. The relation S of Example 2 is a function.

When a function is to be given a name, say f, it is customary to write $y = f(x)$ to specify the functional relationship. The term "$f(x)$" is to be read "f of x" and it signifies the point in the range with which the member x of the domain is associated by the function f. If S is the function of Example 2, then we may specify the function by $y = S(x) = 2x$. Then $S(6) = 12$, $S(-\sqrt{2}) = -2\sqrt{2}$, and $S(\pi) = 2\pi$.

EXAMPLE 3 ● Let T be the relation whose domain is the set of all real numbers and with the property that (x, y) is in T provided that

$$y = |x|.$$

Is T a function? What is the range of T?

SOLUTION. We see that $(2, 2)$ and $(-2, 2)$ are both in T, but that this does *not* contradict the definition of function. The issue is this: Are there two ordered pairs in T with the same first element and different second elements? Can a real number have two *distinct* absolute values? The answer, according to Definition 1.1, is "No." The relation T is a function. The range of T is the set of all results we can get by taking the absolute value of each real number. Hence the range of T is the set of all nonnegative real numbers. ●

A relation or a function may be completely determined by specifying a domain and by an equation relating the elements of the ordered pairs. It frequently happens that an equation is given without specifying a domain. When this occurs, it is understood that the domain is to consist of the largest set of real numbers x for which the equation yields real numbers $y = f(x)$.

EXAMPLE 4 ● Find the domain and range of the function specified by $y = 1/x$.

SOLUTION. For any nonzero real number x the equation yields a real number $y \neq 0$. If either x or y is 0, the equation is a false statement; hence the domain is the set of all nonzero real numbers, as is the range. ●

EXAMPLE 5 ● Find the domain and range of $y = \sqrt{9 - x^2}$.

SOLUTION. The domain consists of all real numbers x for which $9 - x^2 \geq 0$, for if the radicand is negative, y will not be a real number. From algebra we know that $9 \geq x^2$ if and only if $-3 \leq x \leq 3$. Furthermore, as x is allowed to vary from -3

to 3, y varies from 0 to 3 and back to 0. The range is the set of numbers from 0 to 3, inclusive. ●

The functions sine and cosine of trigonometry each have as their domain the set of real numbers and as their range the set of real numbers from -1 to 1, inclusive. We shall encounter the trigonometric functions again later in the book.

It may happen that a function is best defined by different recipes on different intervals. For instance, we might have

$$f(x) = \begin{cases} x & \text{if} & 0 \le x < 1, \\ x^2 & \text{if} & 1 \le x < 3, \\ 2x + 3 & \text{if} & 3 \le x. \end{cases}$$

Then $f(\frac{1}{2}) = \frac{1}{2}$, $f(2) = 4$, and $f(12) = 27$.

Graphs of Relations and Functions

Earlier in the chapter we alluded to the fact that analytic geometry is the marriage of algebra and geometry. Indeed, we are now in a position to see some of the power of analytic geometry. The strength of the discipline is this: It gives us a way to visualize algebraic expressions and, on the other hand, it gives us ways to associate with a geometric figure an algebraic expression. We may be able, then, to manipulate the algebraic expression in order to learn more about the geometric figure.

Likewise, given an algebraic expression, we may associate a geometric figure with it so that examination of the figure reveals properties of the algebraic expression that were not previously apparent.

We turn our attention in this section to the problem of associating a geometric figure with an algebraic or analytic expression (or equation).

DEFINITION 1.8 ● *The **graph** of a relation or function consists of all points in the coordinate plane whose coordinates satisfy the relation or function. If the relation or function is defined by an equation, the graph of the equation is the same as the graph of the relation or function.*

The graph of the relation R from Example 1 of this section is Fig. 1.6 in Section 1.1. It is a discrete graph, consisting of only seven points. The graphs of the functions in Examples 2 through 4 are left as exercises. The graph of $y = \sqrt{9 - x^2}$ of Example 5 will be an exercise in Section 2 of Chapter 5.

EXAMPLE 6 ● Draw the graph of the function defined by the equation

$$2x + 3y = 6.$$

SOLUTION. To draw the graph we assign values to x and find the corresponding values of y. The resulting ordered pairs are shown in the following table. We plot

x	-3	-1.5	0	3	4.5	6
y	4	3	2	0	-1	-2

each of the pairs as the abscissa and ordinate of a point. The points thus obtained appear to lie on a straight line (Fig. 1.31). The variables x and y are of the first degree in the given equation and therefore the equation is said to be **linear**. In the next chapter we shall prove that the graph of a linear equation in two variables is a straight line.

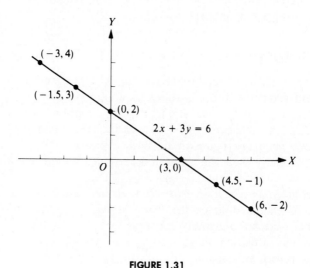

FIGURE 1.31

EXAMPLE 7 ● Construct the graph of the equation

$$y = x^2 - 3x - 3.$$

SOLUTION. Any pair of numbers for x and y that satisfy the equation is called a **solution** of the equation. If a value is assigned to x, the corresponding value of y may be computed. Thus setting $x = -2$, we find $y = 7$. Several values of x and the corresponding values of y are shown in the table. These pairs of values, each constituting a solution, furnish a picture of the relation of x and y. A better representation is had, however, by plotting each value of x and the corresponding value of y as the abscissa and ordinate of a point and then drawing a smooth curve through the points thus obtained. This process is called **graphing the**

equation, and the curve is called the **graph** of the equation.

x	-2	-1	0	1	1.5	2	3	4	5
y	7	1	-3	-5	-5.25	-5	-3	1	7

The plotted points (Fig. 1.32) extend from $x = -2$ to $x = 5$. Points corresponding to smaller and larger values of x could be plotted, and also any number of intermediate points could be located. But the plotted points show approximately where the intermediate points would be. Hence we can use a few points to draw a curve that is reasonably accurate. The curve shown here is called a **parabola**. We can, of course, draw only a part of the graph, since the complete graph extends indefinitely far into the first and second quadrants. ●

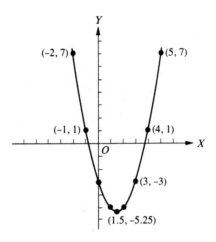

FIGURE 1.32

EXAMPLE 8 ● Construct the graph of the relation defined by the equation

$$4x^2 + 9y^2 = 36.$$

SOLUTION. We solve the equation for y to obtain a suitable form for making a table of values. Thus, we get

$$y = \pm \tfrac{2}{3}\sqrt{9 - x^2}.$$

We now see that x can take values only from -3 to 3; other values for x would yield imaginary values for y. The number pairs in the following table yield points of the graph. The curve drawn through the points (Fig. 1.33) is called an **ellipse**.

x	-3	-2	-1	0	1	2	3
y	0	± 1.5	± 1.9	± 2	± 1.9	± 1.5	0

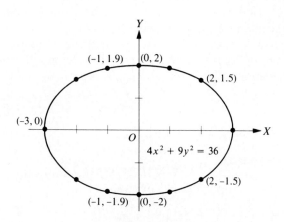

FIGURE 1.33

Observe the graphs in Fig. 1.6 and in Figs. 1.31 through 1.33. We see that it is relatively easy to tell from a graph whether a relation is a function by the test: *If any vertical line crosses or touches the graph of a relation in more than one point, then the relation is not a function.* For then we would have two points (x, y) and (x, z) on the graph, with $y \neq z$. The relations that are graphed in Figs. 1.6 and 1.33 are not functions, while the relations graphed in Figs. 1.31 and 1.32 are readily seen to be functions.

DEFINITION 1.9 ● *The abscissa of a point where a curve touches or crosses the x axis is called an **x intercept**, and the ordinate of a point where a curve touches or crosses the y axis is called a **y intercept**.*

To find the x intercepts, if any, of the graph of an equation, we set $y = 0$ and solve the resulting equation for x. Similarly, we set $x = 0$ and solve for y to find the y intercepts. Thus, the x intercepts of the equation

$$y + x^2 - 2x - 3 = 0$$

are -1 and 3, and the y intercept is 3. The intercepts will be helpful in drawing the graphs of the equations in the following exercises.

Exercises

1. Does the relation $\{(1,5), (2,5), (3,5)\}$ form a function? Why?

2. Does the relation $\{(1,2), (2,3), (1,1)\}$ form a function? Why?

Plot a few points and draw the graph of each equation in Exercises 3 through 16. Use a hand calculator or Table I of the Appendix to find square roots. Decide which represent functions.

3. $y = 2x$

4. $y = -3x$

5. $y = f(x) = \begin{cases} x & \text{if} & 0 \le x < 1 \\ x^2 & \text{if} & 1 \le x < 3 \\ 2x + 3 & \text{if} & 3 \le x \end{cases}$

6. $3x - 5y = 15$

7. $y = |x|$

8. $y = x^2$

9. $y = x^2 - 4x + 2$

10. $x = y^2$

11. $y = \begin{cases} 1 & \text{if} & 0 \le x \le 1 \\ 2 & \text{if} & 1 < x \le 2 \\ 3 & \text{if} & 2 < x \le 3 \end{cases}$

12. $y = |x - 1|$

13. $y = 1/x$

14. $x^2 + y^2 = 1$

15. $y = -\sqrt{x}$

16. $2.158x - 3.804y = 7.116$

17. Is the relation defined by the following equation a function? That is, is the set of all pairs (x, y) a function if x and y are related by the equation?

a) $x = y^2$
b) $y = x^2$
c) $y = \begin{cases} x & \text{if} & x \ge 0 \\ 0 & \text{if} & x < 0 \end{cases}$

d) $y = \pm\sqrt{x}$
e) $y = \pm|x|$

18. If f is a function with all real numbers in its domain, and if $f(x) = x^2 + 1$, what is $f(-1)$, $f(1 + h)$, $f(1 - h)$, $f(0)$? Is there an x for which $f(x) = 0$?

19. Recent research in sociology describes the relationship between the age x one first marries and the years y of education the person completed, by a model of the form

$$y = \begin{cases} ax + b, & \text{if} & 14 \le x \le 22 \\ c, & \text{if} & 22 < x, \end{cases}$$

where the parameters a, b, and c are constants to be found empirically. Graph the particular model (equation)

$$y = \begin{cases} 1 + 1/2x & \text{if} & 14 \le x \le 22 \\ 12, & \text{if} & x > 22. \end{cases}$$

1.6

THE EQUATION OF A GRAPH

Having obtained graphs of equations, we naturally surmise that a graph may have a corresponding equation. We shall consider the problem of writing the equation of a graph all of whose points are definitely fixed by given geometric conditions. This problem is the inverse of drawing the graph of an equation.

DEFINITION 1.10 ● *An equation in x and y that is satisfied by the coordinates of all points of a graph and only those points is said to be an equation of the graph.**

The procedure for finding the equation of a graph is straightforward. Each point $P(x, y)$ of the graph must satisfy the specified conditions. The desired equation can be written by requiring P to obey the conditions. The following examples illustrate the method.

EXAMPLE 1 ● A line passes through the point $(-3, 1)$ with slope 3/2. Find the equation of the line.

SOLUTION. We first draw the line through $(-3, 1)$ with the given slope. Then we apply the formula for the slope of a line through two points (Section 1.2). Thus the slope m through $P(x, y)$ and $(-3, 1)$ is

$$m = \frac{y - 1}{x - (-3)} = \frac{y - 1}{x + 3}.$$

We equate this expression to the given slope. Hence

$$\frac{y - 1}{x + 3} = \frac{3}{2},$$

or, simplifying,

$$3x - 2y + 11 = 0.$$

The graph of this equation is the line in Fig. 1.34. ●

EXAMPLE 2 ● Find the equation of the line that passes through the point $(5, -2)$ with slope $-4/3$.

SOLUTION. We now have

$$m = \frac{y - (-2)}{x - 5} = \frac{y + 2}{x - 5}.$$

---

*A graph may be represented by more than one equation. For instance, the graph of $(x^2 + 1)(x + y) = 0$ and $x + y = 0$ is the same straight line. Sometimes, however, we shall speak of "the" equation when really we mean the simplest obtainable equation.

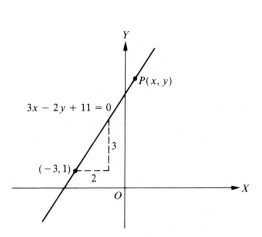

FIGURE 1.34

Hence

$$\frac{y + 2}{x - 5} = -\frac{4}{3}.$$

Simplifying this equation, we obtain the desired equation

$$4x + 3y - 14 = 0.$$

The graph of this equation is in Fig. 1.35. ●

FIGURE 1.35

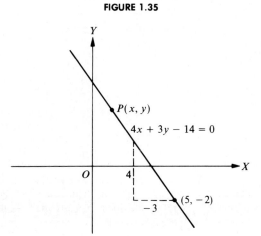

EXAMPLE 3 • Find the equation of the set of all points equally distant from the y axis and $(4, 0)$.

FIGURE 1.36

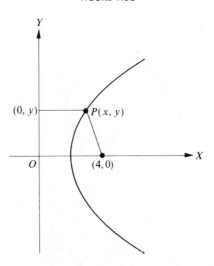

SOLUTION. We take a point $P(x, y)$ of the graph (Fig. 1.36). Then, referring to the distance formula (Section 1.1), we find the distance of P from the y axis to be the abscissa x, and the distance from the point $(4, 0)$ to be

$$\sqrt{(x - 4)^2 + y^2}.$$

Equating the two distances, we obtain

$$\sqrt{(x - 4)^2 + y^2} = x.$$

By squaring both sides and simplifying, we get

$$y^2 - 8x + 16 = 0. \quad \bullet$$

EXAMPLE 4 • Find the equation of the set of all points that are twice as far from $(4, 4)$ as from $(1, 1)$.

SOLUTION. We apply the distance formula to find the distance of a point $P(x, y)$ from each of the given points. Thus we obtain the expressions

$$\sqrt{(x - 1)^2 + (y - 1)^2} \quad \text{and} \quad \sqrt{(x - 4)^2 + (y - 4)^2}.$$

Since the second distance is twice the first, we have the equation

$$2\sqrt{(x-1)^2+(y-1)^2} = \sqrt{(x-4)^2+(y-4)^2}.$$

Simplifying, we get

$$4(x^2-2x+1+y^2-2y+1) = x^2-8x+16+y^2-8y+16$$

or

$$x^2+y^2 = 8.$$

The graph of the equation appears in Fig. 1.37. ●

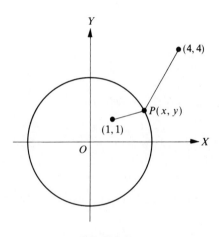

FIGURE 1.37

EXAMPLE 5 ● Find the equation of the set of all points $P(x, y)$ such that the sum of the distances of P from $(-5, 0)$ and $(5, 0)$ is equal to 14.

SOLUTION. Referring to Fig. 1.38, we get the equation

$$\sqrt{(x+5)^2+y^2} + \sqrt{(x-5)^2+y^2} = 14.$$

By transposing the second radical, squaring, and simplifying, we obtain the equation

$$7\sqrt{(x-5)^2+y^2} = 49 - 5x.$$

Squaring again and simplifying, we have the equation

$$24x^2+49y^2 = 1176.$$

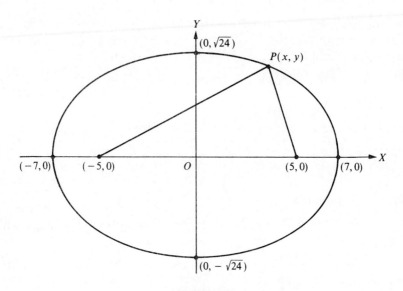

FIGURE 1.38

As shown in the figure, the x intercepts of the graph of this equation are $(-7, 0)$ and $(7, 0)$, and the y intercepts are $(0, -\sqrt{24})$ and $(0, \sqrt{24})$. ●

Exercises

In each of Exercises 1 through 10, draw the line that satisfies the given conditions. Then find the equation of the line.

1. The line passing through $(4, 2)$ with slope 1.

2. The line passing through the origin with slope -2.

3. The line passing through $(-1, 2)$ with slope $\frac{1}{2}$.

4. The line passing through $(5, 7)$ with slope $-\frac{3}{2}$.

ⓒ 5. The line passing through $(-1.8059, 2.1643)$ with slope -3.1786.

6. The horizontal line passing through $(-2, 4)$.

7. The vertical line passing through $(3, -1)$.

8. The line 2 units above the x axis.

9. The line passing through $(2, -3)$ with slope 0.

10. The line 4 units to the left of the y axis.

In each of Exercises 11 through 26, find the equation of the set of all points $P(x, y)$ that satisfy the given conditions. From the equation sketch the graph, if the instructor so requests.

11. $P(x, y)$ is equidistant from $(-2, 4)$ and $(1, -5)$.

12. $P(x, y)$ is equidistant from $(-3, 0)$ and $(3, -5)$.

13. $P(x, y)$ is equidistant from the y axis and $(4, 0)$.

14. $P(x, y)$ is equidistant from $(4, 0)$ and the line $x = -4$.

15. $P(x, y)$ is twice as far from $(4, -4)$ as from $(1, -1)$.

16. $P(x, y)$ is twice as far from $(-8, 8)$ as from $(-2, 2)$.

17. $P(x, y)$ forms with $(0, 3)$ and $(0, -3)$ the vertices of a right triangle with P the vertex of the right angle.

18. $P(x, y)$ forms with $(4, 0)$ and $(-4, 0)$ the vertices of a right triangle with P the vertex of the right angle.

19. The sum of the distances of $P(x, y)$ from $(-4, 0)$ and $(4, 0)$ is equal to 12.

20. The sum of the distances of $P(x, y)$ from $(0, -3)$ and $(0, 3)$ is equal to 10.

21. The difference of the distances of $P(x, y)$ from $(-3, 0)$ and $(3, 0)$ is 2.

22. The difference of the distances of $P(x, y)$ from $(0, -3)$ and $(0, 3)$ is 1.

23. The distance of $P(x, y)$ from $(3, 4)$ is 5.

24. The sum of the squares of the distances of $P(x, y)$ from $(0, 3)$ and $(0, -3)$ is 50.

25. The product of the distances of $P(x, y)$ from the coordinate axes is 5.

26. The product of the distances of $P(x, y)$ from $(0, 4)$ and the x axis is 4.

REVIEW EXERCISES

1. Define the following terms: Directed line, ordered pair, inclination of a line, slope of a line, relation, function, graph of a function, real number line.

2. The points $A(1, 2)$, $B(4, 3)$, and $C(6, 0)$ are vertices of a triangle. Find the lengths of the sides of the triangle.

3. Show that the points $A(-10, 2)$, $B(4, -2)$, $C(16, 2)$, and $D(2, 6)$ are the vertices of the parallelogram $ABCD$.

4. Find the tangents of the angles of the triangle whose vertices are $A(-2,1)$, $B(1,3)$, $C(6, -7)$. Find also each angle to the nearest degree.

5. The points $A(-4,1)$ and $B(2,7)$ determine a line segment. Find (a) the coordinates of the midpoint of the segment, and (b) the coordinates of the point $\frac{1}{3}$ of the way from A to B.

6. A lines passes through $A(2, -2)$ and $B(-4,3)$. Find the coordinates of the point on the line twice as far from A as from B (two cases).

7. Find the equation of the set of all points $P(x, y)$ that are equidistant from $A(-4,3)$ and $B(2, -2)$. Draw the graph of the equation.

8. Construct the graph of the equation $y = x^2 - 4$.

9. Find the equation of the set of all points $P(x, y)$ if the sum of the distances of P from $(-4,0)$ and $(4,0)$ is equal to 12. Draw the graph of the equation.

2

The
Straight Line
and the Circle

2.1

LINES AND FIRST-DEGREE EQUATIONS

We begin our program of discovering the correspondence between geometric curves and equations with the simplest case, the straight line. The equation of a line is, as we shall see, also rather simple. We will find that as the complexity of the geometric curve increases, so will the complexity of the analytic equation. Despite its simplicity, the line is a vital concept of mathematics and enters into our daily experiences in numerous interesting and useful ways. In our study of the straight line, we shall first discover a correspondence between a line and a first-degree equation in x and y. The equation will represent the line, and the line will be the graph of the equation. With respect to this relationship, we state the following theorem.

THEOREM 2.1 ● *The equation of every straight line is expressible in terms of the first degree. Conversely, the graph of a first-degree equation is a straight line.*

PROOF. Suppose we start with a fixed line in the coordinate plane. The line may be parallel to the y axis or not parallel to the y axis. First, let us consider a line L parallel to the y axis and at a distance a from the axis (Fig. 2.1). The abscissas of all points on the line are equal to a. So we see at once that an equation of the line is

$$\boxed{x = a.} \tag{2.1}$$

Conversely, the equation $x = a$ is satisfied by the coordinates of every point on the line at a distance a from the y axis. Hence the line L is the graph of the equation.

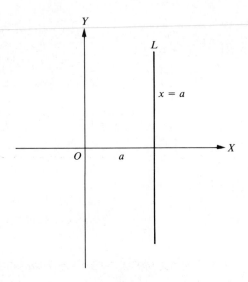

FIGURE 2.1

Next, we consider a line not parallel to the y axis (Fig. 2.2). Such a line has a slope, and the line intersects the y axis at a point whose abscissa is zero. We let m stand for the slope and b for the ordinate of the intersection point. Then if (x, y) are the coordinates of any other point of the line, we apply the formula for the slope of a line through two points and have the equation

$$\frac{y - b}{x - 0} = m$$

which reduces to

$$\boxed{y = mx + b.} \tag{2.2}$$

FIGURE 2.2

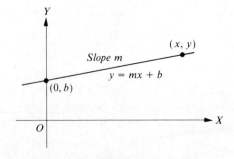

This equation makes evident the slope and y intercept of the line it represents, and is said to be in the **slope–intercept form**.

We started with a fixed line and obtained Eq. (2.2). Now suppose we start with the equation and determine its graph. Clearly the equation is satisfied by $x = 0$ and $y = b$. Let (x, y) be a point other than $(0, b)$ on the graph. This means that x and y satisfy Eq. (2.2) and consequently the equivalent equation.

$$\frac{y - b}{x - 0} = m.$$

This equation tells us that the point (x, y) of the graph must be on the line through $(0, b)$ with slope m. Hence any point whose coordinates satisfy Eq. (2.2) is on the line.

We now have shown that the equations of all lines can be expressed in the forms (2.1) and (2.2) and, conversely, that the graphs of such equations are straight lines. To complete the proof of the theorem, we need only to point out that the general first-degree equation

$$Ax + By + C = 0 \tag{2.3}$$

can be put in the form (2.1) or (2.2). We assume that A, B, and C are constants with A and B not both zero. Accordingly, if $B = 0$, then $A \neq 0$, and Eq. (2.2) reduces to

$$x = -\frac{C}{A}. \tag{2.4}$$

Next, if $B \neq 0$, we can solve Eq. (2.3) for y. Thus, we get

$$y = -\frac{A}{B}x - \frac{C}{B}. \tag{2.5}$$

Equation (2.4) is in the form of Eq. (2.1) with $a = -(C/A)$, and Eq. (2.5) is in the form of Eq. (2.2) with $m = -(A/B)$ and $b = -(C/B)$. This completes the proof of the theorem.

We note that Eq. (2.3), with $B \neq 0$, yields for each value of x a unique corresponding value for y. This means that the equation defines a function (Definition 1.7) whose graph is a straight line not parallel to the y axis.

EXAMPLE 1 ● Write the equation of the line with slope -3 and y intercept 4.

SOLUTION. In the slope–intercept form (2.2), we substitute -3 for m and 4 for b. This gives at once the equation $y = -3x + 4$. ●

EXAMPLE 2 ● Express the equation $4x - 3y - 11 = 0$ in the slope–intercept form.

SOLUTION. The given equation is in the general form (2.3) and, when solved for y, reduces to an equation whose coefficient of x is the slope of the graph and whose constant term is the y intercept. So we solve for y and obtain the desired equation

$$y = \tfrac{4}{3}x - \tfrac{11}{3}. \ \bullet$$

The slope–intercept form (2.2) represents a line that passes through the point $(0, b)$. The equation can be altered slightly to focus the attention on any other point of the line. If (x_1, y_1) is another point of the line, we can replace x by x_1 and y by y_1 in the form (2.2). This gives

$$y_1 = mx_1 + b \qquad \text{or} \qquad b = y_1 - mx_1.$$

Substituting $y_1 - mx_1$ for b in Eq. (2.2), we have

$$y = mx + y_1 - mx_1$$

or, equivalently,

$$\boxed{y - y_1 = m(x - x_1).} \tag{2.6}$$

This equation tells us that the point (x_1, y_1) is on the line and that the slope is m. Accordingly, the equation is said to be in the **point–slope form**.

EXAMPLE 3 ● Find the equation of the line that passes through $(2, -3)$ and has a slope of 5.

SOLUTION. Using the point–slope form (2.6) with $x_1 = 2$, $y_1 = -3$, and $m = 5$, we get

$$y + 3 = 5(x - 2) \qquad \text{or} \qquad 5x - y = 13.$$

We recall from geometry that two points determine a line. Suppose then that we are given two points (x_1, y_1) and (x_2, y_2) and wish to determine the equation of the line through them. We may use the point–slope form of the line with either of the points (x_1, y_1) or (x_2, y_2) as the chosen point. By Theorem 1.3, the slope of the line must be

$$m = \frac{y_2 - y_1}{x_2 - x_1}.$$

Consequently the equation of the line through (x_1, y_1) and (x_2, y_2) is

$$\boxed{y - y_1 = \left(\frac{y_2 - y_1}{x_2 - x_1} \right)(x - x_1)} \ \bullet \tag{2.7}$$

Equation 2.7 is the **two-point form** of a straight line.

EXAMPLE 4 ● Find the equation of the line determined by the points $(3, -3)$ and $(2, 4)$.

SOLUTION. First, we find the slope m of the line through the given points. Thus by Theorem 1.3, we have

$$m = \frac{y_2 - y_1}{x_2 - x_1} = \frac{4 + 3}{2 - 3} = -7.$$

Now we use the point–slope formula (2.6) with $y_1 = -3$ and $x_1 = 3$. Thus we get

$$y + 3 = -7(x - 3) \qquad \text{or} \qquad 7x + y - 18 = 0. \ \bullet$$

We could also arrive at this solution by substituting ($y_1 = 4$ and $x_1 = 2$) in formula (2.7).

EXAMPLE 5 ● Suppose a manufacturer finds that the total cost to manufacture 1000 units of his product is $8500, while the total cost to manufacture 2000 units is $11,500. He assumes that this relationship between cost and number of units made is linear. Find the relationship, graph the equation, and interpret the graph. What is the total cost to produce 2500 units?

SOLUTION. We use the two-point form of the line with x equal to the number of units made and y equal to the total cost to manufacture x units. The line passes through (1000, 8500) and (2000, 11,500) and has the equation

$$y - 8500 = 3(x - 1000)$$

or

$$3x - y = -5500.$$

If $x = 2500$, then $y = 13000$, so the total cost is $13,000 to make 2500 units. The graph is given in Fig. 2.3.

FIGURE 2.3

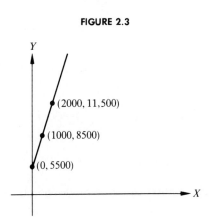

The y intercept is 5500; this means that, if no units are manufactured, the fixed cost (rent, equipment, etc.) is \$5500. The slope is 3, which means that it costs \$3 to make each unit. Note that negative values of x and y are meaningless in this application. ●

Exercises

By inspection, give the slope and intercepts of each line represented by the equations in Exercises 1 through 15. Reduce each equation to the slope–intercept form.

1. $x - 4y = 8$ 2. $x + y = 6$ 3. $x - y + 2 = 0$

4. $9x + 4y = 36$ 5. $4x - 3y = 12$ 6. $3x - 6y + 10 = 0$

7. $7x + y = 11$ 8. $3x + 2y = 14$ 9. $3x + 7y - 6 = 0$

10. $8x - 2y = 5$ 11. $8x - 3y = 4$ 12. $5x + 5y - 1 = 0$

13. $6x + 7y = 12$ ⓒ14. $4.0213x - 8.7631y = 11.3331$ 15. $3x + 3y + 4 = 0$

In each of Exercises 16 through 29, write the equation of the line determined by the slope m and y intercept b.

16. $m = 2$, $b = -3$ 17. $m = -4$, $b = 7$

18. $m = 5$, $b = 0$ 19. $m = -\frac{2}{3}$, $b = 2$

20. $m = \frac{3}{2}$, $b = -4$ 21. $m = 0$, $b = 11$

22. $m = \frac{4}{3}$, $b = 3$ 23. $m = 7$, $b = 0$

24. $m = -3$, $b = -4$ 25. $m = 9$, $b = 11$

26. $m = -8$, $b = 7$ 27. $m = -\frac{3}{4}$, $b = 5$

28. $m = \frac{5}{3}$, $b = -4$ ⓒ 29. $m = 7.0132$, $b = -4.9875$

In each of Exercises 30 through 39, find the equation of the line that passes through A with slope m. Draw the lines.

30. $A(1, 3)$, $m = 2$ 31. $A(-5, 3)$, $m = 1$

32. $A(0, -2)$, $m = 0$ 33. $A(-3, 0)$, $m = \frac{3}{2}$

34. $A(-6, -3)$, $m = 1$ 35. $A(-2, 5)$, $m = -\frac{1}{2}$

36. $A(4, 0)$, $m = -3$ 37. $A(0, 0)$, $m = \frac{4}{3}$

ⓒ 38. $A(0.01157, 6.23981)$, $m = -4.90032$ 39. $A(-3, 6)$, $m = 2$

Find the equation of the line passing through the points A and B in each of the following exercises.

40. $A(-1,3)$, $B(5,-4)$

41. $A(5,1)$, $B(1,4)$

42. $A(2,0)$, $B(-6,4)$

43. $A(-2,3)$, $B(7,4)$

44. $A(0,0)$, $B(-4,3)$

45. $A(5,\frac{1}{2})$, $B(-1,\frac{3}{4})$

46. $A(-1,-6)$, $B(-1,4)$

47. $A(\frac{1}{3},4)$, $B(0,-\frac{2}{3})$

48. $A(0,0)$, $B(3,1)$

49. $A(4,1)$, $B(-2,-5)$

Ⓒ 50. $A(11.211567,-3.70014)$, $B(2.59172,13.00154)$ 51. $A(1,1)$, $B(4,3)$

52. On the Fahrenheit scale F, water freezes at $32°$ and boils at $212°$. On the *Celsius* scale C, water freezes at $0°$ and boils at $100°$. So 100 *Celsius* degrees is equal to 180 Fahrenheit degrees. Now suppose we have a reading of C degrees on the *Celsius* scale and wish to find the corresponding reading F on the Fahrenheit scale. The reading C is measured from the ice point. And $F - 32$ is the number of Fahrenheit degrees measured from the ice point. Hence, we have the proportion $C:(F-32) = 100:180$.

   In a proportion, the product of the extremes is equal to the product of the means. Hence

$$180C = 100(F-32),$$
$$C = \tfrac{5}{9}(F-32),$$

and solving for F, we get

$$F = \tfrac{9}{5}C + 32.$$

Graph this expression, with the Fahrenheit axis horizontal and the Celsius axis vertical.

53. Reduce (a) $68°F$ to a Celsius reading, (b) $70°C$ to a Fahrenheit reading.

54. At what temperature will C and F readings be the same?

55. Let R be the electrical resistance in ohms of a piece of copper wire of fixed diameter and length at a temperature of $T°C$. If $R = 0.0170$ ohm when $T = 0°$ and $R = 0.0245$ ohm when $T = 100°$ and if the relationship between R and T is linear, obtain the equation expressing the relationship between R and T.

56. If the curved surface of a uniform rod is perfectly insulated (no escape of heat) and its ends are kept at constant temperature, then the temperature at any point within the rod is a linear function of the distance from one end. Suppose one end of such a rod, 8 ft long, is kept at $20°C$ while the other end is kept at $140°C$. Determine the equation for the temperature in terms of the distance from the $20°$ end. Now write the equation for the temperature in terms of the distance from the $140°$ end. Do the results agree?

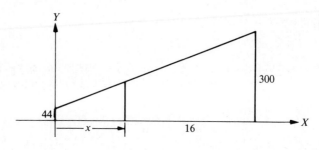

FIGURE 2.4

57. A horizontal beam weighs 80 lbs/ft of length. A 16-ft beam supports a load of dirt distributed as indicated in Fig. 2.4; that is, starting at 44 lbs/ft at the left end and increasing to 300 lbs/ft at the right end.

a) Obtain an equation for the weight of the beam alone for the first x ft from the left end.
b) Find an equation of the straight line along the top of the dirt.
c) Find the weight of the dirt above the first x ft of the beam and the total weight of both beam and dirt for the first 10 ft.

58. A manufacturer finds that it cost $2,790 to make 2000 units of his product each month, while his fixed costs are an additional $2500 per month. (Total cost of 2000 units is $5290.) Assume a linear relationship to find the variable cost per unit to make his product. What is the total cost to make 1000 units?

59. (For students who know determinants.) Show that the equation of the line through (x_1, y_1) and (x_2, y_2) is

$$\begin{vmatrix} x & y & 1 \\ x_1 & y_1 & 1 \\ x_2 & y_2 & 1 \end{vmatrix} = 0.$$

2.2

OTHER FORMS OF FIRST-DEGREE EQUATIONS

The equation of a straight line may be expressed in several different forms such that each form exhibits certain geometric properties of the line. As we have seen, the slope–intercept form brings into focus the slope and y intercept and the point–slope form brings into focus the slope and a point of the line. In this section we shall derive two additional forms, each having its advantages in certain situations.

We first introduce an alternative form for the point–slope equation. To obtain this form, we substitute $-(A/B)$ for m in Eq. (2.6) and have

$$y - y_1 = -\frac{A}{B}(x - x_1).$$

Multiplying by B and transposing terms yields

$$\boxed{Ax + By = Ax_1 + By_1.}$$ (2.8)

EXAMPLE 1 ● Find the equation of the line of slope $2/5$ and passing through the point $(-3, 4)$.

SOLUTION. Since the slope, $-(A/B)$, is equal to $2/5$, we let $A = 2$ and $B = -5$. These values along with $x_1 = -3$ and $y_1 = 4$ may be substituted in Eq. (2.8). Thus we have

$$2x - 5y = 2(-3) - 5(4) = -26$$

or, as a final result,

$$2x - 5y + 26 = 0. \; ●$$

Note that this equation is more directly obtainable by formula (2.8) than by the point–slope formula (2.6).

Suppose the x intercept of a line is a and the y intercept is b, where $a \neq 0$ and $b \neq 0$ (Fig. 2.5). Then the line passes through the point $(a, 0)$ and $(0, b)$, and consequently the slope is $-(b/a)$. Then, applying the point–slope formula, we

FIGURE 2.5

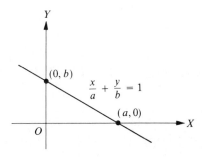

find the equation

$$y - b = -\frac{b}{a}(x - 0),$$

which can be reduced to

$$\boxed{\frac{x}{a} + \frac{y}{b} = 1.}$$

(2.9)

This is called the **intercept form** because the intercepts are exhibited in the denominators.

EXAMPLE 2 ● Write the equation of the line whose x intercept is 3 and whose y intercept is -5. We replace a by 3 and b by -5 in formula (2.9) and get at once

$$\frac{x}{3} + \frac{y}{-5} = 1 \qquad \text{or} \qquad 5x - 3y = 15. ●$$

EXAMPLE 3 ● Write the equation $4x - 9y = -36$ in the intercept form.

SOLUTION. In order to have an equation with the right member equal to unity, as in formula (2.9), we divide the members of the given equation by -36. This gives the desired form

$$\frac{x}{-9} + \frac{y}{4} = 1. ●$$

We could, of course, obtain this result by first finding the intercepts from the given equation and then substituting in formula (2.9).

EXAMPLE 4 ● Find the equation of the line through the point $(2, -3)$ and perpendicular to the line defined by the equation $4x + 5y + 7 = 0$.

SOLUTION. We let $5x - 4y$ be the left side of our equation. This will provide for the necessary slope of the perpendicular line. In order that the line shall pass through the given point, we apply formula (2.8) with $x_1 = 2$ and $y_1 = -3$. Thus we get

$$5x - 4y = 5(2) - 4(-3) \qquad \text{or} \qquad 5x - 4y = 22. ●$$

EXAMPLE 5 ● The ends of a line segment are at $C(7, -2)$ and $D(1, 6)$. Find the equation of the perpendicular bisector of the segment CD.

SOLUTION. The slope of CD is $-4/3$, and the midpoint is at $(4, 2)$. The perpendicular bisector therefore has a slope of $3/4$ and passes through $(4, 2)$. Employing formula (2.8) with $A = 3$, $B = -4$, $x_1 = 4$, and $y_1 = 2$, we obtain

$$3x - 4y = 3(4) - 4(2) \qquad \text{or} \qquad 3x - 4y = 4. ●$$

Exercises

In Exercises 1 through 9, find the equation of the line through the indicated point with slope m. Use formula (2.8).

1. $(4, 1)$, $m = 3/2$

2. $(3, -5)$, $m = -4/5$

3. $(5, 0)$, $m = -2/3$

4. $(2, 2)$, $m = 2/5$

5. $(0, 4)$, $m = -2$

6. $(-1, 2)$, $m = -3$

Ⓒ 7. $(2.1841, -3.0144)$, $m = 0.7642$

8. $(2, -3)$, $m = -3/4$,

9. $(0, 0)$, $m = 4/3$

In Exercises 10 through 15, write the equation of the line with x intercept a and y intercept b.

10. $a = 3$, $b = 2$

11. $a = 1$, $b = 4$

12. $a = 3$, $b = -4$

13. $a = 4/3$, $b = -2$

14. $a = -3/5$, $b = -2/3$

15. $a = 4/5$, $b = 3/4$

Express each of the following equations in intercept form.

16. $4x - y = 8$

17. $3x + 2y = 6$

18. $4y = 9x + 36$

19. $2x/3 + y/2 = 5/6$

20. $3x/5 - 12y/5 = 4/3$

21. $3x/4 - 5y/2 = 2/3$

In the following exercises, find the equation of two lines through the given point, one line parallel and the other perpendicular to the given line.

22. $(1, 4)$, $3x - 2y + 40 = 0$

23. $(2, -1)$, $x - 2y - 5 = 0$

24. $(3, -2)$, $x + 8y = -3$

25. $(6, 0)$, $3x - 3y = 1$

26. $(0, 2)$, $1492x + 1776y = 1985$

27. $(1, -1)$, $y = 1$

Find the equation of the perpendicular bisector of the line joining the points A and B in the following exercises.

28. $A(6, 4)$, $B(4, -2)$

29. $A(-1, -3)$, $B(3, 5)$

30. $A(0, -4)$, $B(4, 6)$

Ⓒ 31. $A(1.784, 2.562)$, $B(-3.190, 0.016)$

32. Show that $Ax + By = 5$ and $Bx - Ay = 2$ are equations of perpendicular lines. Assume that A and B are not both equal to zero.

33. The points $A(-2, 3)$, $B(6, -5)$, $C(8, 5)$ are vertices of a triangle. Find the equations of the medians.

34. A car rents for $20 a day and 11 cents a mile. If the car is rented for one day only, write a formula for the rental charges in terms of the mileage driven. Graph the equation.

2.3

INTERSECTION OF LINES

Our main interest in this section will be in lines that intersect. In the discussion we will, for convenience, refer to a linear equation in two variables itself as a line.

EXAMPLE 1 ● Find the coordinates of the intersection point of the lines:

$$4x - 5y = 26, \tag{2.10}$$
$$3x + 7y = -2. \tag{2.11}$$

SOLUTION. If we multiply the members of equation (2.10) by 3 and those of equation (2.11) by 4, we obtain equivalent equations with the coefficients of x the same.

$$12x - 15y = 78, \tag{2.12}$$
$$12x + 28y = -8. \tag{2.13}$$

We now subtract corresponding members of equation (2.12) from equation (2.13) to eliminate x, and solve for y:

$$43y = -86,$$
$$y = -2.$$

If we substitute -2 for y in either equation (2.10) or (2.11), we find $x = 4$. Hence, the intersection of the given lines is $(4, -2)$. The graph of the lines is drawn in Fig. 2.6. ●

There are many problems in business and economics which are solved by finding the intersection of two lines. A typical example involves what is called **break-even analysis.** Suppose a business has a linear cost function that relates the cost (y dollars) of making and selling x units of its product. Then

$$y = vx + b,$$

where b is the fixed cost (rent, etc.) of doing business. Also, the revenue earned by

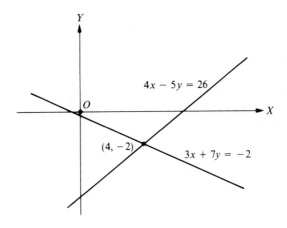

FIGURE 2.6

selling x units at d dollars per unit is given by

$$y = dx.$$

The break-even point is the intersection, (x_1, y_1), of the cost line and the revenue line. The company must then make more than x_1 units to have a profit.

EXAMPLE 2 ● Suppose the Red Double T Corporation has a cost function of

$$y = 500x + 4025,$$

and a revenue function of

$$y = 675x.$$

What is the break-even point for the corporation?

SOLUTION. We solve the two linear equations simultaneously by setting the y values equal to each other, to get

$$675x = 500x + 4025,$$

or

$$175x = 4025.$$

Thus

$$x = 23.$$

The company should make 23 units of its product to break even; hence it should make more than 23 units to have a profit. ●

Exercises

In Exercises 1 through 10, find the intersection point of the pair of lines in each problem.

1. $2x - 5y = 20$, $3x + 2y = 11$

2. $2x - 3y = 6$, $x + y = 3$

3. $4x + 3y = 28$, $2x - 3y = 5$

4. $5x - 4y = 7$, $3x - 2y = 4$

5. $3x + 2y = 30$, $3x - 5y = -19$

6. $3x - 6y = 13$, $4x + 3y = -1$

7. $2x + 3y = 8$, $2x - 3y = 4$

8. $4x - 3y = 8$, $2x + 6y = -1$

9. $3x + 5y = 6$, $5x - y = 10$

10. $5x + 4y = 50$, $5x - 4y = 50$

11. The sides of a triangle are on the lines $x - 2y + 1 = 0$, $3x - y - 13 = 0$, and $2x + y - 13 = 0$. Find the vertices.

12. The sides of a triangle are on the lines $x + y + 6 = 0$, $x + 2y - 4 = 0$, and $3x + y + 2 = 0$. Find the vertices.

13. A person wants to mix super unleaded and unleaded gasoline to obtain 500 gallons of blend valued at $620. If unleaded gasoline costs $1.11 per gallon and super unleaded costs $1.25 per gallon, how many gallons of each should go into the mixture?

14. In marketing research, a demand curve relates the demand (x units) for a product to the price (y dollars) per unit that the market is willing to pay. A supply curve reflects the number of units a manufacturer is willing to supply at a given price. Demand curves have negative slopes, while supply curves have positive slopes. For a certain product, it is found that the demand curve is given by $y = 80 - 3x$, while the supply curve is given by $y = 20 + 2x$. Find the point of market equilibrium for the product (the price and quantity that satisfy both supply and demand).

15. If the demand curve for a product is given by $y = 400 - 0.01x$, while the supply curve is given by $y = 300 + 0.15x$, find the point of equilibrium for the market in the product.

16. *Linear Inequalities.* A line $Ax + By + C = 0$ separates the plane into three regions: the points on one side of the line, the points on the other side, and the points on the line itself. The points on the line satisfy its equation, while the points on one side satisfy the inequality $Ax + By + C > 0$, and those on the other satisfy $Ax + By + C < 0$. For instance, points below the line $x + y = 1$ have ordinates less than ordinates on the line. Hence the shaded region in Fig. 2.7 is described by $y < 1 - x$ or $x + y - 1 < 0$. Graph

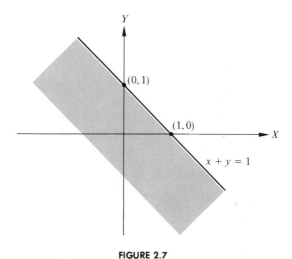

FIGURE 2.7

the region described by

a) $2x - 3y < 6$

b) $2x - 3y \geq 6$

c) $2x - 3y \leq 6$

d) $2x - 3y \leq 6$, $x + y \leq 1$, and $x > 0$.

2.4

DIRECTED DISTANCE FROM A LINE TO A POINT

The directed distance from a line to a point can be found from the equation of the line and the coordinates of the point. For a vertical line and a horizontal line, the distances are easily determined. We notice these special cases first, and then will consider a slant line.

If the point $P(x_1, y_1)$ is to the right of the vertical line $x = a$, then the directed distance from the line to the point is positive and is equal to $x_1 - a$. If $P(x_1, y_1)$ were to the left of the line $x = a$, the directed distance from the line to the point would be negative and it would again equal $x_1 - a$. We see that in each case, the directed distance from the vertical line $x = a$ to the point $P(x_1, y_1)$ is given by

$$d = x_1 - a.$$

A similar argument readily shows that the directed distance from the horizontal line $y = b$ to a point $P(x_1, y_1)$ is given by

$$d = y_1 - b.$$

The general case, that of a slant line L and a point $P_1(x_1, y_1)$ not on L, involves a somewhat more delicate argument than the special cases we have just considered. The reader is urged to follow the argument carefully, for it employs many of the ideas of earlier sections of the book together with some algebraic arguments.

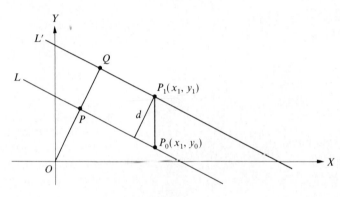

FIGURE 2.8

We next consider a slant line L and a point $P_1(x_1, y_1)$, as shown in Fig. 2.8. We seek now the perpendicular distance from L to the point P_1. The line L' passes through P_1 and is parallel to L. The line through the origin perpendicular to L intersects L and L' at P and Q. We choose the general linear equation to represent the line L. Then we may express the equations of L, L', and the perpendicular line, respectively, by

$$Ax + By + C = 0, \tag{2.14}$$
$$Ax + By + C' = 0, \tag{2.15}$$
$$Bx - Ay = 0. \tag{2.16}$$

The simultaneous solution of Eqs. (2.14) and (2.16) gives the coordinates of P, and the simultaneous solution of Eqs. (2.15) and (2.16) gives the coordinates of Q. These coordinates, as the student may verify, are

$$P\left(\frac{-AC}{A^2 + B^2}, \frac{-BC}{A^2 + B^2}\right) \quad \text{and} \quad Q\left(\frac{-AC'}{A^2 + B^2}, \frac{-BC'}{A^2 + B^2}\right).$$

We let d denote the perpendicular distance from the given line L to the point P_1. Then d is also equal to the distance from P to Q. To find the distance from P to

Q, we use the distance formula of Section 1.1. Thus, we get

$$d^2 = |PQ|^2 = \frac{(C - C')^2 A^2}{(A^2 + B^2)^2} + \frac{(C - C')^2 B^2}{(A^2 + B^2)^2}$$

$$= \frac{(C - C')^2 (A^2 + B^2)}{(A^2 + B^2)^2} = \frac{(C - C')^2}{A^2 + B^2}$$

and

$$d = \frac{C - C'}{\pm \sqrt{A^2 + B^2}}.$$

Since the line of Eq. (2.15) passes through $P_1(x_1, y_1)$, we have

$$Ax_1 + By_1 + C' = 0 \quad \text{and} \quad C' = -Ax_1 - By_1.$$

Hence, substituting for C', we get

$$d = \frac{Ax_1 + By_1 + C}{\pm \sqrt{A^2 + B^2}}. \tag{2.17}$$

To remove the ambiguity of the sign in the denominator, let us agree that d is to be positive if P_1 is above the given line L and negative if P_1 is below the line. We can achieve this end by choosing the sign of the denominator the same as the sign of B. That is, the coefficient of y_1 will be $B/\sqrt{A^2 + B^2}$ when B is positive and will be $B/-\sqrt{A^2 + B^2}$ when B is negative. Therefore, in either case, the coefficient of y_1 will be positive. Referring again to Fig. 2.8, we let $P_0(x_1, y_0)$ be a point on the line L such that $P_0 P_1$ is parallel to the y axis. Since P_0 is on the line L, we have the equation

$$\frac{Ax_1 + By_0 + C}{\pm \sqrt{A^2 + B^2}} = 0.$$

Suppose now that $P_1(x_1, y_1)$ is above the line L. Then y_1 is greater than y_0, hence we have

$$y_1 > y_0$$

and

$$\frac{B}{\pm \sqrt{A^2 + B^2}} y_1 > \frac{B}{\pm \sqrt{A^2 + B^2}} y_0, \tag{2.18}$$

because we agreed to choose the sign in the denominator so that the fraction $B/\pm \sqrt{A^2 + B^2}$ is positive. If we now add $(Ax_1 + C)/\pm \sqrt{A^2 + B^2}$ to both sides of Eq. (2.18), we find that

$$d = \frac{Ax_1 + By_1 + C}{\pm \sqrt{A^2 + B^2}} > \frac{Ax_1 + By_0 + C}{\pm \sqrt{A^2 + B^2}} = 0.$$

So if $P_1(x_1, y_1)$ is above the line L, then d is positive.

Likewise, if $P_1(x_1, y_1)$ is below the line L, then with $P_0(x_1, y_0)$ chosen as above we have

$$y_0 > y_1$$

and

$$\frac{B}{\pm\sqrt{A^2 + B^2}} y_0 > \frac{B}{\pm\sqrt{A^2 + B^2}} y_1.$$

By adding, just as we did above, we find that

$$0 = \frac{Ax_1 + By_0 + C}{\pm\sqrt{A^2 + B^2}} > \frac{Ax_1 + By_1 + C}{\pm\sqrt{A^2 + B^2}} = d.$$

Hence, if $P_1(x_1, y_1)$ is below the line L, d is negative. On the basis of this discussion we state the following theorem.

THEOREM 2.2 ● *The directed distance from the slant line $Ax + By + C = 0$ to the point $P_1(x_1, y_1)$ is given by the formula*

$$\boxed{d = \frac{Ax_1 + By_1 + C}{\pm\sqrt{A^2 + B^2}},} \qquad (2.19)$$

where the denominator is given the sign of B. The distance is positive if the point P_1 is above the line, and negative if P_1 is below the line.

EXAMPLE 1 ● Find the distance from the line $5x = 12y + 26$ to the points $P_1(3, -5)$, $P_2(-4, 1)$, and $P_3(9, 0)$.

SOLUTION. We write the equation in the form $5x - 12y - 26 = 0$, and apply formula (2.19) with the denominator negative. Then referring to Fig. 2.9, where

FIGURE 2.9

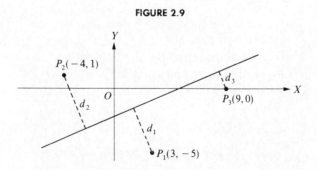

the given line and the given points are plotted, we have

$$d_1 = \frac{5(3) - 12(-5) - 26}{-\sqrt{5^2 + 12^2}} = \frac{49}{-13} = -\frac{49}{13},$$

$$d_2 = \frac{5(-4) - 12(1) - 26}{-13} = \frac{-58}{-13} = \frac{58}{13},$$

$$d_3 = \frac{5(9) - 12(0) - 26}{-13} = \frac{19}{-13} = -\frac{19}{13}. \quad \bullet$$

EXAMPLE 2 ● Find the distance between the parallel lines $15x + 8y + 68 = 0$ and $15x + 8y - 51 = 0$.

SOLUTION. The distance between the lines can be found by computing the distance from each line to a particular point. To minimize the computations, we choose the origin as the particular point. Thus,

$$d_1 = \frac{15(0) + 8(0) + 68}{\sqrt{15^2 + 8^2}} = \frac{68}{17} = 4,$$

$$d_2 = \frac{15(0) + 8(0) - 51}{17} = \frac{-51}{17} = -3.$$

These distances reveal that the origin is 4 units above the first line and 3 units below the second line. Hence the given lines are 7 units apart. ●

An alternative method for this problem would be to find the distance from one of the lines to some point on the other line. The point $(3.4, 0)$ is on the second given line. So using this point and the first equation, we find

$$d = \frac{15(3.4) + 8(0) + 68}{17} = \frac{119}{17} = 7.$$

EXAMPLE 3 ● Find the equation of the bisector of the pair of acute angles formed by the lines $x - 2y + 1 = 0$ and $x + 3y - 3 = 0$.

SOLUTION. We graph the two lines in order to determine which of the two angles is the acute angle between the lines (Fig. 2.10). From the graph we observe that a point $P(x', y')$ is on the bisector of the acute angle if it is above one line and below the other. If the directed distances d_1 and d_2 are as indicated in Fig. 2.10, then, since P is above one line and below the other, we must have $d_1 = -d_2$. By formula 2.19, we have

$$d_1 = \frac{x' + 3y' - 3}{\sqrt{10}} \qquad \text{and} \qquad d_2 = \frac{x' - 2y' + 1}{-\sqrt{5}}.$$

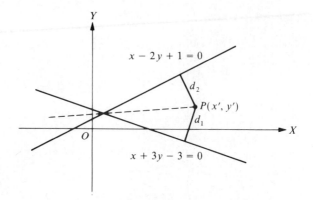

FIGURE 2.10

We equate these distances and simplify the result

$$\frac{x' + 3y' - 3}{\sqrt{10}} = \frac{x' - 2y' + 1}{\sqrt{5}},$$

$$\frac{x' + 3y' - 3}{\sqrt{2}} = x' - 2y' + 1,$$

$$x' + 3y' - 3 = \sqrt{2}\,(x' - 2y' + 1);$$

$$(1 - \sqrt{2})x' + (3 + 2\sqrt{2})\,y' = 3 + \sqrt{2}.$$

Dropping the primes we get the desired equation

$$(1 - \sqrt{2})x + (3 + 2\sqrt{2})y = 3 + \sqrt{2}. \ \bullet$$

EXAMPLE 4 • Find the equation of the bisector of the pair of obtuse angles of Example 3.

SOLUTION. The obtuse angle between the lines, as indicated in Fig. 2.11, is seen to be above both lines (or, if you will, below both lines). Thus if the point $P(x', y')$ is on the bisector of the obtuse angle and if d_1 and d_2 are the directed distances, as indicated in Fig. 2.11, then d_1 and d_2 have the same sign (both positive or both negative). So we get

$$d_1 = \frac{x' - 2y' + 1}{-\sqrt{5}} \qquad \text{and} \qquad d_2 = \frac{x' + 3y' - 3}{\sqrt{10}}.$$

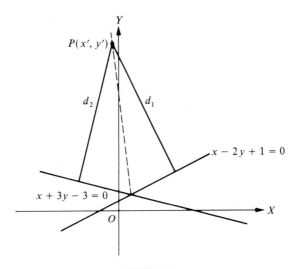

FIGURE 2.11

We equate these distances and simplify the result:

$$\frac{x' - 2y' + 1}{-\sqrt{5}} = \frac{x' + 3y' - 3}{\sqrt{10}},$$

$$\frac{x' - 2y' + 1}{-1} = \frac{x' + 3y' - 3}{\sqrt{2}},$$

$$\sqrt{2}(x' - 2y' + 1) = -x' - 3y' + 3,$$

$$(1 + \sqrt{2})x' + (3 - 2\sqrt{2})y' + \sqrt{2} - 3 = 0.$$

Dropping the primes we get the desired equation.

$$(1 + \sqrt{2})x + (3 - 2\sqrt{2})y - 3 + \sqrt{2} = 0. \ \bullet$$

Exercises

Find the directed distance from the line to the point in each of Exercises 1 through 12.

1. $5x - 12y + 3 = 0$; $(-2, 1)$ 2. $4x + 3y = 5$; $(2, -5)$

3. $12x + 5y - 6 = 0$; $(4, -6)$ 4. $3x - 4y = 12$; $(-2, -1)$

5. $x + y + 4 = 0$; $(5, 4)$ 6. $3x - y = 10$; $(0, 8)$

7. $2x + 3y - 4 = 0$; $(0, 4)$ 8. $x + 2y = 6$; $(2, 2)$

9. $x + 5y + 3 = 0$; $(3, 3)$ 10. $4x - 3y = 7$; $(0, 0)$

11. $y - 6 = 0$; $(5, 3)$ 12. $x + 3 = 0$; $(-1, -4)$

Find the distance between the two parallel lines in Exercises 13 through 25.

13. $3x - 4y - 9 = 0$
    $3x - 4y + 3 = 0$
    14. $12x + 5y = 15$
    $12x + 5y = 12$

15. $15x + 8y + 30 = 0$
    $15x + 8y + 20 = 0$
    16. $x - y + 7 = 0$
    $x - y + 11 = 0$

17. $10x + 24y = 14$
    $5x + 12y = -12$
    18. $2x + 3y - 6 = 0$
    $2x + 3y + 2 = 0$

19. $4x + y = 6$
    $12x + 3y = 14$
    20. $x + 2y + 5 = 0$
    $x + 2y - 4 = 0$

21. $x - y = 16$
    $2x - 2y = 15$
    22. $3x + 2y - 6 = 0$
    $3x + 2y - 4 = 0$

Ⓒ 23. $2.371849x - 15.913565y = 7.109436$
    $2.371849x - 15.913565y = 5.298325$

24. $2x - 5y = 4$
    $4x - 10y = 14$
    25. $3x - 2y + 7 = 0$
    $3x - 2y - 7 = 0$

26. Find the equation of the line that bisects the first quandrant.

27. A circle has its center at $(-4, 2)$ and is tangent to the line $3x + 4y - 5 = 0$. What is the radius of the circle? What is the equation of the diameter that is perpendicular to the line?

28. Find the equation of the bisector of the pair of acute angles formed by the lines $4x - 3y = 8$ and $12x - 5y = 10$.

29. Find the equation of the bisector of the acute angles and also the equation of the bisector of the obtuse angles formed by the lines $7x - 24y = 8$ and $3x + 4y = 12$.

30. Find the equation of the bisector of the acute angles and also the equation of the bisector of the obtuse angles formed by the lines $x + y = 2$ and $x + 2y = 3$.

31. Find the equation of the bisector of the acute angles and also the equation of the bisector of the obtuse angles formed by the lines $3x - y - 5 = 0$ and $3x + 4y - 12 = 0$.

32. The vertices of a triangle are at $A(2, 4)$, $B(-1, 2)$, and $C(7, 1)$. Find the length of the altitude from the vertex B and the length of side AC. Then compute the area of the triangle.

33. The vertices of a triangle are at $A(-3, -2)$, $B(2, 1)$, and $C(6, 5)$. Find the length of the altitude from the vertex C and the length of side AB. Then compute the area of the triangle.

34. The equation of a gas line is $2x + y = 2$. A factory located at the point $(6, 7)$ will connect with the gas line perpendicularly. Find the equation of the connecting line and the length of pipe required if the units are miles.

35. A cylindrical tank, 6 ft in radius, lies on its side parallel to and against the side of a warehouse. A ladder, which leans against the building, passes over and just touches the tank, and has a slope of $-\frac{3}{4}$. Find an equation for the line of the ladder and the length of the ladder. (See Fig. 2.12.)

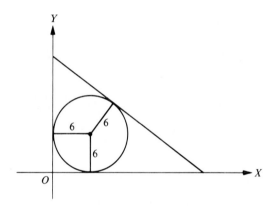

FIGURE 2.12

2.5

FAMILIES OF LINES

We have expressed equations of lines in various forms. Among these are the equations

$$y = mx + b \qquad \text{and} \qquad \frac{x}{a} + \frac{y}{b} = 1.$$

Each of these equations has two constants that have geometrical significance. The constants of the first equation are m and b. When definite values are assigned to these letters, a line is completely determined. Other values for these constants, of course, determine other lines. Thus the quantities m and b are fixed for any particular line but change from line to line. These letters are called **parameters**. In the second equation a and b would be called the parameters.

A linear equation with one parameter represents lines all with a particular property. For example, the equation $y = 3x + b$ represents a line with slope 3 and y intercept b. We consider b a parameter that may assume any real value. Since the slope is the same for all values of b, the equation represents a set of parallel lines. The totality of lines thus determined is called a **family** of lines. There are, of course, infinitely many lines in the family. In fact, a line of the family passes through each point of the coordinate plane.

EXAMPLE 1 ● Write the equation of the family of lines with slope $\frac{2}{3}$.

SOLUTION. We choose the equation

$$2x - 3y = k$$

to represent the family of lines. The graph of this equation is a line of slope $\frac{2}{3}$ for any particular value of the parameter k. Hence, with k allowed to vary, it represents a family of parallel lines. Figure 2.13 shows a few lines of the family corresponding to the indicated values of k. ●

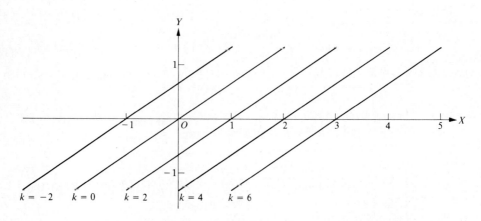

FIGURE 2.13

EXAMPLE 2 ● Discuss the family of lines represented by the equation

$$y - 2 = m(x - 4).$$

SOLUTION. This is an equation of a family of lines passing through the point $(4, 2)$. The family consists of all lines through this point except the vertical line. There is no value of the parameter m that will yield a vertical line. A few lines of the family are drawn in Fig. 2.14. ●

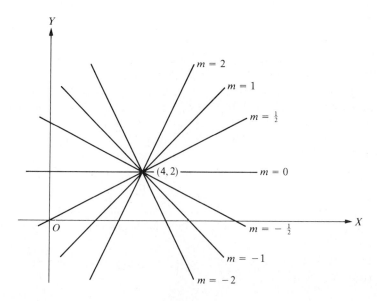

FIGURE 2.14

EXAMPLE 3 ● Write an equation of the family of lines perpendicular to the line
$$3x - 2y = 5.$$

SOLUTION. The slope of the given line is $\frac{3}{2}$. So we desire a family of lines of slope $-\frac{2}{3}$. For this purpose we select the equation
$$2x + 3y = k. \ ●$$

EXAMPLE 4 ● Write an equation of the family of lines having the product of the intercepts equal to 4.

SOLUTION. Using a as a parameter, we take for the intercepts a and $4/a$. Thus we write
$$\frac{x}{a} + \frac{y}{4/a} = 1.$$

The family may also be expressed by
$$4x + a^2y = 4a, \qquad a \neq 0. \ ●$$

EXAMPLE 5 ● Write the equation of the family of lines that are parallel to the line $5x + 12y + 7 = 0$. Find the equations of the members of the family that are 3 units from the point $(2, 1)$.

SOLUTION. Each member of the family $5x + 12y + C = 0$ is parallel to the given line. We seek values of the parameter C that will yield lines 3 units from the point $(2, 1)$, one above the point and the other below the point. Using the formula for the distance from a line to a point, we obtain the equations

$$\frac{5(2) + 12(1) + C}{13} = 3 \quad \text{and} \quad \frac{5(2) + 12(1) + C}{13} = -3.$$

The roots of these equations are $C = 17$ and $C = -61$. Hence the required equations are

$$5x + 12y + 17 = 0 \quad \text{and} \quad 5x + 12y - 61 = 0. \bullet$$

EXAMPLE 6 ● The equation of the family of lines passing through the intersection of two given lines can be written readily. To illustrate, we consider the two intersecting lines

$$2x - 3y + 5 = 0 \quad \text{and} \quad 4x + y - 11 = 0.$$

From the left members of these equations we form the equation

$$(2x - 3y + 5) + k(4x + y - 11) = 0, \tag{2.20}$$

where k is a parameter. This equation is of the first degree in x and y for any value of k. Hence it represents a family of lines. Furthermore, each line of the family goes through the intersection of the given lines. We verify this statement by actual substitution. The given lines intersect at $(2, 3)$. Then, using these values for x and y, we get

$$(4 - 9 + 5) + k(8 + 3 - 11) = 0,$$
$$0 + k(0) = 0,$$
$$0 = 0.$$

This result demonstrates that Eq. (2.20) is satisfied by the coordinates $(2, 3)$, regardless of the value of k. Hence the equation defines a family of lines passing through the intersection of the given lines. ●

More generally, let the equations

$$A_1x + B_1y + C_1 = 0$$
$$A_2x + B_2y + C_2 = 0$$

define two intersecting lines. Then the equation

$$(A_1x + B_1y + C_1) + k(A_2x + B_2y + C_2) = 0$$

represents a family of lines through the intersection of the given lines. To verify this statement, we first observe that the equation is linear for any value of k. Next we note that the coordinates of the intersection point reduce each of the parts in parentheses to zero, and hence satisfy the equation for any value of k.

EXAMPLE 7 ● Write the equation of the family of lines that pass through the intersection of $2x - y - 1 = 0$ and $3x + 2y - 12 = 0$. Find the member of the family that passes through $(-2, 1)$.

SOLUTION. The equation of the family of lines passing through the intersection of the given lines is

$$(2x - y - 1) + k(3x + 2y - 12) = 0. \tag{2.21}$$

To find the member of the family passing through $(-2, 1)$, we replace x by -2 and y by 1. This gives

$$(-4 - 1 - 1) + k(-6 + 2 - 12) = 0,$$
$$k = -\tfrac{3}{8}.$$

Replacing k by $-\tfrac{3}{8}$ in Eq. (2.21), we have

$$(2x - y - 1) - \tfrac{3}{8}(3x + 2y - 12) = 0,$$

or, on simplifying,

$$x - 2y + 4 = 0. \ ●$$

EXAMPLE 8 ● Write the equation of the family of lines through the intersection of $x - 7y + 3 = 0$ and $4x + 2y - 5 = 0$. Find the member of the family that has the slope 3.

SOLUTION. The equation of the family of lines passing through the intersection of the given lines is

$$(x - 7y + 3) + k(4x + 2y - 5) = 0$$

or, collecting terms,

$$(1 + 4k)x + (-7 + 2k)y + 3 - 5k = 0.$$

The slope of each member of this family, except for the vertical line, is $-(1 + 4k)/(2k - 7)$. Equating this fraction to the required slope gives

$$-\frac{1 + 4k}{2k - 7} = 3 \quad \text{and} \quad k = 2.$$

The member of the system for $k = 2$ is $9x - 3y - 7 = 0$. ●

Exercises

Tell what geometric property is possessed by all lines of each family in Exercises 1 through 9. The letters other than x and y are parameters.

1. $y = mx - 4$ 2. $y = 3x + b$ 3. $2x - 9y = k$

4. $y - 4 = m(x + 3)$ 5. $ax - 3y = a$ 6. $x + ay = 2a$

7. $3x - ay = 3a$ 8. $ax - 2y = 4a$ 9. $ax + ay = 3$

Write the equation of the family of lines possessing the given property in Exercises 10 through 17. In each case assign three values to the parameter and graph the corresponding lines.

10. The lines are parallel to $4x - 7y = 3$.

11. The lines pass through $(3, -4)$.

12. The y intercept is twice the x intercept.

13. The lines are perpendicular to $3x - 4y = 5$.

14. The x intercept is equal to 3.

15. The sum of the intercepts is equal to 5.

16. The product of the intercepts is equal to 8.

17. The product of the intercepts is 28.

18. Write the equation of the family of lines with slope -2, and find the equations of the two lines that are 4 units from the origin.

19. Write the equation of the family of lines parallel to $4x - 3y + 5 = 0$, and find the equations of the two lines that are 5 units from the point $(2, -3)$.

20. Write the equation of the family of lines that are parallel to the line $3x + 4y + 5 = 0$. Find the equations of the two members of the family that are 3 units from the point $(1, 2)$.

21. Write the equation of the family of lines that are parallel to $5x + 12y + 2 = 0$. Find the equations of the two members of the family that are 2 units from the point $(2, 2)$.

22. The line $2x - 3y + 2 = 0$ is midway between two parallel lines 6 units apart. Find the equations of the two lines.

23. The line $3x + 2y + 4 = 0$ is midway between two parallel lines that are 4 units apart. Find the equations of the two lines.

24. The line $3x - 4y = 4$ is midway between two parallel lines that are 8 units apart. Find the equations of the two lines.

In Exercises 25 through 31, find the equation of the line that passes through the intersection of the given pair of lines and satisfies the other given condition.

25. $x - 5y - 4 = 0$, $2x + 3y + 2 = 0$; passes through $(2, 5)$.

26. $3x + 5y - 2 = 0$, $x + y + 2 = 0$; passes through $(-4, -2)$.

27. $2x + 5y + 7 = 0$, $4x - 2y + 3 = 0$; the x intercept is 4.

28. $x - y - 4 = 0$, $2x + y - 5 = 0$; passes through $(0, 0)$.

29. $3x + 4y - 2 = 0$, $3x - 4y + 1 = 0$; the intercepts are equal.

30. $5x - 3y + 2 = 0$, $x + y - 2 = 0$; $m = 3$.

31. $3x + y - 4 = 0$, $x + 11y = 0$;   a vertical line.

32. The sides of a triangle are on the lines $2x + 3y + 4 = 0$, $x - y + 3 = 0$, and $5x + 4y - 20 = 0$. Without solving for the vertices, find the equations of the altitudes.

33. The sides of a triangle are on the lines $3x - 5y + 2 = 0$, $x + y - 2 = 0$, and $4x - 3y - 3 = 0$. Without solving for the vertices, find the equations of the altitudes.

34. Find the equation of the family of lines each member of which forms with the coordinate axes a triangle of area 17 square units.

2.6

CIRCLES

We have seen how to write the equation of a line whose position on the coordinate plane is specified. We shall discover that it is equally easy to write the equation of a circle if the location of its center and the radius are known. First, however, we give an explicit definition of a circle.

DEFINITION 1.1 ● *A circle is the set of all points on a plane that are equidistant from a fixed point on the plane. The fixed point is called the* **center**, *and the distance from the center to any point of the circle is called the* **radius**.

Let the center of a circle be at the fixed point $C(h, k)$ and let the radius be equal to r. Then if $P(x, y)$ is any point of the circle, the distance from C to P is equal to r (Fig. 2.15). This condition requires that

$$\sqrt{(x - h)^2 + (y - k)^2} = r$$

FIGURE 2.15

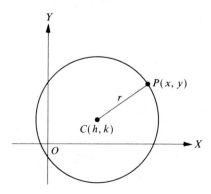

and, by squaring,

$$(x - h)^2 + (y - k)^2 = r^2. \tag{2.22}$$

This formula exhibits the coordinates of the center and the length of the radius and, consequently, is sometimes called the **center–radius form** of the equation of a circle.

Conversely, the graph of an equation of the form (2.22) is a circle with center at (h, k) and radius equal to r. This fact is evident since the equation is satisfied by, and only by, points whose distance from (h, k) is r. Hence it is an easy task to write the equation of a circle whose center and radius are known, or to draw the circle whose equation is expressed in the form (2.22).

If the center of a circle is at the origin ($h = 0, k = 0$) and the radius is r, its equation is

$$x^2 + y^2 = r^2. \tag{2.23}$$

EXAMPLE 1 ● If the center of a circle is at $(3, -2)$ and the radius is 4, the equation of the circle is

$$(x - 3)^2 + (y + 2)^2 = 16$$

or

$$x^2 + y^2 - 6x + 4y - 3 = 0. ●$$

Equation (2.22) can be presented in another form by squaring the binomials and collecting terms. Thus

$$x^2 - 2hx + h^2 + y^2 - 2ky + k^2 = r^2,$$

$$x^2 + y^2 - 2hx - 2ky + h^2 + k^2 - r^2 = 0.$$

The last equation is of the form

$$x^2 + y^2 + Dx + Ey + F = 0. \tag{2.24}$$

This is called the **general form** of the equation of a circle.

Conversely, an equation of the form (2.24) can be reduced to the form (2.22) by the simple expedient of completing the squares in the x terms and the y terms. We illustrate the procedure.

EXAMPLE 2 ● Let us change the general form

$$x^2 + y^2 + Dx + Ey + F = 0$$

to the center–radius form. For this purpose we separate the x terms and the y terms on the left side and put the constant term on the right. Thus

$$x^2 + Dx + y^2 + Ey = -F.$$

Next, we add the square of half the coefficient of x to the x terms and the square of half the coefficient of y to the y terms and add the same amount on the right side. This gives

$$x^2 + Dx + \frac{D^2}{4} + y^2 + Ey + \frac{E^2}{4} = \frac{D^2}{4} + \frac{E^2}{4} - F.$$

The x terms and y terms are now perfect squares. Hence we write

$$\left(x + \frac{D}{2}\right)^2 + \left(y + \frac{E}{2}\right)^2 = \frac{D^2}{4} + \frac{E^2}{4} - F.$$

This equation is in the center–radius form. The equation will have a graph, which is a circle, if

$$\frac{D^2}{4} + \frac{E^2}{4} - F$$

is positive. It will also have a graph if the right member is equal to zero. Then the graph will consist of the single point $(-D/2, -E/2)$. In this case the graph is sometimes called a **point circle**. Clearly, no real values of x and y satisfy the equation when the right member is negative. ●

EXAMPLE 3 ● Change the equation

$$2x^2 + 2y^2 - 8x + 5y - 80 = 0$$

to form (2.22).

SOLUTION. We first divide by 2 to reduce the equation to the general form (2.24). Thus

$$x^2 + y^2 - 4x + \tfrac{5}{2}y - 40 = 0.$$

Next, leaving spaces for the terms to be added to complete squares, we have

$$x^2 - 4x \quad + y^2 + \tfrac{5}{2}y \quad = 40.$$

The square of half the coefficient of x goes in the first space and the square of half the coefficient of y goes in the second space. Then we have

$$(x^2 - 4x + 4) + \left(y^2 + \tfrac{5}{2}y + \tfrac{25}{16}\right) = 40 + 4 + \tfrac{25}{16},$$
$$(x - 2)^2 + \left(y + \tfrac{5}{4}\right)^2 = \left(\tfrac{27}{4}\right)^2.$$

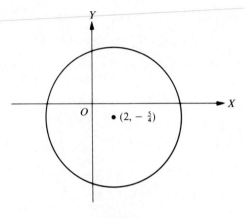

FIGURE 2.16

This equation is in the required form, and it reveals that the given equation is that of a circle with center at $(2, -\frac{5}{4})$ and the radius is equal to 27/4. The graph is in Fig. 2.16. ●

EXAMPLE 4 ● What is the graph, if any, of the equation

$$x^2 + y^2 + 4x - 6y + 14 = 0?$$

SOLUTION. Upon completing the squares, we find

$$(x + 2)^2 + (y - 3)^2 = -1.$$

Clearly the left member of this equation cannot be negative for any real values of x and y. Hence the equation has no graph. ●

Circles Determined By Geometric Conditions

We have seen how to write the equation of a line from certain information that fixes the position of the line in the coordinate plane. We consider now a similar problem concerning the circle. Both the center–radius form and the general form of the equation of a circle will be useful in this connection. There are innumerable geometric conditions that determine a circle. It will be recalled, for example, that a circle can be passed through three points that are not on a straight line. We illustrate this case first.

EXAMPLE 5 ● Find the equation of the circle that passes through the points $P(1, -2)$, $Q(5, 4)$, and $R(10, 5)$.

SOLUTION. The equation of the circle can be expressed in the form

$$x^2 + y^2 + Dx + Ey + F = 0.$$

Our problem is to find values for D, E, and F so that the equation is satisfied by the coordinates of each of the given points. Hence we substitute for x and y the coordinates of these points. This gives the system

$$\begin{aligned}
1 + 4 + \quad D - 2E + F &= 0, \\
25 + 16 + \quad 5D + 4E + F &= 0, \\
100 + 25 + 10D + 5E + F &= 0.
\end{aligned}$$

The solution of these equations is $D = -18$, $E = 6$, and $F = 25$. Therefore the required equation is

$$x^2 + y^2 - 18x + 6y + 25 = 0.$$

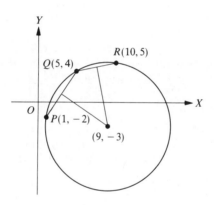

FIGURE 2.17

Alternatively, this problem can be solved by applying the fact that the perpendicular bisectors of two nonparallel chords of a circle intersect at the center. Thus the equations of the perpendicular bisectors of PQ and QR (Fig. 2.17) are

$$2x + 3y = 9 \quad \text{and} \quad 5x + y = 42.$$

The solution of these equations is $x = 9$, $y = -3$. These are the coordinates of the center. The radius is the distance from the center to either of the given points. Hence the resulting equation, in center–radius form, is

$$(x - 9)^2 + (y + 3)^2 = 65. \; \bullet$$

EXAMPLE 6 ● A circle is tangent to the line $2x - y + 1 = 0$ at the point $(2, 5)$, and the center is on the line $x + y = 9$. Find the equation of the circle.

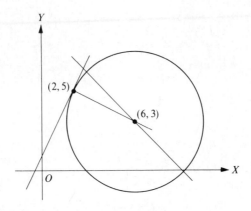

FIGURE 2.18

SOLUTION. The line through $(2,5)$ and perpendicular to the line $2x - y + 1 = 0$ passes through the center of the circle (Fig. 2.18). The equation of this line is $x + 2y = 12$. Hence the solution of the system

$$x + 2y = 12,$$
$$x + y = 9,$$

yields the coordinates of the center. Accordingly, the center is at $(6,3)$. The distance from this point to $(2,5)$ is $\sqrt{20}$. The equation of the circle, therefore, is

$$(x - 6)^2 + (y - 3)^2 = 20. \; \bullet$$

EXAMPLE 7 $\bullet$ A triangle has its sides on the lines $x + 2y - 5 = 0$, $2x - y - 10 = 0$, and $2x + y + 2 = 0$. Find the equation of the circle inscribed in the triangle.

SOLUTION. From geometry we know that the center of the inscribed circle is the point of intersection of the bisectors of the angles of the triangle. Referring to Fig. 2.19, we indicate the meeting point of the bisectors of the angle A and the angle B by $P(x', y')$. The distances from the sides of the triangle to P are indicated by d_1, d_2, and d_3. The distances d_1 and d_2 are positive but d_3 is negative. So $d_1 = d_2$ and $d_2 = -d_3$. We now write:

$$\frac{2x' + y' + 2}{\sqrt{5}} = \frac{2x' - y' - 10}{-\sqrt{5}} \qquad \text{or} \qquad x' = 2,$$

$$\frac{2x' - y' - 10}{-\sqrt{5}} = -\frac{x' + 2y' - 5}{\sqrt{5}} \qquad \text{or} \qquad x' - 3y' = 5.$$

Dropping the primes gives $x = 2$ and $x - 3y = 5$, which are the equations of the

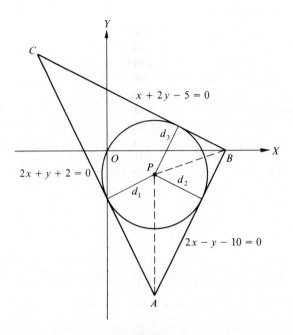

FIGURE 2.19

bisectors of angles A and B. The solution of this pair of equations is $x = 2$ and $y = -1$. Therefore the point $(2, -1)$ is the center of the inscribed circle. The radius is the distance of $(2, -1)$ from each of the three sides, which is $\sqrt{5}$. Hence the desired equation is

$$(x - 2)^2 + (y + 1)^2 = 5. \ \bullet$$

Exercises

In Exercises 1 through 16, write the equation of the circle that satisfies the given conditions.

1. Center $(3, -4)$, radius 6

2. Center $(0, 8)$, radius 5

3. Center $(5, -12)$, radius 13

4. Center $(0, 3)$, radius 7

5. Center $(5, -3)$, radius $\sqrt{6}$

6. Center $(-1, -6)$, radius 8

7. Center $(3, -\frac{1}{2})$, radius $\sqrt{17}$

8. Center $(\frac{3}{5}, \frac{1}{3})$, radius $\sqrt{10}$

9. The line segment joining $A(0,0)$ and $B(6, -8)$ is a diameter.

10. The line segment joining $A(-1, 5)$ and $B(-5, -7)$ is a diameter.

11. The line segment joining $A(-3, -4)$ and $B(4, 3)$ is a diameter.

12. The center is at $(1, -3)$ and the circle passes through $(-3, 5)$.

13. The circle is tangent to the y axis, and the center is at $(5, 3)$.

14. The circle is tangent to the x axis, and the center is at $(-3, -4)$.

15. The circle is tangent to the line $3x + 4y = 16$ and the center is at $(-3, -4)$.

16. The circle is tangent to the line $5x - 12y = 24$ and the center is at $(5, -5)$.

Reduce each equation in Exercises 17 through 26 to the center–radius form and draw the circle.

17. $x^2 + y^2 + 6x - 4y - 12 = 0$ 18. $x^2 + y^2 + 4x + 12y + 36 = 0$

19. $x^2 + y^2 - 8x - 2y + 1 = 0$ 20. $x^2 + y^2 - 10x + 4y - 7 = 0$

21. $x^2 + y^2 + 8x + 6y - 11 = 0$ 22. $x^2 + y^2 - 10x - 24y + 25 = 0$

23. $x^2 + y^2 + 4x - 12y + 3 = 0$ 24. $x^2 + y^2 - 3x - 4y = 0$

25. $2x^2 + 2y^2 + 12x - 2y - 3 = 0$ 26. $3x^2 + 3y^2 - 6x + 5y = 0$

Determine whether each equation in Exercises 27 through 36 represents a circle, or a point, or has no graph.

27. $x^2 + y^2 - 2x + 4y + 5 = 0$ 28. $x^2 + y^2 - 6x + 2y + 10 = 0$

29. $x^2 + y^2 + 4x - 8y - 5 = 0$ 30. $x^2 + y^2 + 10y = 0$

31. $x^2 + y^2 - x = 0$ 32. $x^2 + y^2 + 8x + 15 = 0$

33. $x^2 + y^2 - 4x - 4y + 9 = 0$ 34. $x^2 + y^2 + 7x + 5y + 16 = 0$

35. $x^2 + y^2 - 5x - 3y + \frac{34}{4} = 0$ 36. $x^2 + \frac{1}{2}x + y^2 + \frac{1}{3}y + \frac{13}{144} = 0$

Find the equation of the circle described in each of Exercises 37 through 46.

37. The circle is tangent to the line $x - y = 2$ at the point $(4, 2)$ and the center is on the x axis.

38. The circle is tangent to the line $x + 2y = 3$ at the point $(-1, 2)$ and the center is on the y axis.

39. The circle is tangent to the line $4x + 3y = 4$ at the point $(4, -4)$, and the center is on the line $x - y = 7$.

40. The circle is tangent to the line $5x + y = 3$ at the point $(2, -7)$ and the center is on the line $x - 2y = 19$.

41. The circle is tangent to the line $3x - 4y = 34$ at the point $(10, -1)$ and also tangent to the line $4x + 3y = 12$ at the point $(3, 0)$.

42. The circle is tangent to both coordinate axes and contains the point $(6, 3)$.

43. The circle passes through the points $(0, 3)$, $(2, 4)$, and $(1, 0)$.

44. The circle passes through the points $(0, 0)$, $(0, 5)$, and $(3, 3)$.

45. The circle is circumscribed about the triangle whose vertices are $(3, -2)$, $(2, 5)$, and $(-1, 6)$.

46. The circle is circumscribed about the triangle whose vertices are $(-1, -3)$, $(-2, 4)$, and $(2, 1)$.

47. The sides of a triangle are along the lines $x - 2y = 0$, $5x - 2y = 8$, and $3x + 2y = 24$. Find the equation of the circle circumscribed about the triangle.

48. The sides of a triangle lie on the lines $3x + 4y + 8 = 0$, $3x - 4y - 32 = 0$, and $x = 8$. Find the equation of the circle inscribed in the triangle.

49. The sides of a triangle are on the lines $3x - y - 5 = 0$, $x + 3y - 1 = 0$, and $x - 3y + 7 = 0$. Find the equation of the circle inscribed in the triangle.

50. The sides of a triangle are on the lines $6x + 7y + 11 = 0$, $2x - 9y + 11 = 0$, and $9x + 2y - 11 = 0$. Find the equation of the circle inscribed in the triangle.

2.7

FAMILIES OF CIRCLES

In Section 2.5 we learned how to find the equation of the family of lines passing through the intersection of two lines. The method used there will also serve for finding the equation of families of circles passing through the intersections of two circles. For this purpose, let us consider the equations

$$x^2 + y^2 + D_1x + E_1y + F_1 = 0, \qquad (2.25)$$
$$x^2 + y^2 + D_2x + E_2y + F_2 = 0, \qquad (2.26)$$

and, taking k as a parameter, the equation

$$\left(x^2 + y^2 + D_1x + E_1y + F_1\right) + k\left(x^2 + y^2 + D_2x + E_2y + F_2\right) = 0.$$
$$(2.27)$$

Suppose now that Eqs. (2.25) and (2.26) represent circles that intersect in two points. Then if k is a parameter, with k different from -1, Eq. (2.27) represents a family of circles passing through the intersection points of the given equations. This is true because the coordinates of an intersection point, when substituted for x and y, reduce Eq. (2.27) to $0 + k0 = 0$. If the given circles are tangent to each other, Eq. (2.27) represents the family of circles passing through the point of tangency. We shall give examples of circles intersecting at two points and of circles intersecting at only one point.

EXAMPLE 1 ● Write the equation of the family of circles C_3 all members of which pass through the intersection of the circles C_1 and C_2 represented by the equations

$$C_1: \quad x^2 + y^2 - 6x + 2y + 5 = 0,$$
$$C_2: \quad x^2 + y^2 - 12x - 2y + 29 = 0.$$

Find the member of the family C_3 that passes through the point $(7, 0)$.

SOLUTION. Letting k be a parameter, we express the family of circles by the equation

$$(x^2 + y^2 - 6x + 2y + 5) + k(x^2 + y^2 - 12x - 2y + 29) = 0. \quad (2.28)$$

Replacing x by 7 and y by 0 in this equation, we find

$$(49 - 42 + 5) + k(49 - 84 + 29) = 0,$$
$$12 + k(-6) = 0,$$
$$k = 2.$$

For this value of k, Eq. (2.28) reduces to

$$3x^2 + 3y^2 - 30x - 2y + 63 = 0$$

or, in center–radius form,

$$C_3: \quad (x - 5)^2 + \left(y - \tfrac{1}{3}\right)^2 = \tfrac{37}{9}.$$

Hence the required member of the family of circles has its center at $(5, \tfrac{1}{3})$ and radius equal to $\sqrt{37}/3$, which is approximately equal to 2.03. This circle and the two given circles are constructed in Fig. 2.20. ●

FIGURE 2.20

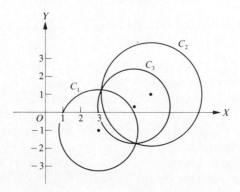

EXAMPLE 2 ● Graph the circles C_1 and C_2 whose equations are

$$C_1: \quad x^2 + y^2 - 12x - 9y + 50 = 0,$$
$$C_2: \quad x^2 + y^2 - 25 = 0.$$

Also graph the member C_3 of the family of circles

$$\left(x^2 + y^2 - 12x - 9y + 50\right) + k\left(x^2 + y^2 - 25\right) = 0,$$

for which $k = 1$.

SOLUTION. Replacing k by 1, we obtain, in center–radius form, the equation

$$C_3: \quad (x - 3)^2 + \left(y - \tfrac{9}{4}\right)^2 = \tfrac{25}{16}.$$

The center of the circle is at $(3, \tfrac{9}{4})$ and the radius is $\tfrac{5}{4}$. As shown in Fig. 2.21, the given circles intersect in only one point and the third circle passes through the point of tangency. ●

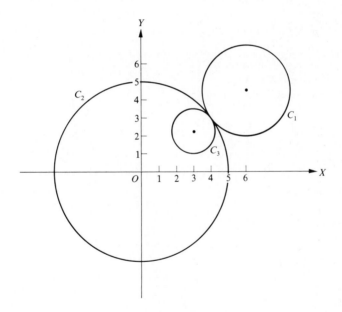

FIGURE 2.21

Equation (2.27), as we saw, represents a circle if k is assigned any real number except $k = -1$. If, however, we let $k = -1$, Eq. (2.27) reduces to the linear equation

$$(D_1 - D_2)x + (E_1 - E_2)y + F_1 - F_2 = 0.$$

The graph of this equation is a straight line called the **radical axis** of the two given circles. If the given circles intersect in two points, the radical axis passes through the intersection points; if the given circles are tangent, the radical axis is tangent to the circles at their point of tangency. If the given circles have no common point, the radical axis is between the circles. In each of the three possibilities, the radical axis is perpendicular to the line joining the centers of the given circles. We leave the proof of this statement to the student.

EXAMPLE 3 ● Draw the graph of the equations

$$x^2 + y^2 - 4x - 6y - 3 = 0 \quad \text{and} \quad x^2 + y^2 - 12x - 14y + 65 = 0.$$

Then find the equation of the radical axis and draw the axis.

SOLUTION. Subtracting the first given equation from the second given equation, we get $-8x - 8y + 68 = 0$, which is equal to $2x + 2y - 17 = 0$. This is the equation of the radical axis. The graphs of the given equations and the radical axis are shown in Fig. 2.22. ●

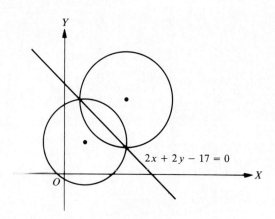

$$2x + 2y - 17 = 0$$

FIGURE 2.22

Exercises

1. Write an equation of the family of circles passing through the intersection of the circles

$$x^2 + y^2 - 2x - 24 = 0 \quad \text{and} \quad x^2 + y^2 + 10x + 6y - 2 = 0.$$

Find the member of the family for which $k = 1$. Construct the three circles on the same coordinate system.

2. Write an equation of the family of circles passing through the intersection of the circles

$$x^2 + y^2 + 2x + 4y - 4 = 0 \quad \text{and} \quad x^2 + y^2 + 6x + 2y + 6 = 0.$$

Find the member of the family for which $k = 2$.

3. Write an equation of the family of circles passing through the intersection of the circles

$$x^2 + y^2 + 2x - 4y = 4 \quad \text{and} \quad x^2 + y^2 - 4x + 6y = 3.$$

Find the member of the family that passes through $(1, 2)$.

4. Find the equation of the line passing through the points of intersection of the circles

$$x^2 + y^2 - 4x + 2y - 4 = 0 \quad \text{and} \quad x^2 + y^2 + 4x = 0.$$

5. Find the equation of the radical axis represented by the circles

$$x^2 + y^2 + 14x + 38 = 0 \quad \text{and} \quad x^2 + y^2 - 4 = 0.$$

6. Draw the graph of the equations

$$x^2 + y^2 + 4x + 6y - 3 = 0 \quad \text{and} \quad x^2 + y^2 + 12x + 14y + 60 = 0.$$

Then find the equation of the radical axis and draw the axis.

7. Draw the circles

$$x^2 + y^2 - 14x + 40 = 0 \quad \text{and} \quad x^2 + y^2 - 4 = 0.$$

Find the equation of the radical axis and draw the axis.

8. Draw the circles

$$x^2 + y^2 + 2x - 5y - 4 = 0 \quad \text{and} \quad x^2 + y^2 + 12x + 8y + 36 = 0.$$

Find the equation of the radical axis and draw the axis.

9. Prove that the radical axis of two circles is perpendicular to the line segment connecting the centers of the circles.

10. Write the equation of the family of circles passing through the intersection points of

$$x^2 + y^2 + 16x + 10y + 24 = 0 \quad \text{and} \quad x^2 + y^2 + 4x - 8y - 6 = 0.$$

Find the member of the family that passes through the origin. Construct the three circles.

2.8

TRANSLATION OF AXES

The equation of a circle of radius r has the simple form $x^2 + y^2 = r^2$ if the origin of coordinates is at the center of the circle. If the origin is not at the center, the corresponding equation may be expressed in either of the less simple forms (2.22) or (2.24), Section 2.6. This is an illustration of the fact that the simplicity of the equation of a curve depends on the relative positions of the curve and the axes.

Suppose we have a curve in the coordinate plane and the equation of the curve. Let us consider the problem of writing the equation of the same curve with respect to another pair of axes. The process of changing from one pair of axes to another is called a **transformation of coordinates**. The most general transformation is one in which the new axes are not parallel to the old axes and the origins are different. Just now, however, we shall consider transformations in which the new axes are parallel to the original axes and similarly directed. A transformation of this kind is called a **translation of axes**.

The coordinates of each point of the plane are changed under a translation of axes. To see how the coordinates are changed, examine Fig. 2.23. The new axes $O'X'$ and $O'Y'$ are parallel, respectively, to the old axes OX and OY. The coordinates of the origin O', referred to the original axes, are denoted by (h, k). Hence the new axes can be obtained by shifting the old axes h units horizontally and k units vertically while keeping their directions unchanged. Let x and y stand for the coordinates of any point P when referred to the old axes, and x' and y' the coordinates of P with respect to the new axes. It is evident from the figure that

$$x = \overline{ON} = \overline{OM} + \overline{O'Q} = h + x',$$
$$y = \overline{NP} = \overline{MO'} + \overline{QP} = k + y'.$$

FIGURE 2.23

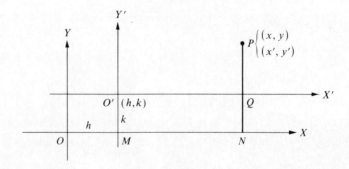

Hence

$$x = x' + h, \qquad y = y' + k.$$

These formulas give the relations of the old and new coordinates. They hold for all points of the plane, where the new origin O' is any point of the plane. Consequently, the substitutions $x' + h$ for x and $y' + k$ for y in the equation of a curve referred to the original axes yield the equation of the same curve referred to the translated axes. It is imperative that each set of axes be properly labeled. Otherwise, a graph becomes a confusion of lines.

EXAMPLE 1 ● Find the new coordinates of the point $P(4, -2)$ if the origin is moved to $(-2, 3)$ by a translation.

SOLUTION. Since we are to find the new coordinates of the given point, we write the translation formulas as $x' = x - h$ and $y' = y - k$. The original coordinates of the given point are $x = -2$, $y = 3$. Making the proper substitutions, we find

$$x' = 4 - (-2) = 6, \qquad y' = -2 - 3 = -5.$$

The new coordinates of P are $(6, -5)$. This result can be obtained directly from Fig. 2.24. ●

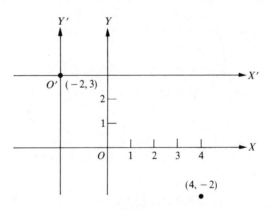

FIGURE 2.24

EXAMPLE 2 ● Find the new equation of the circle

$$x^2 + y^2 - 6x + 4y - 3 = 0$$

after a translation that moves the origin to the point $(3, -2)$.

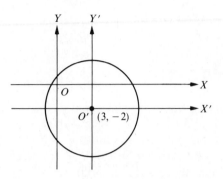

FIGURE 2.25

SOLUTION. The translation formulas here become $x = x' + 3$ and $y = y' - 2$. These substitutions for x and y in the given equation yield

$$(x' + 3)^2 + (y' - 2)^2 - 6(x' + 3) + 4(y' - 2) - 3 = 0$$

or, by simplification,

$$x'^2 + y'^2 = 16.$$

Both sets of axes and the graph are drawn in Fig. 2.25. ●

EXAMPLE 3 ● Translate the axes so that no first-degree terms will appear in the transformed equation of the circle

$$x^2 + y^2 + 6x - 10y + 12 = 0.$$

SOLUTION. We first express the equation in the center–radius form and have

$$(x + 3)^2 + (y - 5)^2 = 22.$$

If we choose the new origin at the center of the circle, $(-3, 5)$, the translation formulas are $x = x' - 3$ and $y = y' + 5$. These substitutions for x and y give

$$x'^2 + y'^2 = 22$$

as the equation of the circle referred to the translated axes. ●

Exercises

Determine the new coordinates of the points in Exercises 1 through 6 if the axes are translated so that the new origin is at the given point O'. Draw both sets of axes and verify your result from a figure.

1. $(3, 2)$, $(3, -2)$, $(-3, -2)$, $(-3, 2)$; $O'(4, 1)$.

2. $(7, 1)$, $(4, 2)$, $(4, -3)$, $(2, 3)$; $O'(2, 3)$.

3. $(4, -3)$, $(5, -1)$, $(-3, -2)$, $(3, 4)$; $O'(-3, 1)$.

4. $(6, 3)$, $(2, -3)$, $(7, 4)$, $(-4, -2)$; $O'(-5, -2)$.

5. $(5, 3)$, $(2, -3)$, $(7, 4)$, $(-4, -2)$; $O'(-2, 2)$.

6. $(3, 1)$, $(4, 2)$, $(5, 3)$, $(4, 6)$, $(-2, -3)$; $O'(-2, 3)$.

Find the new equation in Exercises 7 through 17 if the origin is moved to the given point O' by a translation of axes. Draw both sets of axes and the graph.

7. $2x + y - 6 = 0$; $O'(-2, 2)$ 8. $x - 2y - 4 = 0$; $O'(3, -2)$

9. $3x + 2y + 5 = 0$; $O'(-1, 2)$ 10. $4x - 3y = 0$; $O'(3, 3)$

11. $x^2 + y^2 - 2x + 4y = 0$; $O'(2, 2)$ 12. $x^2 + y^2 + 2x = 0$; $O'(0, 1)$

13. $x^2 + y^2 - 4x + 4y - 2 = 0$; $O'(1, 3)$

14. $x^2 + y^2 + 6x - 8y + 5 = 0$; $O'(1, -4)$

15. $x^2 + y^2 - 16x + 6y + 8 = 0$; $O'(-6, 3)$

16. $x^2 + y^2 - 8x + 12y - 7 = 0$; $O'(3, -8)$

Ⓒ17. $x^2 + y^2 + 14x - 10y + 20 = 0$; $O'(4.015, 2.193)$

In Exercises 18 through 23 find the point to which the origin must be translated in order that the transformed equation will have no first-degree term. Give the transformed equation.

18. $x^2 + y^2 + 6x + 4y + 8 = 0$ 19. $x^2 + y^2 - 4x + 2y = 5$

20. $x^2 + y^2 + 10x - 12y + 3 = 0$ 21. $4x^2 + y^2 + 16x - 6y = 3$

22. $x^2 + 4y^2 - 8x - 8y + 5 = 0$ 23. $x^2 + y^2 + 10x - 14y = 7$

Review Exercises

1. Find the equation of the line through $(2, -3)$ and perpendicular to the line defined by the equation $4x + 5y + 6 = 0$.

2. Find the directed distance from the line $5x - 2y - 26 = 0$ to the points $P_1(4, -5)$, $P_2(-4, 2)$, and $P_3(9, 1)$.

3. Find the equation of the bisector of the acute angles formed by the lines $x - 2y + 6 = 0$ and $x + 2y - 4 = 0$.

4. Write the equation of the family of lines that are parallel to the line $5x + 12y + 6 = 0$. Find the member of the family that is 3 units from the point $(2, 1)$.

5. Write the equation of the family of lines passing through the point $(5, -2)$. Find the member that passes through $(-3, 4)$.

6. The line segment joining $(5, -1)$ and $(-7, -5)$ is a diameter of a circle. Find the equation of the circle.

7. A circle is tangent to the line $3x - 4y - 4 = 0$ at the point $(-4, -4)$ and the center is on the line $x + y + 7 = 0$. Find the equation of the circle.

8. Reduce the circle $x^2 + y^2 - 10x - 7y - 7 = 0$ to the center–radius form.

9. Write the equation of the family of circles passing through the intersection of the circles

$$x^2 + y^2 - 2x - 24 = 0 \quad \text{and} \quad x^2 + y^2 + 10x - 9y + 39 = 0.$$

Find the member of the family for which $k = 1$. Construct the three circles on the same coordinate axes.

10. Construct the graphs of the circles

$$x^2 + y^2 + 4x = 0 \quad \text{and} \quad x^2 + y^2 - 4 = 0.$$

Find, in center–radius form, the equation of the member of the family of circles

$$\left(x^2 + y^2 + 4x\right) + k\left(x^2 + y^2 - 4\right) = 0$$

for which $k = 1$. Draw the graph of the two given circles and the circle that you found. Put the three circles on the same coordinate axes.

11. Find the new equation of the circle

$$x^2 + y^2 - 2x - 6y + 4 = 0$$

if the origin is moved to $O'(2, 3)$.

12. Find the point to which the origin must be translated in order for the transformed equation of

$$x^2 + y^2 + 10x - 12y - 3 = 0$$

to have no first-degree terms.

3

Conics

In the preceding chapter, we defined a circle in terms of a set of points. In this chapter, we shall give names to other sets of points, or curves, and derive the corresponding equations. As in the case of a circle, the equations will be of the second degree, or quadratic, in two variables. All the equations, with a few exceptions, define relations some of which are also functions (Definitions 1.6 and 1.7).

The general quadratic equation in x and y may be expressed in the form

$$Ax^2 + Bxy + Cy^2 + Dx + Ey + F = 0. \qquad (3.1)$$

The graph of a second-degree equation in the coordinates x and y is called a **conic section** or, more simply, a **conic**. This designation comes from the fact that the curve can be obtained as the intersection of a right circular cone and a plane.*

The Greek mathematician Apollonius (262 B.C.–200 B.C.) wrote the definitive treatise *Conic Sections* on this subject. It superseded the works of earlier Greek geometers and formed the cornerstone of thought on the subject for well over a thousand years. Indeed, eighteen centuries passed before Descartes wrote his *La Géométrie*.

The conic sections are of more than historic or academic interest; they have many interesting and important applications in science, engineering, and industry. While we cannot examine each application in detail, we can indicate the rich variety of known applications of the conics. Furthermore, we have faith that new applications will be uncovered in the future, just as they have for the past twenty-two centuries. Many of today's applications could not have been imagined just fifty or one hundred years ago.

Obviously, different kinds of conic sections are possible. A plane not passing through the vertex of a cone may cut all the elements of one nappe and make a closed curve (Fig. 3.1). If the plane is parallel to an element, the intersection

---

*Let P be a point on a fixed line L. Then the surface composed of all lines through P and forming a constant angle with L is called a **right circular cone**. The line L is the **axis** of the cone, the point P the **vertex**, and each line composing the surface of the cone is called an **element**. The vertex separates the cone into two parts called **nappes**.

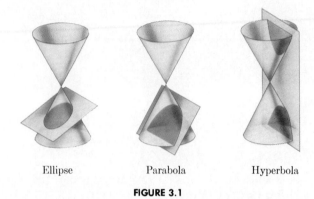

| Ellipse | Parabola | Hyperbola |

FIGURE 3.1

extends indefinitely far along one nappe but does not cut the other nappe. The plane may cut both nappes and make a section of two parts, each extending indefinitely far along a nappe. In addition to those sections, the plane may pass through the vertex of the cone and determine a point, a line, or two intersecting lines. An intersection of each of these kinds is sometimes called a **degenerate conic**.

3.1

THE PARABOLA

We now describe and name a curve that, like the circle, is an important concept in mathematics and is used frequently in applied problems.

DEFINITION 3.1 ● *A **parabola** is the set of all points in a plane that are equidistant from a fixed point and a fixed line of the plane. The fixed point is called the **focus** and the fixed line the **directrix**.*

In Fig. 3.2 the point F is the focus and the line D the directrix. The point V, midway between the focus and the directrix, must belong to the parabola (Definition 3.1). This point is called the **vertex**.

Other points of the parabola can be located in the following way. Draw a line L parallel to the directrix (Fig. 3.2). With F as a center and radius equal to the distance between the lines D and L, describe arcs cutting L at P and P'. Each of these points, being equidistant from the focus and directrix, is on the parabola. The curve can be sketched by determining a few points in this manner. The line VF through the vertex and focus is the perpendicular bisector of PP' and of all other chords similarly drawn. For this reason the line is called the **axis** of the parabola, and the parabola is said to be **symmetric** with respect to its axis.

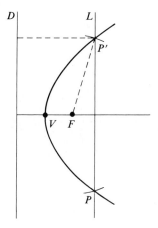

FIGURE 3.2

Although points of a parabola can be located by a direct application of the definition of a parabola, it is easier to obtain them from an equation of the curve. The simplest equation of a parabola can be written if the coordinate axes are placed in a special position relative to the directrix and focus. Let the x axis be on the line through the focus and perpendicular to the directrix and let the origin be at the vertex. Then, choosing $a > 0$, we denote the coordinates of the focus by $F(a, 0)$, and the equation of the directrix by $x = -a$ (Fig. 3.3). Since any point

FIGURE 3.3

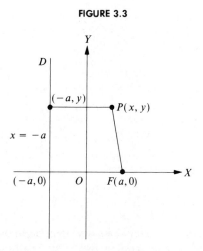

$P(x, y)$ of the parabola is the same distance from the focus and directrix, we have

$$\sqrt{(x - a)^2 + y^2} = x + a.$$

We next square the binomials in this equation and collect terms. Thus

$$(x - a)^2 + y^2 = (x + a)^2,$$

$$x^2 - 2ax + a^2 + y^2 = x^2 + 2ax + a^2,$$

$$y^2 = 4ax.$$

This is the equation of a parabola with the vertex at the origin and focus at $(a, 0)$. Since $a > 0$, x may have any positive value or zero, but no negative value. Hence the graph extends indefinitely far into the first and fourth quadrants, and the axis of the parabola is the positive x axis (Fig. 3.4). It is evident from the equation that the parabola is symmetric with respect to its axis because $y = \pm 2\sqrt{ax}$.

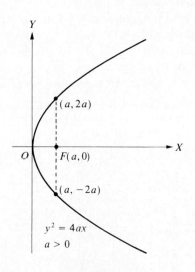

FIGURE 3.4

The chord drawn through the focus and perpendicular to the axis of the parabola is given the Latin name **latus rectum**. The length of the latus rectum can be determined from the coordinates of its endpoints. By substituting a for x in the equation $y^2 = 4ax$, we find

$$y^2 = 4a^2 \qquad \text{and} \qquad y = \pm 2a.$$

Hence the endpoints are $(a, -2a)$ and $(a, 2a)$. This makes the length of the latus rectum equal to $4a$. The vertex and the extremities of the latus rectum are sufficient for drawing a rough sketch of the parabola.

We can, of course, have the focus of a parabola to the left of the origin. For this case, we choose $a < 0$, and let $F(a, 0)$ denote the focus and $x = -a$ the directrix (Fig. 3.5). Then the positive measurement from a point $P(x, y)$ of the parabola to the directrix is $-a - x$. Hence

$$\sqrt{(x - a)^2 + y^2} = -a - x = |a + x|,$$

and this equation, as before, reduces to

$$y^2 = 4ax.$$

Since $a < 0$, the variable x can take only negative values and zero, as shown in Fig. 3.5.

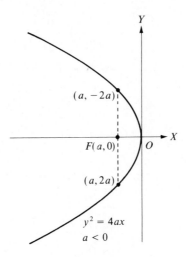

$(a, -2a)$

$F(a, 0)$

$(a, 2a)$

$y^2 = 4ax$

$a < 0$

FIGURE 3.5

In the preceding discussion, we placed the x axis on the line through the focus and perpendicular to the directrix. By choosing this position for the y axis, we would be interchanging the roles of x and y. Hence the equation of the

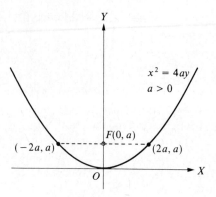

FIGURE 3.6

parabola would then be

$$x^2 = 4ay.$$

The graph of this equation when $a > 0$ is in Fig. 3.6 and when $a < 0$ in Fig. 3.7.

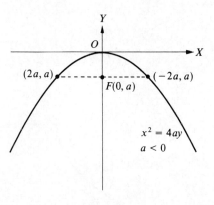

FIGURE 3.7

Summarizing, we make the following statements.

THEOREM 3.1 ● *The equation of a parabola with vertex at the origin and focus at* $(a, 0)$ *is*

$$y^2 = 4ax.$$

(3.2)

The parabola opens to the right if $a > 0$ *and opens to the left if* $a < 0$.

The equation of a parabola with vertex at the origin and focus at $(0, a)$ is

$$x^2 = 4ay.$$ (3.3)

The parabola opens upward if $a > 0$ and opens downward if $a < 0$.

Equations (3.2) and (3.3) can be applied to find the equations of parabolas that satisfy specified conditions. We illustrate their use in some examples.

EXAMPLE 1 ● Write the equation of the parabola with vertex at the origin and the focus at $(0, 4)$.

SOLUTION. Equation (3.3) applies here. The distance from the vertex to the focus is 4, and hence $a = 4$. Substituting this value for a, we get

$$x^2 = 16y. ●$$

EXAMPLE 2 ● A parabola has its vertex at the origin, its axis along the x axis, and passes through the point $(-3, 6)$. Find its equation.

SOLUTION. The equation of the parabola is of the form $y^2 = 4ax$. To determine the value of $4a$, we substitute the coordinates of the given point in this equation. Thus we obtain

$$36 = 4a(-3) \quad \text{and} \quad 4a = -12.$$

The required equation is $y^2 = -12x$. The focus is at $(-3, 0)$, and the given point is the upper end of the latus rectum. The graph is constructed in Fig. 3.8. ●

FIGURE 3.8

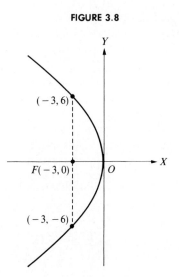

EXAMPLE 3 ● The equation of a parabola is $x^2 = -6y$. Find the coordinates of the focus, the equation of the directrix, and the length of the latus rectum.

SOLUTION. The equation is of the form (3.3), where a is negative. Hence the focus is on the negative y axis and the parabola opens downward. From the equation $4a = -6$, we find $a = -\frac{3}{2}$. Therefore the coordinates of the focus are $(0, -\frac{3}{2})$ and the directrix is $y = \frac{3}{2}$. The length of the latus rectum is equal to the absolute value of $4a$, and in this case is 6. The latus rectum extends 3 units to the left and 3 units to the right of the focus. The graph may be sketched by drawing through the vertex and the ends of the latus rectum. For more accurate graphing a few additional points could be plotted. (See Fig. 3.9.)

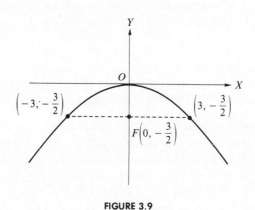

FIGURE 3.9

Exercises

In Exercises 1 through 9 find the coordinates of the focus, the length of the latus rectum, and the coordinates of its endpoints for each of the given parabolas. Find also the equation of the directrix of each parabola. Sketch each curve.

1. $y^2 = 4x$
2. $y^2 = -16x$
3. $x^2 = 4y$

4. $x^2 = -10y$
5. $y^2 + 3x = 0$
6. $x^2 - 8y = 0$

7. $2y^2 = 7x$
8. $2y^2 = -3x$
9. $x^2 - 7y = 0$

Write the equation of the parabola with vertex at the origin that satisfies the given conditions in Exercises 10 through 19.

10. Focus at $(3, 0)$
11. Focus at $(0, 3)$

12. Focus at $(-4, 0)$
13. Focus at $(0, -3)$

14. Directrix is $x + 4 = 0$
15. Directrix is $y - 4 = 0$

16. The length of the latus rectum is 10 and the parabola opens to the right.

17. The length of the latus rectum is 8 and the parabola opens upward.

18. The focus is on the x axis and the parabola passes through the point $(3, 4)$.

19. The parabola opens to the left and passes through the point $(-3, 4)$.

20. On the same axes, sketch the parabolas $x^2 = 4ay$ for $a = \frac{1}{4}, \frac{1}{2}, 2, 4$, and note the effect that changing the value of a has on the shape of the parabola.

21. A cable suspended from supports that are the same height and 600 ft apart has a sag of 100 ft. If the cable hangs in the form of a parabola, find its equation, taking the origin at the lowest point.

22. Find the width of the cable of Exercise 21 at a height 50 ft above the lowest point.

23. Find the equation of the set of all points in the coordinate plane that are equidistant from the point $(0, a)$ and the line $y = -a$, and thus derive Eq. (3.3) in Section 3.1.

3.2

PARABOLA WITH VERTEX AT (h, k)

We now consider a parabola whose axis is parallel to, but not on, a coordinate axis. In Fig. 3.10, the vertex is at (h, k) and the focus at $(h + a, k)$. We introduce another pair of axes by a translation to the point (h, k). Since the distance from

FIGURE 3.10

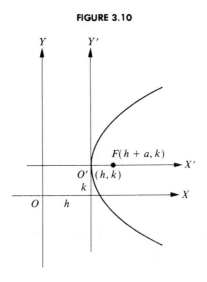

the vertex to the focus is a, we have at once the equation

$$y'^2 = 4ax'.$$

To write the equation of the parabola with respect to the original axes, we apply the translation formulas of Section 2.8, and thus obtain

$$(y - k)^2 = 4a(x - h).$$

We observe from this equation, and also from the figure, that when $a > 0$, the factor $x - h$ of the right member must be greater than or equal to zero. Hence the parabola opens to the right. For $a < 0$, the factor $x - h$ must be less than or equal to zero, and therefore the parabola would open to the left. The axis of the parabola is on the line $y - k = 0$. The length of the latus rectum is equal to the absolute value of $4a$, and hence the endpoints can be easily located.

A similar discussion can be made if the axis of a parabola is parallel to the y axis. Consequently, we make the following statements.

THEOREM 3.2 ● *The equation of a parabola with vertex at (h, k) and focus at $(h + a, k)$ is*

$$(y - k)^2 = 4a(x \quad h).$$
(3.4)

The parabola opens to the right if $a > 0$ and opens to the left if $a < 0$.

The equation of a parabola with vertex at (h, k) and focus at $(h, k + a)$ is

$$(x - h)^2 = 4a(y - k).$$
(3.5)

The parabola opens upwards if $a > 0$ and opens downwards if $a < 0$.

Each of Eqs. (3.4) and (3.5) is said to be in **standard form**. When $h = 0$ and $k = 0$, they reduce to the simpler equations of the preceding section. If the equation of a parabola is in standard form, its graph can be quickly sketched. The vertex and the ends of the latus rectum are sufficient for a rough sketch. The plotting of a few additional points would, of course, improve the accuracy.

We note that each of Eqs. (3.4) and (3.5) is quadratic in one variable and linear in the other variable. This fact can be expressed more vividly if we perform the indicated squares and transpose terms to obtain the general forms

$$x^2 + Dx + Ey + F = 0,$$
(3.6)

$$y^2 + Dx + Ey + F = 0.$$
(3.7)

Conversely, an equation in the form (3.6) or (3.7) can be presented in a standard form, provided $E \neq 0$ in (3.6) and $D \neq 0$ in (3.7).

EXAMPLE 1 ● Draw the graph of the equation

$$y^2 + 8x - 6y + 25 = 0.$$

SOLUTION. The equation represents a parabola because y appears quadratically and x linearly. The graph can be more readily drawn if we first reduce the equation to a standard form. Thus, completing the square, we get

$$y^2 - 6y + 9 = -8x - 25 + 9,$$
$$(y - 3)^2 = -8(x + 2).$$

The vertex is at $(-2, 3)$. Since $4a = -8$ and $a = -2$, the focus is two units to the left of the vertex. The length of the latus rectum, equal to the absolute value of $4a$, is 8. Hence the latus rectum extends 4 units above and below the focus. The graph is constructed in Fig. 3.11. ●

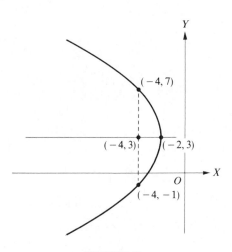

FIGURE 3.11

EXAMPLE 2 ● Construct the graph of the equation

$$x^2 - 6x - 12y - 51 = 0.$$

SOLUTION. The given equation represents a parabola since x appears quadratically and y linearly. We first express the equation in a standard form.

$$x^2 - 6x + 9 = 12y + 51 + 9,$$
$$(x - 3)^2 = 12(y + 5).$$

The vertex is at $(3, -5)$. Since $4a = 12$, $a = 3$. So the focus is 3 units above the vertex, or at $(3, -2)$. The length of the latus rectum is 12, and this makes the

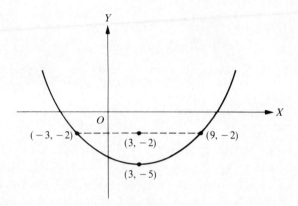

FIGURE 3.12

coordinates of its ends $(-3, -2)$ and $(9, -2)$. The graph is constructed in Fig. 3.12. •

EXAMPLE 3 • A parabola whose axis is parallel to the y axis passes through the points $(1, 1)$, $(2, 2)$, and $(-1, 5)$. Find its equation.

SOLUTION. Since the axis of the parabola is parallel to the y axis, the equation must be quadratic in x and linear in y. Hence we start with the general form

$$x^2 + Dx + Ey + F = 0.$$

This equation is to be satisfied by the coordinates of each of the given points. Substituting the coordinates of each point, in turn, we have the system of equations:

$$1 + D + E + F = 0,$$
$$4 + 2D + 2E + F = 0,$$
$$1 - D + 5E + F = 0.$$

The simultaneous solution of these equations is $D = -2$, $E = -1$, and $F = 2$. Hence the equation of the parabola is $x^2 - 2x - y + 2 = 0$. •

Symmetry

We have observed that the axis of a parabola bisects all chords of the parabola that are perpendicular to the axis. For this reason a parabola is said to be symmetric with respect to its axis. The fact that many other curves possess the property of symmetry leads us to the following discussion.

DEFINITION 3.2 ● *Two points A and B are said to be* **symmetric** *with respect to a line if the line is the perpendicular bisector of the line segment AB. A curve is symmetric with respect to a line if each of its points is one of a pair of points symmetric with respect to the line.*

Two points A and B are symmetric with respect to a point O if O is the midpoint of the line segment AB. A curve is symmetric with respect to a point O if each of its points is one of a pair of points symmetric with respect to O.

Symmetry of a curve with respect to a coordinate axis or the origin is of special interest. Hence we make the following observations. The points (x, y) and $(x, -y)$ are symmetric with respect to the x axis. Accordingly, a curve is symmetric with respect to the x axis if, for each point (x, y) of the curve, the point $(x, -y)$ also belongs to the curve. Similarly, a curve is symmetric with respect to the y axis if, for each point (x, y) of the curve, the point $(-x, y)$ also belongs to the curve. The points (x, y) and $(-x, -y)$ are symmetric with respect to the origin. Hence a curve is symmetric with respect to the origin if, for each point (x, y) of the curve, the point $(-x, -y)$ also belongs to the curve. (See Fig. 3.13.)

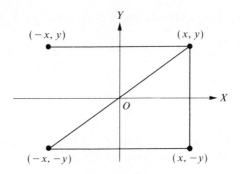

FIGURE 3.13

An equation can easily be tested to determine whether its graph is symmetric with respect to either coordinate axis or the origin. Consider, for example, the equation $x^2 = 4y + 6$. If x is replaced by $-x$, the equation is not altered. This means that if x is given a value and then the negative of that value, the corresponding values of y are the same. Hence, for each point (x, y) of the graph, there is also the point $(-x, y)$ on the graph. Therefore the graph is symmetric with respect to the y axis. On the other hand, the assigning of equal absolute values to y, one positive and the other negative, leads to different corresponding values of x. Hence the graph is not symmetric with respect to the x axis. Similarly, the graph is not symmetric with respect to the origin.

From the definition of symmetry, we formulate the following tests:

1. *If an equation is unchanged when y is replaced by* $-y$, *then the graph of the equation is symmetric with respect to the x axis.*
2. *If an equation is unchanged when x is replaced by* $-x$, *then the graph of the equation is symmetric with respect to the y axis.*
3. *If an equation is unchanged when x is replaced by* $-x$ *and y by* $-y$, *then the graph of the equation is symmetric with respect to the origin.*

We have three kinds of symmetries, and it is easy to see that a graph that possesses two of the symmetries also possesses the third symmetry. Suppose, for example, that the point (x, y) is on the graph, which is symmetric with respect to both the x axis and the y axis. The symmetry with respect to the y axis means that the point $(-x, y)$ is on the graph. Then the symmetry with respect to the x axis means the point $(-x, -y)$ is on the graph. Hence the graph is symmetric with respect to the origin. Clearly, also, a graph that is symmetric with respect to one coordinate axis and the origin is symmetric with respect to the other axis.

EXAMPLE 4 ● The graph of the equation $xy = 1$ is symmetric with respect to the origin. This is true because a point (x_1, y_1) that satisfies the given equation means that the product $x_1 y_1 = 1$. Consequently, the product $(-x_1)(-y_1)$ is also equal to 1.

The equation was graphed in Exercise 13 on page 41. It will reappear in the next chapter. ●

EXAMPLE 5 ● Construct the graph of the equation

$$x^4 - 36y^2 = 0.$$

SOLUTION. Clearly, the graph possesses three symmetries. To obtain the graph, we first factor the left side of the given equation. Thus we get

$$(x^2 + 6y)(x^2 - 6y) = 0.$$

Then, setting each factor equal to 0, we have

$$x^2 + 6y = 0 \qquad \text{and} \qquad x^2 - 6y = 0.$$

Hence the desired graph consists of two parabolas, each with the vertex at the origin, one opening upward and the other downward. We will draw the graph of $x^2 = 6y$ and then obtain the graph of the other parabola by the use of symmetry. For the parabola opening upward, we have $4a = 6$ and $a = \frac{3}{2}$. Hence the focus is at the point $(0, \frac{3}{2})$ and the ends of the latus rectum are at $(3, \frac{3}{2})$ and $(-3, \frac{3}{2})$. By plotting a few additional points, we can obtain a good drawing. The complete graph is shown in Fig. 3.14. ●

A projectile (for example, a ball or bullet) travels in a path that is approximately a **parabola** (this fact is discussed in greater detail in Sec. 8.2). But the parabola has a very important feature that has made it useful for a wide variety of

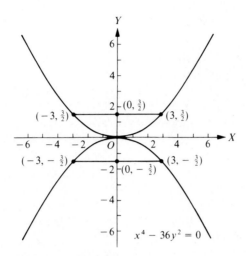

FIGURE 3.14

applications: The parabola has the reflecting or focusing property. The two angles, θ and ϕ in Fig. 3.15, formed by a line parallel to the axis and a tangent to the parabola at a point, and by the line from the focus F to the point, are equal. If the parabola is a reflecting surface, then light rays traveling parallel to the axis are reflected to the focus. Hence, a paraboloid of revolution (surface formed by

FIGURE 3.15

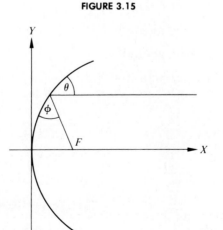

rotating a parabola about its axis) is the ideal shape for reflecting telescopes, headlights on cars, radar and microwave antennas, home TV-satellite dishes, and some solar generators of electricity.

In this vein it should be mentioned that the Crosbyton (Texas) Solar Project employs a fixed hemisphere, rather than a paraboloid, to focus the sun's heat along a diameter rather than at a point. This avoids the problem of aiming the axis at the moving sun and allows a larger solar collector. In this case, the conic is a **circle** and the heat is concentrated along a segment of a diameter rather than at the center.

Exercises

Express the equation, in standard form, of the parabola that satisfies the given condition in each of Exercises 1 through 13.

1. Vertex at $(3, 2)$,   focus at $(3, 4)$.

2. Vertex at $(3, -2)$,   focus at $(3, -8)$.

3. Vertex at $(-6, -4)$,   focus at $(0, -4)$.

4. Vertex at $(4, 1)$, $x = 2$ as directrix.

5. Vertex at $(4, 1)$, $y = -3$ as directrix.

6. Vertex at $(4, -2)$,   latus rectum 8;   opens to the right.

7. Vertex at $(1, 2)$,   latus rectum 8;   opens downward.

8. Vertex at $(3, -2)$,   ends of latus rectum $(-2, \frac{1}{2})$, $(8, \frac{1}{2})$.

9. Vertex $(2, 1)$,   ends of latus rectum $(-1, -5)$, $(-1, 7)$.

10. Focus at $(2, -3)$, $x = 6$ as directrix.

11. Focus at $(5, 0)$, $x = -1$ as directrix.

12. Focus at $(-2, 2)$, $y = 4$ as directrix.

13. Focus at $(2, 0)$, $y = -3$ as directrix.

Express each equation in Exercises 14 through 27 in standard form. Give the coordinates of the vertex, the focus, and the ends of the latus rectum. Sketch the graph.

14. $y^2 + 8x + 8 = 0$

15. $x^2 + 4y + 8 = 0$

16. $y^2 - 12x - 48 = 0$

17. $x^2 + 16y - 32 = 0$

18. $x^2 + 4x + 16y + 4 = 0$

19. $y^2 - 6y - 4x + 9 = 0$

20. $y^2 + 8y + 6x + 16 = 0$

21. $x^2 + 10x - 20y + 25 = 0$

22. $y^2 - 4y + 8x - 28 = 0$

23. $x^2 + 2x + 12y + 37 = 0$

24. $x^2 - 8x - 6y - 8 = 0$ 25. $y^2 + 6y + 10x - 1 = 0$

c 26. $y^2 - 12.63y - 21.49x + 120.09 = 0$ 27. $x^2 - 12x - 16y - 60 = 0$

Find the equation of the parabola in each of Exercises 28 through 35.

28. Vertex at $(3, -4)$, axis horizontal;   passes through $(2, -5)$.

29. Vertex at $(-1, -2)$, axis vertical;   passes through $(3, 6)$.

30. Axis vertical;   passes through $(0, 0)$, $(3, 0)$ and $(-1, 4)$.

31. Axis horizontal;   passes through $(1, 1)$, $(1, -3)$, and $(-2, 0)$.

32. Axis vertical;   passes through $(-1, 0)$, $(5, 0)$, and $(1, 8)$.

33. Axis horizontal;   passes through $(0, 4)$, $(0, -1)$, and $(6, 1)$.

34. Axis horizontal;   passes through $(-1, 1)$, $(3, 4)$, and $(2, -2)$.

35. Axis vertical;   passes through $(-1, -3)$, $(1, -2)$, and $(2, 1)$.

36. Derive Eq. (3.4), by finding the equation of the set of all points equally distant from the focus $(h + a, k)$ and the directrix $x = h - a$.

37. Derive Eq. (3.5), by finding the equation of the graph of a point equally distant from the focus $(h, k + a)$ and the directrix $y = k - a$.

38. Draw the graph of $y^4 - 36x^2 = 0$.

39. Prove that a curve that is symmetric with respect to both coordinate axes is symmetric with respect to the origin.

40. Prove that a curve that is symmetric with respect to a coordinate axis and the origin is symmetric with respect to the other coordinate axis.

41. Figure 3.16 shows an arched truss 80 ft long with heights as indicated. The vertical "stringers" are 10 ft apart. If both the top and bottom of the arch are arcs of parabolas, find to the nearest foot the sum of the lengths of the vertical and inclined stringers.

FIGURE 3.16

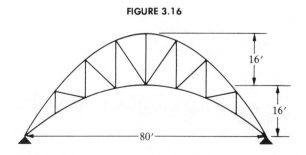

42. A TV-satellite dish is parabolic and has its receiver 24 inches from its vertex. Find the equation of the parabolic cross section of the dish (locate the vertex at the origin).

3.3

THE ELLIPSE

We obtained second-degree equations for the circle and parabola; hence these curves are conics. We come now to another kind of curve, which, like the circle but unlike the parabola, is a closed curve. The new curve is a conic because its equation, as we shall demonstrate, is of the second degree in x and y.

DEFINITION 3.3 ● *An **ellipse** is the set of all points P in a plane such that the sum of the distances of P from two fixed points F' and F on the plane is constant.*

Each of the fixed points is called a **focus** (plural **foci**). Figure 3.17 shows how the foci can be used in drawing an ellipse. The ends of a string are fastened at the foci F' and F. As the pencil at P moves, with the string taut, the curve traced is an ellipse.

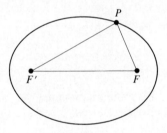

FIGURE 3.17

To find the equation of an ellipse, we take the origin of coordinates midway between the foci and one of the coordinate axes on the line through the foci (Fig. 3.18). We denote the distance between the foci by $2c$, and, accordingly, label the

FIGURE 3.18

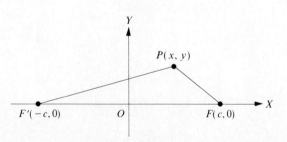

foci as $F'(-c,0)$ and $F(c,0)$. Now if we let the sum of the distances from a point $P(x, y)$ of the ellipse to the foci be $2a$, we obtain

$$PF' + PF = 2a,$$

$$\sqrt{(x + c)^2 + y^2} + \sqrt{(x - c)^2 + y^2} = 2a.$$

By transposing the second radical, squaring, and simplifying, we get

$$cx - a^2 = -a\sqrt{(x - c)^2 + y^2}.$$

By squaring again and simplifying, we find

$$(a^2 - c^2)x^2 + a^2y^2 = a^2(a^2 - c^2).$$

We observe from the figure that the length of one side of triangle $F'PF$ is $2c$, and the sum of the lengths of the other sides is $2a$. Hence $2a > 2c$, and, consequently, $a^2 - c^2 > 0$. Letting $b^2 = a^2 - c^2$ and dividing by the nonzero quantity a^2b^2, we obtain the final form

$$\boxed{\frac{x^2}{a^2} + \frac{y^2}{b^2} = 1.}$$ (3.8)

We first observe that the graph of Eq. (3.8) is symmetric with respect to both coordinate axes. When $y = 0$, then $x = \pm a$, and when $x = 0$, $y = \pm b$. Hence the ellipse cuts the x axis at $V'(-a,0)$ and $V(a,0)$ and cuts the y axis at $B'(0, -b)$ and $B(0, b)$ as in Fig. 3.19. The segment $V'V(= 2a)$ is called the **major axis** of the ellipse and the segment $B'B(= 2b)$ the **minor axis**. The ends of the

FIGURE 3.19

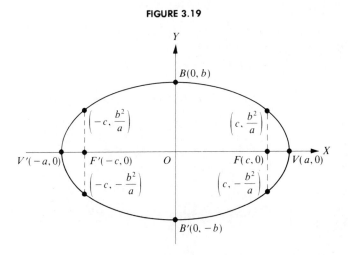

major axis are called **vertices**. The intersection of the axes of the ellipse is the **center**. The chord through a focus and perpendicular to the major axis is called a **latus rectum**. Substituting $x = c$ in Eq. (3.8) and using the relation $c^2 = a^2 - b^2$, we find the points $(c, -b^2/a)$ and $(c, b^2/a)$ to be the ends of one latus rectum. The ends of the other latus rectum are at $(-c, -b^2/a)$ and $(-c, b^2/a)$. These results show that the length of each latus rectum is $2b^2/a$. The major axis is longer than the minor axis. This is true because $b^2 = a^2 - c^2 < a^2$, and therefore $b < a$. We note also that the foci are on the major axis. The ellipse and the important points are shown in Fig. 3.19.

If we take the foci of an ellipse on the y axis at $(0, -c)$ and $(0, c)$, we can obtain, by steps similar to those above, the equation

$$\frac{y^2}{a^2} + \frac{x^2}{b^2} = 1. \tag{3.9}$$

In this case the major axis is on the y axis and the minor axis on the x axis (Fig. 3.20).

FIGURE 3.20

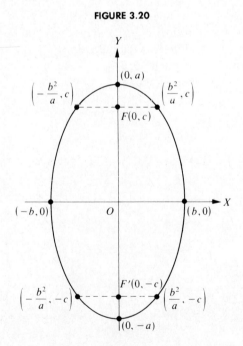

Extent of Ellipse

From the definition of an ellipse, we know that its points do not extend indefinitely far from the foci. The fact that an ellipse is limited in extent can be deduced from its equation. Thus solving Eq. (3.8) for x and y in turn, we get

$$x = \pm \frac{a}{b}\sqrt{b^2 - y^2} \qquad \text{and} \qquad y = \pm \frac{b}{a}\sqrt{a^2 - x^2}.$$

These equations reveal that y^2 must not exceed b^2, and x^2 must not exceed a^2. In other words, the permissible values are

$$-b \le y \le b \qquad \text{and} \qquad -a \le x \le a.$$

Hence no point of the ellipse is outside the rectangle formed by the horizontal lines $y = -b$, $y = b$ and the vertical lines $x = -a$, $x = a$.

In many equations the extent of the graph can be readily determined by solving for each variable in terms of the other. The concept of extent, as well as that of symmetry, is often an aid in drawing a graph.

EXAMPLE 1 ● Find the equation of an ellipse with foci at $(0, \pm 4)$ and a vertex at $(0, 6)$.

SOLUTION. The location of the foci shows that the center of the ellipse is at the origin, that $c = 4$, and that the desired equation may be expressed in the form (3.9). The given vertex, 6 units from the origin, makes $a = 6$. Using the relation $b^2 = a^2 - c^2$, we find $b^2 = 20$. Hence we obtain the equation

$$\frac{y^2}{36} + \frac{x^2}{20} = 1. \ ●$$

EXAMPLE 2 ● Sketch the ellipse $9x^2 + 25y^2 = 225$.

SOLUTION. Dividing by 225 yields the form

$$\frac{x^2}{25} + \frac{y^2}{9} = 1.$$

Since the denominator of x^2 is greater than the denominator of y^2, the major axis is along the x axis. We see also that $a^2 = 25$, $b^2 = 9$, and $c = \sqrt{a^2 - b^2} = 4$. Hence the vertices are at $(\pm 5, 0)$, the ends of the minor axis at $(0, \pm 3)$, and the foci at $(\pm 4, 0)$. The length of a latus rectum is $2b^2/a = 18/5$. The locations of the ends of the axes and the ends of each latus rectum are sufficient for making a sketch of the ellipse. Figure 3.21 shows the curve with several important points indicated. ●

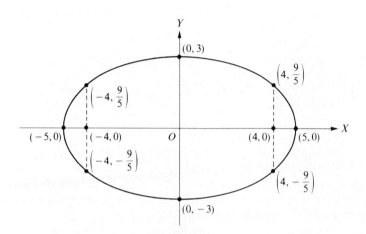

FIGURE 3.21

Focus–Directrix Property of an Ellipse

We defined a parabola in terms of a focus and directrix, but we made no use of a directrix in defining an ellipse. It turns out, however, that an ellipse has a directrix. In deriving the equation of an ellipse [Eq. (3.8)], we arrived at the equation

$$-a\sqrt{(x-c)^2 + y^2} = cx - a^2.$$

This equation implies the existence of a directrix that is evident after a slight modification. Thus, dividing by $-a$ and factoring the right member, we obtain

$$\sqrt{(x-c)^2 + y^2} = \frac{c}{a}\left(\frac{a^2}{c} - x\right).$$

The left member of this equation is the distance from a point (x, y) of the ellipse to the focus $(c, 0)$. The factor $(a^2/c) - x$ of the right member is the distance from

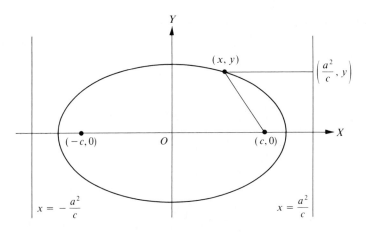

FIGURE 3.22

the point of the ellipse to the line $x = a^2/c$ (Fig. 3.22), and the factor c/a is a constant between zero and unity. Hence we have a proof, based on our definition of an ellipse, of the following theorem.

THEOREM 3.3 ● *An ellipse is the set of points in a plane such that the distance of each point of the set from a fixed point of the plane is equal to a constant (between 0 and 1) times its distance from a fixed line of the plane.*

Sometimes an ellipse is defined in terms of a focus and directrix. When so defined, it is possible to prove the statement contained in our definition of an ellipse (Definition 3.3).

The line $x = a^2/c$ is the directrix corresponding to the focus $(c, 0)$. It would be easy to show that the point $(-c, 0)$ and the line $x = -a^2/c$ constitute another focus and directrix. This fact, however, is geometrically evident from considerations of symmetry.

We wish to discuss further the quantity c/a. This ratio is called the **eccentricity** e of the ellipse. The shape of an ellipse depends on the value of its eccentricity. Suppose, for example, we visualize an ellipse in which the major axis remains constant while e starts at zero and approaches unity. If $e = 0$, the equations $e = c/a$ and $b^2 = a^2 - c^2$ show that $c = 0$ and $a = b$. The two foci are then coincident at the center, and the ellipse is a circle. As e increases, the foci separate, each receding from the center, and b decreases. As e approaches 1, c approaches a, and b approaches 0. Hence the ellipse, starting as a circle, becomes narrow and narrower. If $e = 1$, or $c = a$, then $b = 0$. Equation (3.8) would not then apply because b^2 is a denominator. But, if $e = 1$, the definition of an ellipse requires the graph to be the line segment connecting the foci.

Summarizing, we have an actual ellipse if e is between 0 and 1. When e is slightly more than 0, the ellipse is somewhat like a circle; when e is slightly less than 1, the ellipse is relatively long and narrow.

Ellipses with the same major axis and different eccentricities are constructed in Fig. 3.23.

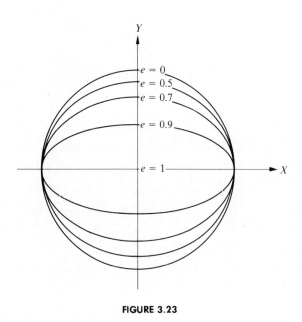

FIGURE 3.23

EXAMPLE 3 ● The moon's orbit is an ellipse with the earth at one focus. The length of the major axis is 478,000 miles and the eccentricity $e = 0.0549$. Find the least and greatest distances from the earth to the moon.

SOLUTION. The ellipse in Fig. 3.24 represents the path of the moon as it revolves about the earth. The major axis of the ellipse extends from $V'(-a, 0)$ to $V(a, 0)$, and $F(c, 0)$ represents the position of the earth, where $a = 239{,}000$, $c = ae = 13{,}000$ miles, rounded off. Then the least distance from the earth to the moon is $239{,}000 - 13{,}000 = 226{,}000$ miles. And the greatest distance is $239{,}000 + 13{,}000 = 252{,}000$ miles. ●

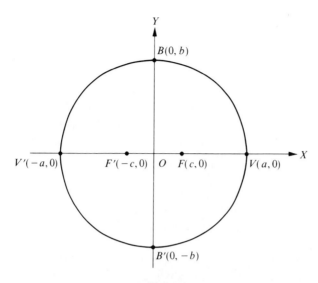

FIGURE 3.24

Ellipse with Center at (h, k)

If the axes of an ellipse are parallel to the coordinate axes and the center is at (h, k), we can obtain its equation by applying the translation formulas of Section 2.8. We draw a new pair of coordinate axes along the axes of the ellipse (Fig. 3.25).

The equation of the ellipse referred to the new axes is

$$\frac{x'^2}{a^2} + \frac{y'^2}{b^2} = 1.$$

FIGURE 3.25

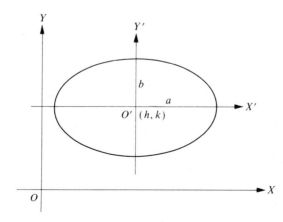

The substitutions $x' = x - h$ and $y' = y - k$ yield

$$\frac{(x-h)^2}{a^2} + \frac{(y-k)^2}{b^2} = 1.$$
(3.10)

Similarly, when the major axis is parallel to the y-axis, we have

$$\frac{(y-k)^2}{a^2} + \frac{(x-h)^2}{b^2} = 1.$$
(3.11)

These are the **standard forms** of equations of ellipses. They reduce to Eqs. (3.8) and (3.9) when $h = k = 0$. The quantities a, b, and c have the same meaning whether or not the center of the ellipse is at the origin. Therefore constructing the graph of an equation in either of the forms of Eqs. (3.10) or (3.11) presents no greater difficulty than drawing the graph of one of the simpler Eqs. (3.8) or (3.9).

EXAMPLE 4 ● Find the equation of the ellipse with foci at $(4, -2)$ and $(10, -2)$, and a vertex at $(12, -2)$.

SOLUTION. The center, midway between the foci, is at $(7, -2)$. The distance between the foci is 6 and the given vertex is 5 units from the center; hence $c = 3$ and $a = 5$. Then $b^2 = a^2 - c^2 = 16$. Since the major axis is parallel to the x axis, we substitute in Eq. (3.10) and get

$$\frac{(x-7)^2}{25} + \frac{(y+2)^2}{16} = 1.$$

The graph of this ellipse is constructed in Fig. 3.26. ●

FIGURE 3.26

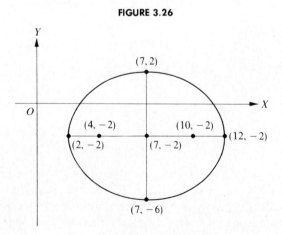

EXAMPLE 5 ● Reduce to standard form the equation

$$4y^2 + 9x^2 - 24y - 72x + 144 = 0.$$

SOLUTION. The essential steps in the process follow.

$$4(y^2 - 6y) + 9(x^2 - 8x) = -144,$$

$$4(y^2 - 6y + 9) + 9(x^2 - 8x + 16) = -144 + 36 + 144,$$

$$4(y - 3)^2 + 9(x - 4)^2 = 36,$$

$$\frac{(y - 3)^2}{9} + \frac{(x - 4)^2}{4} = 1.$$

The coordinates of the center are $(4, 3)$; $a = 3$, $b = 2$, and

$$c = \sqrt{a^2 - b^2} = \sqrt{9 - 4} = \sqrt{5}.$$

The vertices at $(4, 0)$ and $(4, 6)$, and the ends of the minor axis are at $(2, 3)$ and $(6, 3)$. The coordinates of the foci are $(4, 3 - \sqrt{5})$ and $(4, 3 + \sqrt{5})$. The graph is constructed in Fig. 3.27. ●

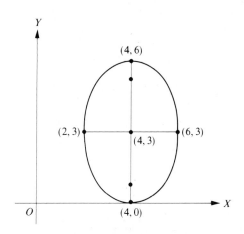

FIGURE 3.27

Conic sections have been used to describe natural and astronomical data. Kepler's Laws of Planetary Motion state that (1) the planets have **elliptic** paths with the sun at a focus; (2) the line connecting the sun with a planet sweeps over equal areas in equal time intervals; and (3) the square of the length of time for a complete revolution of a planet is proportional to the cube of one-half the length of the major axis, that is, a^3. In industry, much use is made of elliptic-shaped gears and of semi-elliptic springs. The ellipse has a reflective property not totally unlike the parabola: A light or sound wave emanating from one focus would be reflected by an elliptic surface to arrive at the other focus. So-called "whispering galleries" utilize this principle: A whisper at one focus is audible at the other.

Exercises

For each of Exercises 1 through 16, find the coordinates of the foci, the ends of the major and minor axes, and the ends of each latus rectum. From this information sketch the curve.

1. $\dfrac{x^2}{25} + \dfrac{y^2}{9} = 1$

2. $\dfrac{y^2}{25} + \dfrac{x^2}{9} = 1$

3. $\dfrac{x^2}{169} + \dfrac{y^2}{144} = 1$

4. $\dfrac{y^2}{9} + \dfrac{x^2}{4} = 1$

5. $\dfrac{x^2}{49} + \dfrac{y^2}{25} = 1$

6. $\dfrac{y^2}{25} + \dfrac{x^2}{16} = 1$

7. $\dfrac{y^2}{169} + \dfrac{x^2}{144} = 1$

8. $\dfrac{y^2}{25} + \dfrac{x^2}{4} = 1$

9. $2x^2 + 3y^2 = 12$

10. $x^2 + 4y^2 = 4$

11. $2x^2 + 3y^2 = 18$

12. $16x^2 + y^2 = 16$

13. $\dfrac{(x-3)^2}{16} + \dfrac{(y-2)^2}{9} = 1$

14. $\dfrac{(x-3)^2}{36} + \dfrac{(y-2)^2}{9} = 1$

15. $\dfrac{(x-5)^2}{169} + \dfrac{(y+5)^2}{49} = 1$

16. $\dfrac{(y+3)^2}{36} + \dfrac{(x-6)^2}{16} = 1$

Reduce each equation in Exercises 17 through 24 to standard form. Then find the coordinates of the center, the foci, the ends of the major and minor axes, and the ends of each latus rectum. Sketch the curve.

17. $16x^2 + 25y^2 + 160x + 200y + 400 = 0$

18. $x^2 + 4y^2 + 6x + 16y + 21 = 0$

19. $4x^2 + y^2 + 8x - 4y - 8 = 0$

c 20. $3.104x^2 + 2.078y^2 + 8.219x - 1.734y - 2.413 = 0$

21. $4x^2 + 8y^2 - 4x - 24y - 13 = 0$

22. $25(x+1)^2 + 169(y-2)^2 = 4{,}225$

23. $225(x - 2)^2 + 289(y - 3)^2 = 65{,}025$

24. $169(x - 1)^2 + 144(y - 3)^2 = 24{,}336$

Write the equation of the ellipse that satisfies the given conditions in each of Exercises 25 through 36. Sketch each curve.

25. Center at $(5, 1)$, vertex at $(5, 4)$, end of a minor axis at $(3, 1)$.

26. Vertex at $(6, 3)$, foci at $(-4, 3)$ and $(4, 3)$.

27. Ends of minor axis at $(-1, 2)$ and $(-1, -4)$, focus at $(1, -1)$.

28. Vertices at $(-1, 3)$ and $(5, 3)$, length of minor axis 4.

29. Center at $(0, 0)$, one vertex at $(0, -6)$, end of a minor axis at $(4, 0)$.

30. Vertices at $(-5, 0)$ and $(5, 0)$, length of a latus rectum $\frac{8}{5}$.

31. Center at $(-2, 2)$, a vertex at $(-2, 6)$, a focus at $(-2, 2 + \sqrt{12})$.

32. Foci at $(-4, 2)$ and $(4, 2)$, length of major axis 10.

33. Center at $(5, 4)$, length of major axis 16, length of minor axis 6, major axis parallel to x axis.

34. Foci at $(0, -8)$ and $(0, 8)$, length of major axis 34.

35. Foci at $(-5, 0)$ and $(5, 0)$, length of minor axis 8.

36. Center at $(3, 2)$, a focus at $(3, 7)$, and a vertex at $(3, -5)$.

37. Find the intersection of the two ellipses

$$\frac{x^2}{9} + \frac{y^2}{4} = 1 \quad \text{and} \quad \frac{x^2}{4} + \frac{y^2}{9} = 1.$$

38. The perimeter of a triangle is 30 and the points $(0, -5)$ and $(0, 5)$ are two of the vertices. Find the graph of the third vertex.

39. Find the equation of the graph of the midpoints of the ordinates of the circle $x^2 + y^2 = 36$.

40. The ordinates of a curve are k times the ordinates of the circle $x^2 + y^2 = a^2$. Show that the curve is an ellipse if k is a positive number different from 1.

41. A line segment of length 12 moves with its ends always touching the coordinate axes. Find the equation of the graph of the point on the segment that is 4 units from the end in contact with the x axis.

42. A rod of length $a + b$ moves with its ends in contact with the coordinate axes. Show that the point at a distance a from the end in contact with the x axis describes an ellipse if $a \neq b$.

43. The earth's orbit is an ellipse with the sun at one focus. The length of the major axis is 186,000,000 miles, and the eccentricity 0.0167. Find the distances from the ends of the major axis to the sun. These are the greatest and least distances from the earth to the sun.

44. The arch of an underpass is a semi-ellipse 60 ft wide and 20 ft high. Find the clearance at the edge of a lane if the edge is 20 ft from the middle.

45. A point moves in the coordinate plane so that the sum of its distances from $(1, 2)$ and $(5, 2)$ is equal to 16. Find the equation of the path of the moving point.

46. Find the equation of the path of a point $P(x, y)$ which moves so that its distance from $(5, 0)$ is one-half its distance from the line $x = 20$.

47. Find the equation of the path of a point $P(x, y)$ which moves so that its distance from $(-4, 0)$ is equal to two-thirds its distance from the line $x = -9$.

48. Derive Eq. (3.9) and Eq. (3.11).

49. If p is a parameter of positive numbers, show that all members of the family of ellipses

$$\frac{x^2}{a^2 + p} + \frac{y^2}{b^2 + p} = 1$$

have the same foci.

50. A family of ellipses with the same foci may be represented by the equation

$$\frac{x^2}{15 + p} + \frac{y^2}{p} = 1, \qquad p > 0.$$

Find the coordinates of the ends of the axes of the ellipses when $p = 1$, 10, 21, and 34. Sketch the four ellipses.

51. Write an equation of the family of ellipses whose common foci are $(-6, 0)$ and $(6, 0)$. Find the member of the family that passes through $(9, 0)$.

52. An arched entrance to a zoo has a cross section as shown on the facing page, with the curve a semi-ellipse.

   a) Find the heights of the arch measured from the ground at every 3-ft distance from the point P.

   b) If it is 10 feet thick, find the number of cubic yards of concrete required for its construction. The area of an ellipse of semi-axes a and b is given by πab.

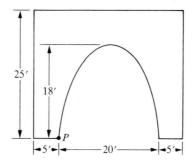

3.4

THE HYPERBOLA

The third type of conic that we shall consider is the hyperbola. The equations of hyperbolas resemble those of ellipses but the properties of these two kinds of conics differ considerably in some respects.

DEFINITION 3.4 ● *A hyperbola is the set of points in a plane such that the difference of the distances of each point of the set from two fixed points (foci) in the plane is constant.*

To derive the equation of a hyperbola, we take the origin midway between the foci and a coordinate axis on the line through the foci (Fig. 3.28). We denote the foci by $F'(-c,0)$ and $F(c,0)$ and the difference of the distances between a point of the hyperbola and the foci by the positive constant $2a$. Then, from the

FIGURE 3.28

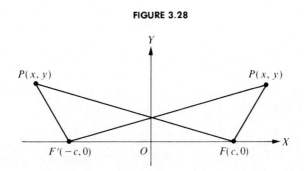

above definition,

$$|F'P| - |FP| = 2a$$

or

$$|F'P| - |FP| = -2a,$$

depending on whether the point $P(x, y)$ of the hyperbola is to the right or left of the y axis. We combine these equations by writing

$$|F'P| - |FP| = \pm 2a$$

or

$$\sqrt{(x + c)^2 + y^2} - \sqrt{(x - c)^2 + y^2} = \pm 2a.$$

Now, transposing the second radical, squaring, and simplifying, we obtain

$$cx - a^2 = \pm a\sqrt{(x - c)^2 + y^2}.$$

Squaring the members of this equation and simplifying, we get

$$(c^2 - a^2)x^2 - a^2 y^2 = a^2(c^2 - a^2).$$

Then letting $b^2 = c^2 - a^2$ and dividing by $a^2 b^2$, we have

$$\boxed{\frac{x^2}{a^2} - \frac{y^2}{b^2} = 1.}$$

(3.12)

The division is allowable if $c^2 - a^2 \neq 0$. In Fig. 3.28, the length of one side of triangle $F'PF$ is $2c$ and the difference of the other two sides is $2a$. Hence

$$c > a \quad \text{and} \quad c^2 - a^2 > 0.$$

The graph of Eq. (3.12) is symmetric with respect to the coordinate axes. The permissible values of x and y become evident where each is expressed in terms of the other. Thus we get

$$x = \pm \frac{a}{b}\sqrt{b^2 + y^2} \quad \text{and} \quad y = \pm \frac{b}{a}\sqrt{x^2 - a^2}.$$

We see, from the first of these equations, that y may have any real value and, from the second, that x may have any real value except those for which $x^2 < a^2$. Hence the hyperbola extends indefinitely far from the axes in each quadrant. But there is no part of the graph between the line $x = -a$ and the line $x = a$. Accordingly, the hyperbola consists of two separate parts, or **branches** (Fig. 3.29). The points $V'(-a, 0)$ and $V(a, 0)$ are called **vertices**, and the segment $V'V$ is called the **transverse axis**. The segment from $B'(0, -b)$ to $B(0, b)$ is called the **conjugate axis**. Although the conjugate axis has no point in common with the

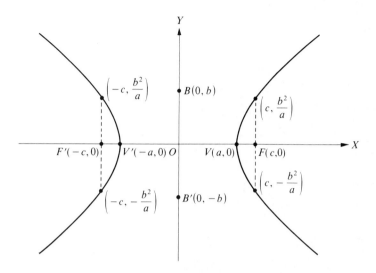

FIGURE 3.29

hyperbola, it has an important relation to the curve, as we shall discover. The intersection of the axes is the **center**. The chord through a focus and perpendicular to the transverse axis is called a **latus rectum**. By substituting $x = c$ in Eq. (3.12) and using the relation $c^2 = a^2 + b^2$, we find the extremities of a latus rectum to be $(c, -b^2/a)$ and $(c, b^2/a)$. Hence the length is $2b^2/a$.

We note that the meanings of the three positive quantities a, b, and c appearing here are analogous to their meanings when used with an ellipse. The quantity a is the distance from the center of the hyperbola to a vertex, b is the distance from the center to an end of the conjugate axis, and c is the distance from the center to a focus. But the relations of a, b, and c for an ellipse are not the same as for a hyperbola. For an ellipse, $a > c$, $c^2 = a^2 - b^2$, and $a > b$. For a hyperbola, $c > a$, $c^2 = a^2 + b^2$, and no restriction is placed on the relative values of a and b.

If the foci are on the y axis at $F'(0, -c)$ and $F(0, c)$, the equation of the hyperbola is

$$\frac{y^2}{a^2} - \frac{x^2}{b^2} = 1. \tag{3.13}$$

The vertices are at $V'(0, -a)$ and $V(0, a)$, and the relations among a, b, and c are unchanged.

The generalized equations of hyperbolas with axes parallel to the coordinate axes and centers at (h, k) are

$$\frac{(x - h)^2}{a^2} - \frac{(y - k)^2}{b^2} = 1, \tag{3.14}$$

$$\frac{(y - k)^2}{a^2} - \frac{(x - h)^2}{b^2} = 1. \tag{3.15}$$

Equations (3.12) through (3.15) are said to be in **standard forms**. The meanings and relations of a, b, and c are the same in all the equations.

The Asymptotes of a Hyperbola

Unlike the other conics, a hyperbola has associated with it two lines bearing an important relation to the curve. These lines are the extended diagonals of the rectangle in Fig. 3.30. One pair of sides of the rectangle pass through the vertices and are perpendicular to the transverse axis. The other pair pass through the ends of the conjugate axis. Suppose we consider the extended diagonal and the part of the hyperbola in the first quadrant. The equations of the diagonal and this part of the hyperbola are, respectively,

$$y = \frac{b}{a}x \quad \text{and} \quad y = \frac{b}{a}\sqrt{x^2 - a^2}.$$

We see that for any $x > a$ the ordinate of the hyperbola is less than the ordinate of the line. If, however, x is many times as large as a, the corresponding ordinates

FIGURE 3.30

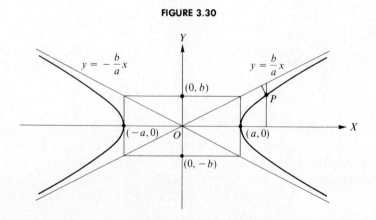

are almost equal. This may be seen more convincingly by examining the difference of the two ordinates. Thus, by subtracting, and then multiplying the numerator and denominator of the resulting fraction by $x + \sqrt{x^2 - a^2}$, we get

$$\frac{b}{a}x - \frac{b}{a}\sqrt{x^2 - a^2} = \frac{b\left(x - \sqrt{x^2 - a^2}\right)}{a}$$

$$= \frac{b\left(x - \sqrt{x^2 - a^2}\right)\left(x + \sqrt{x^2 - a^2}\right)}{a\left(x + \sqrt{x^2 - a^2}\right)}$$

$$= \frac{ab}{x + \sqrt{x^2 - a^2}}.$$

The numerator of the last fraction is constant. The denominator, however, increases as x increases. In fact, we can make the denominator as large as we please by taking a sufficiently large value for x. This means that the fraction, which is the difference of the ordinates of the line and the hyperbola, gets closer and closer to zero as x gets larger and larger. The perpendicular distance from a point P of the hyperbola to the line is less than the fraction. Consequently, we can make the perpendicular distance, though never zero, as near zero as we please by taking x sufficiently large. When the perpendicular distance from a line to a curve approaches zero as the curve extends indefinitely far from the origin, the line is said to be an **asymptote** of the curve. From considerations of symmetry, we conclude that each extended diagonal is an asymptote of the curve. Hence the equations of the asymptotes of the hyperbola represented by Eq. (3.12) are

$$y = \frac{b}{a}x \quad \text{and} \quad y = -\frac{b}{a}x.$$

Similarly, the equations of the asymptotes associated with Eq. (3.13) are

$$y = \frac{a}{b}x \quad \text{and} \quad y = -\frac{a}{b}x.$$

We observe that for each of the hyperbolas, (3.12) through (3.15), the equations of the asymptotes may be obtained by factoring the left member and equating each factor to zero.

The asymptotes of a hyperbola are helpful in sketching a hyperbola. A rough drawing can be made from the associated rectangle and its extended diagonals. The accuracy may be improved considerably, however, by plotting the endpoints of each latus rectum.

If $a = b$, the associated rectangle is a square and the asymptotes are perpendicular to each other. For this case the hyperbola is said to be **equilateral** because its axes are equal, or is said to be **rectangular** because its asymptotes intersect at right angles.

The ratio c/a is called the **eccentricity** e of the hyperbola. The angle of intersection of the asymptotes, and therefore the shape of the hyperbola, depends on the value of e. Since $c > a$, the value of e is greater than 1. If c is just slightly greater than a, so that e is near 1, the relation $c^2 = a^2 + b^2$ shows that b is small compared with a. Then the asymptotes make a pair of small angles. The branches of the hyperbola, enclosed by small angles, diverge slowly. If e increases, the branches are enclosed by larger angles. And the angles can be made near 180° by taking large values for e.

Having found that the eccentricity of an ellipse is a number between 0 and 1 and that the eccentricity of a hyperbola is greater than 1, we naturally ask if there is a conic whose eccentricity is equal to 1. To have an answer, we recall that any point of a parabola is equally distant from the directrix and the focus (Definition 3.1). The ratio of these distances is 1, and consequently this value is the eccentricity of a parabola.

EXAMPLE 1 ● Sketch the curve $36x^2 - 64y^2 = 2304$.

SOLUTION. We divide by 2304 and reduce the equation to

$$\frac{x^2}{64} - \frac{y^2}{36} = 1.$$

The graph is a hyperbola in which $a = 8$, $b = 6$, and $c = \sqrt{a^2 + b^2} = 10$. The vertices therefore are $(\pm 8, 0)$ and the foci $(\pm 10, 0)$. Each latus rectum has a length of $2b^2/a = 9$. The equations of the asymptotes are $3x - 4y = 0$ and $3x + 4y = 0$. From this information the hyperbola can be drawn (Fig. 3.31). ●

FIGURE 3.31

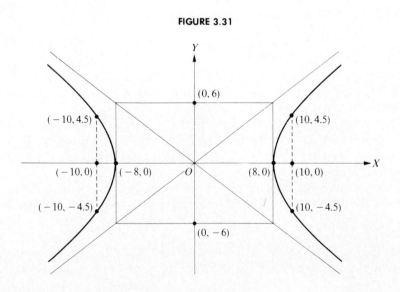

EXAMPLE 2 ● Draw the graph of $12y^2 - 4x^2 + 72y + 16x + 44 = 0$.

SOLUTION. We first reduce the equation to a standard form. Thus

$$12(y^2 + 6y + 9) - 4(x^2 - 4x + 4) = -44 + 108 - 16,$$
$$12(y + 3)^2 - 4(x - 2)^2 = 48, \tag{3.16}$$
$$\frac{(y + 3)^2}{4} - \frac{(x - 2)^2}{12} = 1.$$

We see now that $a = 2$, $b = 2\sqrt{3}$, $c = \sqrt{4 + 12} = 4$, and that the center of the hyperbola is at $(2, -3)$. Hence the ends of the transverse axis are at $(2, -5)$ and $(2, -1)$, and the ends of the conjugate axis are at $(2 - 2\sqrt{3}, -3)$ and $(2 + 2\sqrt{3}, -3)$. The sides of the associated rectangle pass through these points (Fig. 3.32). The extended diagonals of the rectangle are the asymptotes. The coordinates of the foci, 4 units from the center, are $(2, -7)$ and $(2, 1)$. The length of each latus rectum is $2b^2/a = 12$, and therefore the ends of each latus rectum are 6 units from a focus. The vertices of the hyperbola, the ends of each latus rectum, and the asymptotes are sufficient for drawing a reasonably accurate graph. The equations of the asymptotes, though not needed for the drawing, are

$$\frac{y + 3}{2} + \frac{x - 2}{2\sqrt{3}} = 0 \quad \text{and} \quad \frac{y + 3}{2} - \frac{x - 2}{2\sqrt{3}} = 0.$$

We obtain these equations by factoring the left member of Eq. (3.16) and equating each factor to zero. ●

FIGURE 3.32

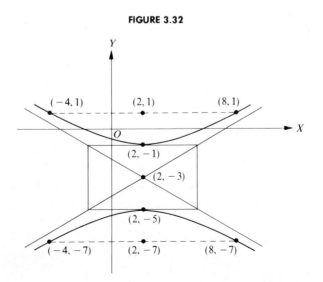

A very interesting and important application of the hyperbola is that of locating the place from which a sound, such as gunfire, emanates. From the difference in the times at which the sound reaches two listening posts, the difference between distances of the posts from the gun can be determined. Then the gun is known to be located on a branch of a hyperbola of which the posts are foci. The position of the gun on this curve can be found by the use of a third listening post. Either of the two posts and the third are foci of a branch of another hyperbola on which the gun is located. Hence the gun is at the intersection of the two branches.

The same principle of the hyperbola is used in modern navagational systems to locate the position of a ship or of an airplane. In this case the vessel receives signals from three known locations and hence is positioned at the intersection of two hyperbolas.

Exercises

In each of Exercises 1 through 12 find the coordinates of the vertices and foci, the length of each latus rectum, and the equations of the asymptotes. Draw the asymptotes and sketch the hyperbola.

1. $\dfrac{x^2}{16} - \dfrac{y^2}{4} = 1$

2. $\dfrac{y^2}{16} - \dfrac{x^2}{4} = 1$

3. $\dfrac{x^2}{36} - \dfrac{y^2}{64} = 1$

4. $\dfrac{y^2}{36} - \dfrac{x^2}{9} = 1$

5. $\dfrac{x^2}{25} - \dfrac{y^2}{9} = 1$

6. $\dfrac{y^2}{25} - \dfrac{x^2}{9} = 1$

7. $\dfrac{x^2}{49} - \dfrac{y^2}{49} = 1$

8. $\dfrac{y^2}{64} - \dfrac{x^2}{64} = 1$

9. $\dfrac{(x+2)^2}{16} - \dfrac{(y-3)^2}{9} = 1$

10. $\dfrac{(y-3)^2}{16} - \dfrac{(x+2)^2}{9} = 1$

11. $\dfrac{(x-4)^2}{25} - \dfrac{(y-5)^2}{25} = 1$

12. $\dfrac{(y-5)^2}{36} - \dfrac{(x+5)^2}{36} = 1$

Reduce each equation, Exercises 13 through 18, to standard form. Then find the coordinates of the center, the vertices, and the foci. Draw the asymptotes and sketch the graph of the equation.

13. $3x^2 - 2y^2 + 4y - 26 = 0$

14. $9x^2 - 4y^2 + 90x + 189 = 0$

15. $9x^2 - 4y^2 + 36x - 16y - 16 = 0$

16. $x^2 - 2y^2 + 6x + 4y + 5 = 0$

17. $49y^2 - 4x^2 + 98y - 48x - 291 = 0$

18. $4y^2 - 9x^2 + 8y - 54x - 81 = 0$

In each of Exercises 19 through 28, find the equation of the hyperbola that satisfies the given conditions.

19. Center at $(2,0)$,   a focus at $(10,0)$,   a vertex at $(6,0)$.

20. Center at $(0,2)$,   a focus at $(0,10)$,   a vertex at $(0,6)$.

21. Center at $(0,0)$,   transverse axis along the x axis,   transverse axis 12,   latus rectum 12.

22. Center at $(0,0)$,   transverse axis along the y axis,   transverse axis 12,   latus rectum 6.

23. Center at $(2,2)$,   a focus at $(10,2)$,   a vertex at $(5,2)$.

24. Center at $(-2,2)$,   a vertex at $(4,2)$,   a focus at $(6,2)$,   transverse axis parallel to the x axis.

© 25. Center at $(4.1025, -2.1374)$,   a vertex at $(4.1025, 1.7144)$,   a focus at $(4.1025, 2.4198)$.

26. Center at $(2, -2)$,   transverse axis parallel to the x axis,   transverse axis 6, conjugate axis 10.

27. Center at $(0,0)$,   transverse axis along the x axis,   passing through the points $(3,5)$, $(2, -3)$.

28. Center at $(6,0)$,   conjugate axis along the x axis,   asymptotes $5x - 6y - 30 = 0$ and $5x + 6y - 30 = 0$.

29. The pair of hyperbolas

$$\frac{x^2}{a^2} - \frac{y^2}{b^2} = 1 \quad \text{and} \quad \frac{y^2}{b^2} - \frac{x^2}{a^2} = 1$$

are called **conjugate hyperbolas**. Show that these hyperbolas have the same asymptotes.

30. Sketch on the same coordinate axes the conjugate hyperbolas

$$\frac{x^2}{25} - \frac{y^2}{9} = 1 \quad \text{and} \quad \frac{y^2}{9} - \frac{x^2}{25} = 1$$

31. Find the equation of the path of a point that moves so that its distance from $(4,0)$ is twice its distance from the line $x = 1$.

32. Find the equation of the path of a point that moves so that its distance from $(5,0)$ is $5/4$ of its distance from the line $x = 16/5$.

33. Show that $(c,0)$ is a focus and $x = (a^2/c)$ is a directrix of the hyperbola

$$\frac{x^2}{a^2} - \frac{y^2}{b^2} = 1.$$

Proceed as in the case of the ellipse (Section 3.3).

34. Listening posts are at A, B, and C. Point A is 2,000 ft north of point B, and point C is 2,000 ft east of B. The sound of a gun reaches A and B simultaneously one second after it reaches C. Show that the coordinates of the gun's position are approximately (860, 1,000), where the x axis passes through B and C and the origin is midway between B and C. Assume that sound travels 1,100 ft/sec.

35. If a and b are constants and p is a parameter such that $p > 0$ and $b^2 - p > 0$, show that the equation

$$\frac{x^2}{a^2 + p} - \frac{y^2}{b^2 - p} = 1$$

represents a family of hyperbolas with common foci on the x axis.

36. Write an equation of the family of hyperbolas whose foci are $(-4, 0)$ and $(4, 0)$. Find the member of the family that passes through $(2, 0)$.

REVIEW EXERCISES

1. Define: *parabola, ellipse, hyperbola*.

2. Derive the equation of the graph of a point that is equidistant from the point $(a, 0)$ and the line $x = -a$.

3. Express, in standard form, the equation

$$x^2 + 16y - 32 = 0.$$

Give the coordinates of the vertex, the focus, and the ends of the latus rectum. Sketch the graph.

4. Express, in standard form, the equation

$$y^2 - 6y - 8x - 7 = 0.$$

Then give the coordinates of the vertex of the parabola, the focus and the ends of the latus rectum. Sketch the graph.

5. Sketch the ellipse

$$16x^2 + 25y^2 = 40.$$

Give the coordinates of the foci, the ends of the major and minor axes, and the ends of each latus rectum.

6. Reduce the equation

$$3x^2 + 2y^2 - 24x + 12y + 60 = 0$$

to standard form. Then find the coordinates of the center, the foci, and the ends of the major and minor axes. Sketch the ellipse.

7. Give the coordinates of the vertices, the foci, the ends of the conjugate axis, the ends of each latus rectum of the hyperbola

$$\frac{x^2}{36} - \frac{y^2}{64} = 1.$$

Give also the equations of the asymptotes. Sketch the graph of the hyperbola.

8. Construct the graph of the equation

$$12x^2 - 4y^2 + 72x + 16y + 44 = 0.$$

Give the coordinates of the center, the vertices, the foci, and the ends of each latus rectum of the hyperbola.

4

Simplification of Equations

4.1

SIMPLIFICATION BY TRANSLATION

In Section 2.8 we discovered that the equation of a circle whose center is not at the origin can be expressed in a simpler form by an appropriate translation of axes. Also, in Chapter 3 we used the idea of translation of axes to express the equations of parabolas, ellipses, and hyperbolas in standard forms. We might surmise, then, that general equations of conics are expressible in simple forms. This is indeed true, as we shall see.

The general equation of the second degree may be presented in the form

$$Ax^2 + Bxy + Cy^2 + Dx + Ey + F = 0. \qquad (4.1)$$

But before taking up the general equation we shall deal with special cases obtained by setting one or more of the coefficients equal to zero. We advise the student to review the derivation of the translation formulas (Section 2.8) and then to study carefully the following examples.

EXAMPLE 1 ● By a translation of axes, simplify the equation

$$x^2 - 6x - 6y - 15 = 0.$$

SOLUTION. We complete the square in the x terms and select the translation that will eliminate the x' term and the constant term in the new equation. Thus, we get

$$x^2 - 6x + 9 = 6y + 15 + 9,$$
$$(x - 3)^2 = 6(y + 4).$$

The vertex of the parabola is at $(3, -4)$. So we put $h = 3$ and $k = -4$ in the translation formulas $x = x' + h$ and $y = y' + k$, and have

$$(x' + 3 - 3)^2 = 6(y' - 4 + 4),$$
$$x'^2 = 6y'.$$

136

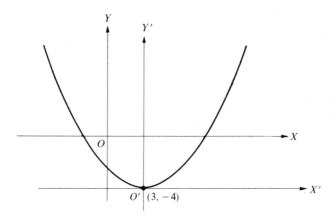

FIGURE 4.1

The parabola in Fig. 4.1 is the graph of the given equation referred to the original axes and also the graph of the new equation referred to the new axes. ●

EXAMPLE 2 ● Translate the axes so as to simplify the equation

$$2x^2 + 3y^2 + 10x - 18y + 26 = 0.$$

FIGURE 4.2

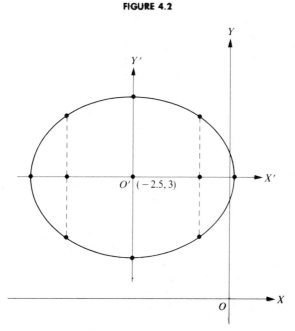

SOLUTION. The simplification can be made by completing the squares in the x and y terms. By this plan, we obtain

$$2(x^2 + 5x) + 3(y^2 - 6y) = -26,$$

$$2\left(x^2 + 5x + \tfrac{25}{4}\right) + 3(y^2 - 6y + 9) = -26 + \tfrac{25}{2} + 27,$$

$$2\left(x + \tfrac{5}{2}\right)^2 + 3(y - 3)^2 = \tfrac{27}{2}.$$

The translation formulas $x = x' - \tfrac{5}{2}$ and $y = y' + 3$ will reduce this equation to one free of first-degree terms. Making these substitutions, we find

$$2\left(x' - \tfrac{5}{2} + \tfrac{5}{2}\right)^2 + 3(y' + 3 - 3)^2 = \tfrac{27}{2},$$

$$4x'^2 + 6y'^2 = 27.$$

The graph is an ellipse with $a = \tfrac{3}{2}\sqrt{3}$, $b = \tfrac{3}{2}\sqrt{2}$, and $c = \tfrac{3}{2}$. Both sets of axes and the graph are in Fig. 4.2. ●

Examples 1 and 2 illustrate the fact that the axes can be translated so as to reduce a second-degree equation of the form

$$Ax^2 + Cy^2 + Dx + Ey + F = 0$$

to one of the forms

$$A'x'^2 + E'y' = 0,$$
$$C'y'^2 + D'x' = 0,$$
$$A'x'^2 + C'y'^2 + F' = 0.$$

In the two examples, we completed squares to locate the origins of the new coordinate systems. That plan would need to be altered when an xy term is present.

EXAMPLE 3 ● Translate the axes to rid the equation $2xy + 3x - 4y = 12$ of first-degree terms.

SOLUTION. For this equation we use the translation formulas with h and k unknown. The equation then becomes

$$2(x' + h)(y' + k) + 3(x' + h) - 4(y' + k) = 12$$

or, by multiplying and collecting terms,

$$2x'y' + (2k + 3)x' + (2h - 4)y' + 2hk + 3h - 4k = 12.$$

To eliminate the x' term and the y' term, we equate their coefficients to zero, thus obtaining $h = 2$ and $k = -\frac{3}{2}$. Using these values for h and k, we get

$$x'y' = 3.$$

We shall learn in Section 4.4 that the graph of this equation, and the given equation, is a hyperbola. Both sets of axes and the hyperbola are drawn in Fig. 4.3. ●

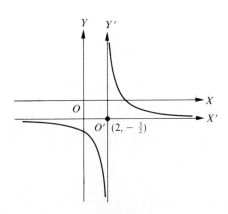

FIGURE 4.3

Exercises

Determine the new equation in each of Exercises 1 through 12 if the origin is translated to the given point.

1. $3x - 2y = 6$, $(4, -3)$

2. $5x + 4y + 3 = 0$, $(1, 2)$

3. $y^2 - 6x + 4y + 22 = 0$, $(3, 2)$

4. $x^2 - 4x + 7y = 0$, $(2, 6)$

5. $3x^2 + 4y^2 + 12x - 8y + 8 = 0$, $(-2, -1)$

6. $9x^2 + y^2 + 36x + 8y + 43 = 0$, $(-2, 4)$

7. $4y^2 - 5x^2 + 8y - 10x - 21 = 0$, $(-1, 1)$

8. $16x^2 - 4y^2 - 160x + 24y + 300 = 0$, $(5, -3)$

9. $xy - x + y - 10 = 0$, $(1, 1)$

10. $3xy - 21x - 6y - 47 = 0$, $(-2, 7)$

Ⓒ11. $3.015x^2 - 2.991x + 0.005y - 0.123 = 0$, $(1.028, 2.314)$

12. $3x^3 - 18x^2 + 36x + 4y - 36 = 0$, $(2, -3)$

In each of Exercises 13 through 22, find the point to which the origin must be translated in order that the transformed equation will have no first-degree term. Find also the new equation.

13. $xy - 2x + 4y - 4 = 0$ 14. $xy + 3x + 3y - 3 = 0$

15. $2x^2 + 2y^2 - 8x + 5 = 0$ 16. $x^2 + 2y^2 + 6x - 4y + 2 = 0$

17. $3x^2 - 2y^2 + 24x + 8y - 34 = 0$ 18. $2y^2 - 3x^2 - 12x - 16y + 14 = 0$

19. $x^2 - xy + y^2 - 9x - 6y - 27 = 0$ 20. $2x^2 - 3xy - y^2 + x - 5y - 3 = 0$

21. $x^3 + 5x^2 + 2xy + 4x - 4y - 4 = 0$

C 22. $xy + 2.8761x - 3.0019y - 9.6338 = 0$

In each of Exercises 23 through 28, eliminate the constant term and one of the first-degree terms.

23. $y^2 + 6y + 4x + 5 = 0$ 24. $x^2 - 2x + 8y - 15 = 0$

25. $y^2 + 10x - 4y + 24 = 0$ 26. $y^2 + 4y - x + 1 = 0$

27. $2x^2 - 20x + 7y + 36 = 0$ 28. $3y^2 + 11x - 6y - 19 = 0$

4.2

ROTATION OF AXES

We now wish to consider a transformation of coordinates where the new axes have the same origin but different directions from the original axes. Since the new axes may be obtained by rotating the original axes through an angle about the origin, the transformation is called a **rotation of axes**.

For a rotation through an angle θ, we will derive transformation formulas which express the old coordinates in terms of θ and the new coordinates. In Fig. 4.4 the coordinates of the point P are (x, y) when referred to the original axes OX and OY, and are (x', y') when referred to the new axes OX' and OY'. We note that

$$x = \overline{OM} \quad \text{and} \quad y = \overline{MP}, \qquad x' = \overline{OS} \quad \text{and} \quad y' = \overline{SP}.$$

The segment $\overrightarrow{RS}$ is drawn parallel to the x axis and $\overrightarrow{NS}$ is parallel to the y axis. Hence we have

$$x = \overline{OM} = \overline{ON} - \overline{MN} = \overline{ON} - \overline{RS} = x'\cos\theta - y'\sin\theta,$$
$$y = \overline{MP} = \overline{MR} + \overline{RP} = \overline{NS} + \overline{RP} = x'\sin\theta + y'\cos\theta.$$

The rotation formulas, therefore, are

$$\boxed{\begin{aligned} x &= x'\cos\theta - y'\sin\theta, \\ y &= x'\sin\theta + y'\cos\theta. \end{aligned}} \qquad (4.2)$$

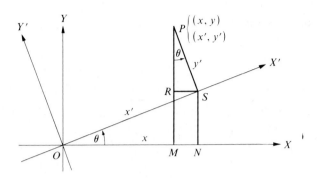

FIGURE 4.4

We have derived these formulas for the special case in which θ is an acute angle and the point P is in the first quadrant of both sets of axes. The formulas hold, however, for any θ and for all positions of P. A proof that the formulas hold generally could be made by observing the proper conventions as to the sign of θ and the signs of all distances involved.

EXAMPLE 1 ● Transform the equation $x^2 - y^2 - 9 = 0$ by rotating the axes through $45°$.

SOLUTION. When $\theta = 45°$, the rotation formulas (4.2) are

$$x = \frac{x'}{\sqrt{2}} - \frac{y'}{\sqrt{2}}, \qquad y = \frac{x'}{\sqrt{2}} + \frac{y'}{\sqrt{2}}.$$

FIGURE 4.5

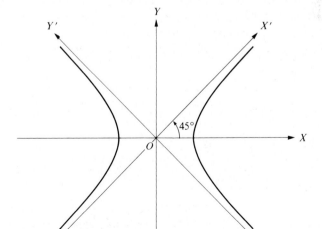

We make the substitutions in the given equation and have

$$\left(\frac{x'}{\sqrt{2}} - \frac{y'}{\sqrt{2}}\right)^2 - \left(\frac{x'}{\sqrt{2}} + \frac{y'}{\sqrt{2}}\right)^2 - 9 = 0,$$

$$\frac{x'^2}{2} - x'y' + \frac{y'^2}{2} - \frac{x'^2}{2} - x'y' - \frac{y'^2}{2} - 9 = 0,$$

$$2x'y' + 9 = 0.$$

The graph and both sets of axes are constructed in Fig. 4.5. ●

EXAMPLE 2 ● Find the acute angle of rotation such that the transformed equation of $2x^2 + \sqrt{3}\,xy + y^2 = 8$ will have no $x'y'$ term.

SOLUTION. We employ the rotation formulas to find the required angle θ. Substituting for x and y, we get

$$2(x'\cos\theta - y'\sin\theta)^2 + \sqrt{3}\,(x'\cos\theta - y'\sin\theta)(x'\sin\theta + y'\cos\theta)$$

$$+ (x'\sin\theta + y'\cos\theta)^2 = 8.$$

We perform the indicated multiplications, collect like terms, and obtain

$$(2\cos^2\theta + \sqrt{3}\sin\theta\cos\theta + \sin^2\theta)x'^2 + (-2\sin\theta\cos\theta + \sqrt{3}\cos^2\theta - \sqrt{3}\sin^2\theta)x'y'$$

$$+ (2\sin^2\theta - \sqrt{3}\sin\theta\cos\theta + \cos^2\theta)y'^2 = 8. \quad (4.3)$$

Since the $x'y'$ term is to vanish, we set its coefficient equal to zero. Thus we have

$$-2\sin\theta\cos\theta + \sqrt{3}\,(\cos^2\theta - \sin^2\theta) = 0.$$

Using the identities $\sin 2\theta = 2\sin\theta\cos\theta$ and $\cos 2\theta = \cos^2\theta - \sin^2\theta$, we obtain the equation in the form

$$-\sin 2\theta + \sqrt{3}\cos 2\theta = 0.$$

Hence

$$\tan 2\theta = \sqrt{3}, \qquad 2\theta = 60°, \qquad \theta = 30°.$$

A rotation of 30° eliminates the $x'y'$ term. This value of θ reduces Eq. (4.3) to

$$5x'^2 + y'^2 = 16.$$

Figure 4.6 shows both sets of axes and the graph. ●

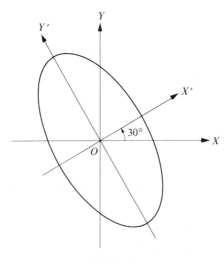

FIGURE 4.6

Exercises

Find the new equation in Exercises 1 through 8 when the axes are rotated through the given angle θ. Sketch the graph of the new equation, showing both sets of axes.

1. $x + y = 6$, $\theta = 45°$

2. $\sqrt{3}\,x - y = 4$, $\theta = 60°$

3. $x^2 + y^2 = a^2$, $\theta = 50°$

4. $xy = 1$, $\theta = 45°$

5. $x^2 - \sqrt{3}\,xy + 2y^2 = 2$, $\theta = 30°$

6. $x^2 + xy + y^2 = 1$, $\theta = 45°$

7. $x^2 + \sqrt{3}\,xy + 2y^2 = 3$, $\theta = \arctan\sqrt{3}$

8. $x^2 - 4xy + 4y^2 - 8\sqrt{5}\,x = 0$, $\theta = \arctan\frac{1}{2}$

Find the angle of rotation in each of Exercises 9 through 12 such that the transformed equation will have no $x'y'$ term.

9. $x^2 - xy + 3 = 0$

10. $3xy + y - 2 = 0$

11. $x^2 + 3xy - x + y = 0$

12. $x^2 - 3xy + 4y^2 + 7 = 0$

4.3

SIMPLIFICATION BY ROTATIONS AND TRANSLATIONS

The general second-degree, or quadratic, equation in x and y is represented by

$$Ax^2 + Bxy + Cy^2 + Dx + Ey + F = 0. \qquad (4.1)$$

At least one of the constants A, B, and C must be different from zero for the equation to be of the second degree. We assume, too, that not all the coefficients of terms involving one of the variables are zero; that is, both x and y appear in the equation.

In Section 4.1 we discussed the procedure for simplifying Eq. (4.1) when $B = 0$. If $B \neq 0$, an essential part of the simplification consists in obtaining a transformed equation lacking the product term $x'y'$. We will show how to determine immediately an angle of rotation that will serve for this purpose. In Eq. (4.1) we substitute the right members of the rotation formulas for x and y. After collecting like terms, we obtain

$$A'x'^2 + B'x'y' + C'y'^2 + D'x' + E'y' + F' = 0,$$

where the new coefficients are

$$A' = A\cos^2\theta + B\sin\theta\cos\theta + C\sin^2\theta,$$
$$B' = B\cos 2\theta - (A - C)\sin 2\theta,$$
$$C' = A\sin^2\theta - B\sin\theta\cos\theta + C\cos^2\theta,$$
$$D' = D\cos\theta + E\sin\theta,$$
$$E' = E\cos\theta - D\sin\theta,$$
$$F' = F.$$

The $x'y'$ term will vanish only if its coefficient is zero. Hence θ must satisfy the equation $B' = B\cos 2\theta - (A - C)\sin 2\theta = 0$. If $A \neq C$, the solution is

$$\tan 2\theta = \frac{B}{A - C}.$$

This formula yields the angle of rotation except when $A = C$. If $A = C$, the coefficient of $x'y'$ is $B\cos 2\theta$. Then the term vanishes by giving θ the value $45°$. Thus we see that an equation of the form (4.1) with an xy term can be transformed into an equation free of the product term $x'y'$.

We summarize the preceding results in the following theorem.

THEOREM 4.1 ● *A second-degree equation*

$$Ax^2 + Bxy + Cy^2 + Dx + Ey + F = 0,$$

in which $B = 0$, can be transformed by a translation of axes into one of the forms

$$A'x'^2 + C'y'^2 + F = 0,$$
$$A'x'^2 + E'y' = 0, \qquad (4.4)$$
$$C'y'^2 + D'x' = 0.$$

If $B \neq 0$, one of these forms can be obtained by a rotation and a translation (if *necessary). The angle of rotation θ (chosen acute) is obtained from the equation*

$$\tan 2\theta = \frac{B}{A - C}, \qquad \text{if } A \neq C,$$

or

$$\theta = 45°, \qquad \text{if } A = C.$$

By this theorem we see how to find the value of $\tan 2\theta$. The rotation formulas, however, contain $\sin \theta$ and $\cos \theta$. These functions can be obtained from the trigonometric identities

$$\sin \theta = \sqrt{\frac{1 - \cos 2\theta}{2}}, \qquad \cos \theta = \sqrt{\frac{1 + \cos 2\theta}{2}}.$$

The positive sign is selected before each radical because we shall restrict θ to an acute angle.

We can interpret geometrically the transformations that reduce the equation of a conic to one of the simplified forms. The rotation orients the coordinate axes in the directions of the axes of an ellipse or hyperbola, and the translation brings the origin to the center of the conic. For a parabola, the rotation makes one coordinate axis parallel to the axis of the parabola, and the translation moves the origin to the vertex.

Although Eqs. (4.4) above usually represent conics, there are exceptional cases, depending on the values of the coefficients. The exceptional cases are not our main interest, but we do observe them. The first of the equations has no graph if A', C', and F' all have the same sign, for then there are no real values of x' and y' for which the terms of the left member add to zero. If $F' = 0$ and A' and C' have the same sign, only the coordinates of the origin satisfy the equation. The values for the coefficients may be selected arbitrarily so that the equation represents two intersecting lines, two parallel lines, or one line. Each of the other equations always has a graph, but the graph is a line if the coefficient of the first-degree term is zero. The point and line graphs of second-degree equations are called **degenerate** conics, as was previously noted.

EXAMPLE 1 ● Express, in one of the form (4.4), the equation

$$3x^2 + 2\sqrt{3}\,xy + y^2 - 2x - 2\sqrt{3}\,y - 2 = 0.$$

SOLUTION. We first rotate the axes so as to eliminate the $x'y'$ terms. Thus, we have

$$\tan 2\theta = \frac{B}{A - C} = \frac{2\sqrt{3}}{3 - 1} = \sqrt{3}.$$

Hence $2\theta = 60°$, $\theta = 30°$, and the formulas for the rotation are

$$x = \frac{\sqrt{3}\,x'}{2} - \frac{y'}{2}, \qquad y = \frac{x'}{2} + \frac{\sqrt{3}\,y'}{2}.$$

Substituting for x and y in the given equation, we obtain

$$3\left(\frac{\sqrt{3}\,x'}{2} - \frac{y'}{2}\right)^2 + 2\sqrt{3}\left(\frac{\sqrt{3}\,x'}{2} - \frac{y'}{2}\right)\left(\frac{x'}{2} + \frac{\sqrt{3}\,y'}{2}\right) + \left(\frac{x'}{2} + \frac{\sqrt{3}\,y'}{2}\right)^2$$

$$-2\left(\frac{\sqrt{3}\,x'}{2} - \frac{y'}{2}\right) - 2\sqrt{3}\left(\frac{x'}{2} + \frac{\sqrt{3}\,y'}{2}\right) - 2 = 0.$$

We now pick out the coefficients of each variable stemming from this equation. Ignoring the $x'y'$ terms that add to zero, we find

$$\left(\tfrac{9}{4} + \tfrac{6}{4} + \tfrac{1}{4}\right)x'^2 + \left(\tfrac{3}{4} - \tfrac{6}{4} + \tfrac{3}{4}\right)y'^2 + \left(-\sqrt{3} - \sqrt{3}\right)x' + (1 - 3)y' - 2 = 0$$

and

$$4x'^2 - 2\sqrt{3}\,x' - 2y' - 2 = 0.$$

FIGURE 4.7

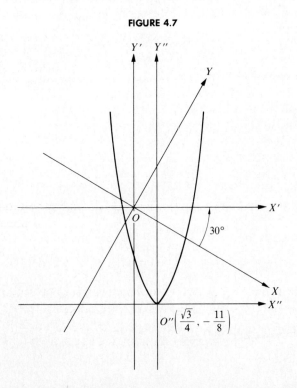

Dividing by 2 and completing the square in the x' terms, we get

$$2\left(x'^2 - \frac{\sqrt{3}}{2}x' + \frac{3}{16}\right) = y' + 1 + \frac{3}{8}$$

or

$$2\left(x' - \frac{\sqrt{3}}{4}\right)^2 = y' + \frac{11}{8}.$$

A translation of axes that moves the origin to the point $(\sqrt{3}/4, -11/8)$ yields the final equation

$$2x''^2 = y''.$$

The graph is the parabola drawn in Fig. 4.7 along with the three sets of axes. ●

EXAMPLE 2 ● Reduce the equation

$$73x^2 - 72xy + 52y^2 + 100x - 200y + 100 = 0$$

to one of the forms (4.4).

SOLUTION. We first transform the equation so that the product term $x'y'$ will disappear. To find the angle of rotation, we use

$$\tan 2\theta = \frac{B}{A - C} = \frac{-72}{73 - 52} = \frac{-24}{7}$$

and

$$\cos 2\theta = \frac{-7}{25}.$$

Hence

$$\sin\theta = \sqrt{\frac{1 - \cos 2\theta}{2}} = \frac{4}{5} \quad \text{and} \quad \cos\theta = \sqrt{\frac{1 + \cos 2\theta}{2}} = \frac{3}{5}.$$

The rotation formulas then become

$$x = \frac{3x' - 4y'}{5} \quad \text{and} \quad y = \frac{4x' + 3y'}{5}.$$

By substituting for x and y in the given equation and simplifying, we get

$$x'^2 + 4y'^2 - 4x' - 8y' + 4 = 0.$$

Completing the squares in the x' and y' terms, we obtain

$$(x' - 2)^2 + 4(y' - 1)^2 - 4 = 0.$$

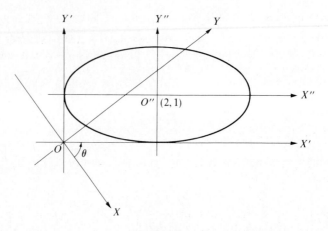

FIGURE 4.8

Finally, a translation to the point $(2, 1)$ yields the desired form

$$x''^2 + 4y''^2 - 4 = 0.$$

It is much easier to draw the graph from this equation than by using the original equation. The graph and the three sets of axes are constructed in Fig. 4.8. ●

Exercises

Reduce each equation to one of the simplified forms (4.4). Draw the graph of the resulting equation, showing all coordinate axes.

1. $x^2 + 4xy + 4y^2 + 2\sqrt{5}\,x - \sqrt{5}\,y = 0$

2. $x^2 - 2xy + y^2 - 8\sqrt{2}\,x - 8 = 0$

3. $x^2 - xy + y^2 - 3 = 0$

4. $xy - x + y = 0$

5. $x^2 + 4xy - 2y^2 + \sqrt{5}\,x + 2\sqrt{5}\,y - \frac{1}{4} = 0$

6. $3x^2 - 10xy + 3y^2 + 22x - 26y + 43 = 0$

7. $7x^2 + 48xy - 7y^2 - 150x - 50y + 100 = 0$

8. $73x^2 - 72xy + 52y^2 + 380x - 160y + 400 = 0$

9. $30x^2 - 12xy + 35y^2 - 60x + 12y - 48 = 0$

Ⓒ 10. $104x^2 + 60xy + 41y^2 - 60x - 82y - 75 = 0$

c 11. $9x^2 - 6xy + y^2 + x + 1 = 0$

12. Prove that a translation of axes does not introduce any second-degree terms in the equation of a conic, and that a rotation does not introduce any first-degree terms.

4.4

IDENTIFICATION OF A CONIC

We have seen that the graph, if any, of the general second-degree equation

$$Ax^2 + Bxy + Cy^2 + Dx + Ey + F = 0$$

can be determined immediately when the equation is reduced to one of the simplified forms (4.4). The determination can also be made from the coefficients of the general equation. To show that this is true, let us first assume that the graph of Eq. (4.1) is a parabola, an ellipse, or a hyperbola. Then applying the rotation formulas (4.2), we obtain

$$A'x'^2 + B'x'y' + C'y'^2 + D'x' + E'y' + F' = 0,$$

where

$$A' = A\cos^2\theta + B\sin\theta\cos\theta + C\sin^2\theta,$$
$$B' = B\cos 2\theta - (A - C)\sin 2\theta,$$
$$C' = A\sin^2\theta - B\sin\theta\cos\theta + C\cos^2\theta.$$

If the expressions for A', B', and C' are substituted in $B'^2 - 4A'C'$ and simplified, the result is

$$B'^2 - 4A'C' = B^2 - 4AC.$$

This relation among the coefficients of the original equation and the transformed equation holds for any rotation. For this reason, $B^2 - 4AC$ is called an **invariant**. By selecting the particular rotation for which $B' = 0$, we have

$$\boxed{-4A'C' = B^2 - 4AC.}$$

With $B' = 0$, the kind of conic represented by the transformed equation, and therefore the original equation, can be determined from the signs of A' and C'. The conic is an ellipse if A' and C' have like signs, and a hyperbola if the signs are different. If either A' or C' is zero, the conic is a parabola. These relations of A' and C', in the order named, would make $-4A'C'$ negative, positive, or zero. Hence we have the following important theorem.

THEOREM 4.2 ● *Let the coefficients of the equation*

$$Ax^2 + Bxy + Cy^2 + Dx + Ey + F = 0 \qquad (4.1)$$

be such that the graph is a nondegenerate conic. Then the graph is

1. *an ellipse if $B^2 - 4AC < 0$;*

2. *a hyperbola if $B^2 - 4AC > 0$;*

3. *a parabola if $B^2 - 4AC = 0$.*

This theorem is based on the condition that the equation has a graph and that the graph is not a degenerate conic. The degenerate conics, as we have already mentioned, consist of two intersecting lines, two parallel lines, one line, and a single point. So, in order to use the theorem with certainty, we need to know how to detect the exceptional cases. A general discussion of this question is a bit tedious; consequently, we will omit it. We shall, however, illustrate a workable plan when A and C are not both zero, and we shall point out the approach for the simpler situation when A and C are both zero and B is different from zero.

EXAMPLE 1 ● Determine the nature of the graph, if any, of the equation

$$2x^2 + 7xy + 3y^2 + x - 7y - 6 = 0.$$

SOLUTION. We find that $B^2 - 4AC = 49 - 4(2)(3) = 25 > 0$. This positive result does not tell us that the graph is a hyperbola. To make the determination, we treat the equation as a quadratic in x and apply the quadratic formula. Thus we get

$$2x^2 + (7y + 1)x + (3y^2 - 7y - 6) = 0;$$

$$x = \frac{-(7y + 1) \pm \sqrt{(7y + 1)^2 - 8(3y^2 - 7y - 6)}}{4}$$

$$= \frac{-7y - 1 \pm \sqrt{25y^2 + 70y + 49}}{4}$$

$$= \frac{-7y - 1 \pm (5y + 7)}{4}.$$

Hence,

$$x = \frac{-y + 3}{2} \quad \text{and} \quad x = -3y - 2$$

are the solutions. This means that the given equation may be presented in the factored form

$$(2x + y - 3)(x + 3y + 2) = 0.$$

If the coordinates of a point make one of these factors equal to zero, they make the product equal to zero and therefore satisfy the original equation. Hence the graph consists of the two lines whose equations are

$$2x + y - 3 = 0 \quad \text{and} \quad x + 3y + 2 = 0. ●$$

EXAMPLE 2 ● Text the equation $y^2 - 4xy + 4x^2 - 2x + y - 12 = 0$.

SOLUTION. Solving the equation for y in terms of x, we find the equations
$$y - 2x - 3 = 0 \quad \text{and} \quad y - 2x + 4 = 0.$$
Hence the graph of the given equation consists of two parallel lines. ●

EXAMPLE 3 ● Test the equation $4x^2 - 12xy + 9y^2 + 20x - 30y + 25 = 0$.

SOLUTION. We can make the test by solving for one variable in terms of the other. Scrutinizing the equation, though, we see that the left member is readily factorable. Thus, we find
$$(4x^2 - 12xy + 9y^2) + 10(2x - 3y) + 25 = 0,$$
$$(2x - 3y)^2 + 10(2x - 3y) + 25 = 0,$$
$$(2x - 3y + 5)(2x - 3y + 5) = 0.$$
The graph is the line $2x - 3y + 5 = 0$. ●

EXAMPLE 4 ● Determine the graph, if any, of the equation
$$x^2 + y^2 - 4x + 2y + 6 = 0.$$

SOLUTION. The student may verify that this equation yields
$$x = 2 \pm \sqrt{-y^2 - 2y - 2}$$
$$= 2 \pm \sqrt{-[(y + 1)^2 + 1]}.$$
It is evident that the radicand is negative for all real values of y, and this means that x is imaginary. Consequently the given equation has no graph. ●

EXAMPLE 5 ● Determine whether the equation
$$x^2 + 3xy + 3y^2 - x + 1 = 0$$
has a graph.

SOLUTION. Solving the equation for x gives
$$x^2 + (3y - 1)x + 3y^2 + 1 = 0,$$
$$x = \frac{1 - 3y \pm \sqrt{(3y - 1)^2 - 12y^2 - 4}}{2}$$
$$= \frac{1 - 3y \pm \sqrt{-3y^2 - 6y - 3}}{2}$$
$$= \frac{1 - 3y \pm (y + 1)\sqrt{-3}}{2}.$$
As we see, x will be imaginary if y has any real value except $y = -1$. For this

value of y, the value of x is 2. Hence the graph of the given equation consists of the single point $(2, -1)$. ●

If the squared terms are missing in a second-degree equation of the form (4.1), we may divide by the coefficient of xy and express the equation as

$$xy + Dx + Ey + F = 0.$$

The procedure for testing in this case is to translate the axes so that the new equation will have no x' or y' term. The transformed equation, as the student may verify, is

$$x'y' + (F - DE) = 0.$$

Clearly, the graph is a hyperbola if $F - DE \neq 0$. And it is equally evident that the graph consists of two intersecting lines if $F - DE = 0$.

We have pointed out the five ways in which Eq. (4.1) will not represent a conic. It is important to understand that the equation represents a conic, provided it does not conform to one of these exceptional cases. The $B^2 - 4AC$ test of Theorem 4.2 is then applicable.

Exercises

Assuming that the equations in Exercises 1 through 8 represent nondegenerate conics, classify each by computing $B^2 - 4AC$.

1. $3x^2 + xy + x - 4 = 0$

2. $2x^2 - 4xy + 8y^2 + 7 = 0$

3. $x^2 + 5xy + 15y^2 = 1$

4. $2xy - x + y - 3 = 0$

5. $x^2 - 2xy + y^2 + 3x = 0$

6. $x^2 - y^2 + 4 = 0$

7. $4x^2 - 3xy + y^2 + 13 = 0$

8. $3x^2 + 6xy + 5y^2 - x + y = 0$

Test each equation in Exercises 9 through 24 and tell whether the graph is a conic, or a degenerate conic, or there is no graph. Give the equation (or equations) of any degenerate graph.

9. $y^2 - 8xy + 17x^2 + 4 = 0$

10. $6x^2 - xy - 2y^2 + 2x + y = 0$

11. $y^2 + 2xy + x^2 - 2x - 2y = 0$

12. $x^2 + 2xy + y^2 - 2x + 1 = 0$

13. $y^2 - 2xy + 2x^2 + 1 = 0$

14. $y^2 - 4xy + 4x^2 + 4x + 1 = 2y$

15. $2x^2 - xy - y^2 + 2x + y = 0$

16. $4y^2 - 12xy + 10x^2 + 2x + 1 = 0$

17. $9x^2 - 6xy + y^2 + 3x - y = 0$

18. $y^2 - 2xy + 2x^2 - 5x = 0$

19. $9x^2 - 6xy + y^2 = 0$

20. $9x^2 - 6xy + y^2 + x + 1 = 0$

21. $xy + 4x - 4y + 20 = 0$

22. $xy - 6x + 5y + 30 = 0$

23. $3x^2 + 3xy - y^2 - x + y = 0$

24. $2x^2 + xy - y^2 + x - y = 0$

Assume, in Exercises 25 through 29, that the graph of the equation

$$Ax^2 + Bxy + Cy^2 + Dx + Ey + F = 0$$

is a nondegenerate conic.

25. If A and C have opposite signs, prove that the graph is a hyperbola.

26. If $B \neq 0$ and either A or C is zero, prove that the graph is a hyperbola.

27. If both D and E are zero, prove that the graph is not a parabola.

28. If both D and E are zero, prove that the center of the conic is at the origin.

29. If the center of the conic is at the origin, prove that both D and E are zero.

30. Work out all steps in showing that $B'^2 - 4A'C' = B^2 - 4AC$. Show also that $A' + C' = A + C$.

REVIEW EXERCISES

Determine the new equation in Exercises 1, 2, and 3 if the origin is translated to the indicated point.

1. $5x - 3y + 2 = 0$, $(2, 1)$

2. $x^2 + 3x + 2y = 0$, $(4, -2)$

3. $9x^2 + 2y^2 - 18x + 14y + 28 = 0$, $(1, -\frac{7}{2})$

Find the new equation in Exercises 4, 5, and 6 if the axes are rotated through the given angle θ.

4. $x^2 - xy + y^2 = 1$, $\theta = 45°$

5. $3x^2 + \sqrt{3}\,xy + 2y^2 = 0$, $\theta = 30°$

6. $x^2 - 4xy + 4y^2 + 6\sqrt{5}\,x - 4\sqrt{5}\,y + 1 = 0$, $\theta = \arctan \frac{1}{2}$

7. Translate the axes to a point such that the equation

$$2x^2 + 3y^2 + 18x - 16y + 61 = 0$$

will be transformed into one without first-degree terms.

8. Find the equation to which

$$x^2 - \sqrt{3}\,xy + 2y^2 = 2$$

will be transformed by a rotation of 30°.

9. If the axes are rotated through 45°, find the new equation corresponding to $x - y = 6$.

10. Simplify the equation

$$2x^2 + xy + 2y^2 - 12\sqrt{2}\,x - 16\sqrt{2}\,y + 12 = 0$$

by making an appropriate rotation followed by a translation.

11. Rotate axes to remove the xy term in $xy = 1$. Graph the equation.

5

Algebraic Curves

We considered first-degree equations in Chapter 2 and second-degree equations in Chapters 3 and 4. In this chapter we attack the problem of drawing the graphs of equations of a degree higher than two. All the equations define relations, and some of the equations define functions (Definitions 1.6 and 1.7). The point-by-point method of constructing a graph is tedious except for simple equations. The task can often be lightened, however, by first discovering certain characteristics of the graph as revealed by the equation. We have already discussed the concepts of intercepts, symmetry, and extent of a curve. These topics occur in Sections 3.1, 3.2, 3.3, and 3.5, and should be restudied. The following section gives another helpful idea which we shall introduce before drawing the graphs of particular equations.

5.1

VERTICAL AND HORIZONTAL ASYMPTOTES

If the distance of a point from a straight line approaches zero as the point moves indefinitely far from the origin along a curve, then the line is called an **asymptote** of the curve. In drawing the graph of an equation, it is well to determine the asymptotes, if any. They are often easily found and will facilitate the graphing. The utility of the asymptotes in drawing a hyperbola has already been demonstrated (Section 3.4). Here, however, we shall deal mainly with curves whose asymptotes, if any, are either horizontal or vertical.

EXAMPLE 1 ● Examine the equation and draw the graph of

$$y = (x^2 - 1)(x - 2)^2.$$

SOLUTION. The three tests for symmetry (Section 3.2) reveal that the graph does not have symmetry with respect to either axis or the origin. The x intercepts are

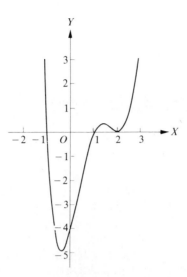

FIGURE 5.1.

−1, 1, and 2, and the y intercept is −4. These points greatly facilitate the graphing. As x increases beyond 2, the y values increase rapidly (Fig. 5.1). If x has a value between 1 and 2, the equation shows that y is positive because both factor $x^2 - 1$ and $(x - 2)^2$ are then positive. If x has a value between −1 and 1, the factor $x^2 - 1$ is negative, and the curve is below the x axis. As x takes values to the left of $x = -1$, the y values become large. With the preceding information we can quickly sketch the curve. The plotting of a few points other than the intercepts makes possible a fairly accurate graph. ●

Example 1 illustrates a technique for graphing a polynomial function in factored form: The graph of

$$y = (x - r_1)^{\alpha_1}(x - r_2)^{\alpha_2} \cdots (x - r_n)^{\alpha_n}$$

crosses the x axis at r_i if α_i is an odd integer, but it touches the x axis without crossing it at r_i if α_i is an even integer. We can get a quick sketch of the graph by utilizing this fact and by choosing a few points to plot. Looking again at Example 1, since $y = (x + 1)(x - 1)(x - 2)^2$, and since the exponents on $x + 1$ and $x - 1$ are the odd integer 1, we see that the curve crosses the axis at −1 and at 1, but touches without crossing at $x = 2$. Finally, since y is −4 if x is 0, we can get a sketch such as in Fig. 5.1.

EXAMPLE 2 ● Sketch a graph of the polynomial function

$$y = (x^2 - 4)(x^2 - 1).$$

SOLUTION. We see that $y = (x - 2)(x - 1)(x + 1)(x + 2)$ and that each exponent on the linear factors is 1. Hence, the curve crosses the axis when x is 2, 1, -1, and -2. When x is 0, y is 4, so the graph is as seen in Fig. 5.2. Note that the graph is symmetric with respect to the y axis since x occurs only to even powers. ●

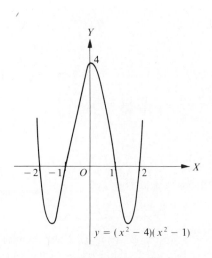

$$y = (x^2 - 4)(x^2 - 1)$$

FIGURE 5.2.

EXAMPLE 3 ● Examine the equation and draw the graph of

$$x^2y - 4y = 8.$$

SOLUTION. The y intercept is -2. But if we set $y = 0$, there is obviously no value of x that will satisfy the equation. Hence there is no x intercept. The graph has symmetry with respect to the y axis but not with respect to the x axis. The part of the graph to the right of the y axis may first be determined and then the other part drawn by the use of symmetry.

Solving the equation for y gives

$$y = \frac{8}{x^2 - 4}. \tag{5.1}$$

We see that the right member of this equation is negative for $-2 < x < 2$, and consequently the graph is below the x axis in this interval. Furthermore, if x has a value slightly less than 2, the denominator is near zero. Then the fraction, which is equal to y, has a large absolute value. As x is assigned values less than but still closer to 2, the corresponding values of $|y|$ increase and can be made greater than any chosen number by taking x close enough to 2. This property of the given equation is indicated by the table of corresponding values of x and y. If, however, x is greater than 2, the value of y is positive. And y can be made to exceed any chosen positive number by letting x be greater than, yet close enough, to 2. Hence the line $x = 2$ is a vertical asymptote of the graph both below and above the x axis. To check for a horizontal asymptote, we think of assigning x larger and larger positive values in Eq. (5.1). It is evident that in this manner we can make y, though positive, as close to 0 as we wish. This means that $y = 0$ is an asymptote. The y intercept, the asymptotes, and the symmetric property aid greatly in constructing the graph of the given equation (Fig. 5.3).

x	0	1	1.5	1.9	1.99	1.999
y	-2	-2.7	-4.6	-20.5	-200	-2000

FIGURE 5.3.

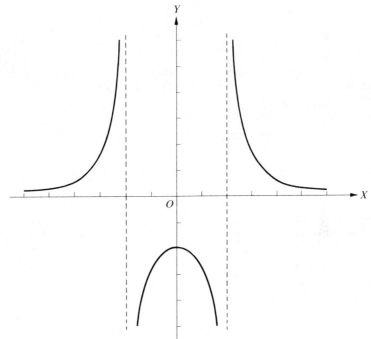

We may also determine the asymptotes in this example by solving the equation for x in terms of y. Thus, taking positive square roots, we find

$$x = 2\sqrt{\frac{y+2}{y}}\,. \tag{5.2}$$

The radicand in this equation must not be negative. Hence the excluded values of y are $-2 < y < 0$. Observing Eq. (5.2), we see that a positive value of y close to 0 makes x large, and that we can make x as large as we please by taking y close enough to 0. Also, by taking larger and larger positive values for y, we can make the corresponding values of x as close to 2 as we wish. We conclude, then, that $y = 0$, $x = -2$, and $x = 2$ are asymptotes of the graph. ●

We next discuss an alternative procedure for determining horizontal asymptotes when y is equal to the quotient of two polynomials in x. Consider, for example, the equation

$$y = \frac{3x^3 - 2x^2 + x - 5}{2x^3 + 4x^2 - 8x - 1},$$

where the numerator and denominator are of the same degree. To determine the behavior of the right member as $|x|$ becomes large, we divide the numerator and denominator by x^3. This gives the equation

$$y = \frac{3 - (2/x) + (1/x^2) - (5/x^3)}{2 + (4/x) - (8/x^2) - (1/x^3)}.$$

If now $|x|$ is assigned a large value, each term of both the numerator and denominator, after the first, is close to zero. Hence, the value of the fraction is close to $\frac{3}{2}$. Furthermore, the value can be made as near $\frac{3}{2}$ as desired by assigning $|x|$ a sufficiently large value. We conclude, then, that $y = \frac{3}{2}$ is an asymptote.

Suppose next that the degree of the numerator is less than the degree of the denominator, as in

$$y = \frac{3x^2 - 5x + 6}{2x^3 + 7x^2 + 5} = \frac{(3/x) - (5/x^2) + (6/x^3)}{2 + (7/x) + (5/x^3)}.$$

Clearly, for a large $|x|$, the numerator is close to 0 and the denominator is close to 2. The fraction can be made arbitrarily close to 0 by taking $|x|$ large enough. Consequently, $y = 0$ is an asymptote.

Finally, we let the degree of the numerator be greater than that of the denominator, as in

$$y \doteq \frac{x^3 - 3x + 1}{3x^2 + 4x + 5} = \frac{x - (3/x) + (1/x^2)}{3 + (4/x) + (5/x^2)}.$$

In this case $|y|$ will be larger than any chosen number when $|x|$ is sufficiently large.

We generalize this discussion by employing the equation

$$y = \frac{Ax^n + (\text{terms of lower degree})}{Bx^m + (\text{terms of lower degree})}.$$

There are three possibilities here, depending on the relative values of m and n.

> If $m = n$, $y = A/B$ is a horizontal asymptote;
> If $m > n$, $y = 0$ is a horizontal asymptote;   (5.3)
> If $m < n$, there is no horizontal asymptote.

EXAMPLE 4 ● Draw the graph of the equation

$$y = \frac{(x + 3)(x - 1)}{(x + 1)(x - 2)}.$$

SOLUTION. The x intercepts are -3 and 1, and the y intercept is $\frac{3}{2}$. The lines $x = -1$ and $x = 2$ are vertical asymptotes. The numerator and denominator of the fraction are quadratic and the coefficient of x^2 in each is unity. Hence $y = 1$ is a horizontal asymptote. Since there are two vertical asymptotes, the graph consists of three separate parts. To aid in sketching, we examine the given equation to determine the signs of y to the right and left of each vertical asymptote and to the right and left of each x intercept. In this examination the notations $(+)$ and $(-)$ indicate the signs of the factors of the numerator and denominator for the specified values of x.

When $x < -3$, the signs are $\dfrac{(-)(-)}{(-)(-)}$, and hence $y > 0$.

When $-3 < x < -1$, the signs are $\dfrac{(+)(-)}{(-)(-)}$, and hence $y < 0$.

When $-1 < x < 1$, the signs are $\dfrac{(+)(-)}{(+)(-)}$, and hence $y > 0$.

When $1 < x < 2$, the signs are $\dfrac{(+)(+)}{(+)(-)}$, and hence $y < 0$.

When $x > 2$, the signs are $\dfrac{(+)(+)}{(+)(+)}$, and hence $y > 0$.

The asymptotes, the signs of y in the various intervals, and just a few plotted points in addition to the intercepts allow us to draw a good graph (Fig. 5.4). ●

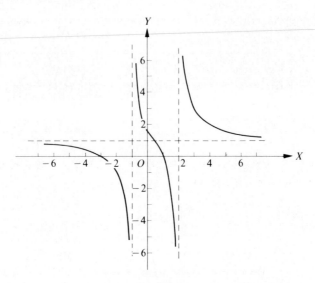

FIGURE 5.4.

Exercises

Find the x intercepts of the graphs of the equations 1 through 10. Determine the sign of y to the left of the smallest intercept, between consecutive intercepts, and to the right of the largest intercept. Then sketch the graph of the given equation.

1. $y = x(x^2 - 9)$

2. $y = (x - 2)^2(x - 4)$

3. $y = x^2(4 - x)$

4. $y = (x - 3)^2(x - 2)$

5. $y = (x + 3)(x^2 - 9)$

6. $y = (x^2 - 4)(x - 3)^2$

7. $y = (x^2 - 9)(x + 1)^2$

8. $y = x(x^2 - 1)(x^2 - 16)$

9. $y = x^2(x + 2)(x - 3)$

10. $y = (x - 4)(x + 1)(x + 3)$

Discuss each of the following equations with regard to intercepts, symmetry, extent of curve, and vertical and horizontal asymptotes. Draw the asymptotes and make at least a rough graph of the given equation.

11. $xy - x + 3 = 0$

12. $x^2y - 9y + 4 = 0$

13. $x^2y + 4y - 3x = 0$

14. $x^2y - y - x = 0$

15. $y = \dfrac{3x + 2}{(x - 1)^2}$

16. $y = \dfrac{x^2 - 4}{x^2 + 4}$

17. $y = \dfrac{x^2 - 9}{x^2 - 4}$

18. $y = \dfrac{x^2 - 1}{x^2 - 4}$

19. $y = \dfrac{(x-1)(x+2)}{x(x^2-4)}$

20. $y = \dfrac{(x+1)(x-2)}{(x+3)(x-1)}$

21. $y = \dfrac{(x+1)(x-2)}{x(x-3)(x+2)}$

22. $y = \dfrac{(2-x)(3-x)}{(1-x^2)(x-4)}$

23. $y = \dfrac{x^2(x-2)(x-4)}{(4-x^2)(25-x^2)}$

24. $y = \dfrac{(x+2)(x+4)(x+6)}{x(x-1)(x+3)}$

Find the equations of the horizontal asymptotes, when they exist, of the graphs of the following equations. Use (5.3).

25. $y = \dfrac{x^{1.1}+2}{x+3}$

26. $y = \dfrac{x^3+x-1}{2x^2-x+3}$

27. $y = \dfrac{5x^4-5}{3x^4+7}$

28. $y = \dfrac{4x^5+x^4+2}{3x^6+x^5-3}$

29. $y = \dfrac{x^{2.4}-x^2+2}{x^{2.5}-x+10}$

30. $y = \dfrac{x+3}{x^{0.9}-3}$

31. $y = \dfrac{x^{3.001}+3}{x^3-3}$

32. $y = \dfrac{x+5}{x^{0.99}+4}$

33. In an ecosystem in which one species is the prey of a fixed number of predators, the number y of prey consumed during a given period depends on the density x of the prey. An equation that may be used to model this situation is

$$y = \frac{ax}{1+abx}, \qquad x \geq 0.$$

Since a predator can consume only a certain number of prey, the curve has a horizontal asymptote. Sketch this curve when $a = 6$ and $b = \frac{1}{2}$, identifying the horizontal asymptote.

5.2
IRRATIONAL EQUATIONS

In this section we consider equations in which y is equal to the square root of a polynomial in x or the square root of the quotient of two polynomials. The procedure for constructing the graphs is like that of the previous section.

EXAMPLE 1 ● Draw the graph of the equation

$$y = \sqrt{x(x^2-16)}.$$

SOLUTION. The x intercepts are 0, ± 4. The permissible values of x are those for which the radicand is not negative. The radicand is positive when $-4 < x < 0$ and also if $x > 4$. The radicand is negative if $x < -4$ and when $0 < x < 4$; hence these values must be excluded. This information and the few plotted points are sufficient for drawing the graph (Fig. 5.5).

x	-3	-2	-1	4.5	5
y (approx.)	4.6	4.9	3.9	4.4	6.7

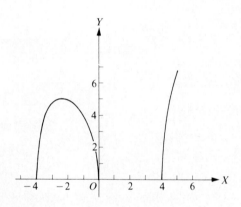

FIGURE 5.5.

EXAMPLE 2 ● Draw the graph of the equation

$$y^2 = \frac{x^2 - 9}{x^2 - 16}.$$

SOLUTION. The graph is symmetric with respect to both axes. Hence we can draw the part in the first quadrant and readily finish the construction by the use of symmetry. The positive x intercept is 3, and the positive y intercept is $\frac{3}{4}$. The line $x = 4$ is a vertical asymptote, and the line $y = 1$ is a horizontal asymptote. We observe that y^2 is positive when $0 \le x < 3$ and when $x > 4$. But y^2 is negative when x has a value between 3 and 4; hence these values must be excluded. From the analysis of the given equation, we sketch the graph (Fig. 5.6). Somewhat greater accuracy can be obtained by plotting the points indicated in the accompanying table.

x	1	2	4.2	5	6	7
y (approx.)	0.7	0.6	2.3	1.3	1.2	1.1

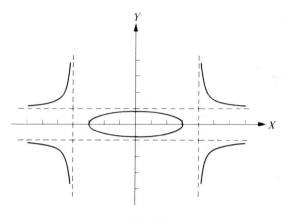

FIGURE 5.6.

Exercises

Sketch the graph of each equation.

1. $y = \sqrt{9 - x^2}$

2. $y = \sqrt{x(x^2 - 25)}$

3. $y = \sqrt{\dfrac{x}{x^2 - 25}}$

4. $y = \sqrt{\dfrac{3x}{x^2 + 1}}$

5. $y^2 = \dfrac{6}{x + 3}$

6. $y^2 = \dfrac{6}{x^2 + 9}$

7. $y^2 = \dfrac{x^2 - 4}{x^2 - 9}$

8. $y^2 = \dfrac{x^2 - 9}{x^2 - 25}$

9. $y^2 = \dfrac{x(x - 1)}{x^2 - 9}$

10. $y^2 = \dfrac{x(x + 3)}{(x^2 - 16)}$

5.3

SLANT ASYMPTOTES

When y is equal to the quotient of two polynomials in x where the degree of the numerator exceeds that of the denominator by unity, the graph will usually have a slant asymptote. The simplest case arises when the numerator is quadratic and the denominator is linear.

EXAMPLE 1 ● Draw the graph of the equation

$$2xy - x^2 + 6x - 4y - 10 = 0.$$

SOLUTION. The $B^2 - 4AC$ test reveals that the graph is either a hyperbola or a degenerate conic (Section 4.4). To determine which of these situations exists, we solve the equation for y. Thus, we get

$$y = \frac{x^2 - 6x + 10}{2x - 4}$$

and, by dividing,

$$y = \frac{1}{2}x - 2 + \frac{1}{x - 2}.$$

For all real values of x, except $x = 2$, this equation yields real values for y. Hence the graph of the equation, and also the given equation, is a hyperbola, not a degenerate conic. Clearly, $x = 2$ is a vertical asymptote. As $|x|$ increases, the ordinate of the hyperbola gets closer and closer to the ordinate of the line $y = \frac{1}{2}x - 2$. And the difference of the two ordinates can be made arbitrarily near zero by taking $|x|$ large enough. Hence the line $y = \frac{1}{2}x - 2$ is an asymptote. The asymptote and just a few plotted points furnish a guide for drawing the graph (Fig. 5.7). ●

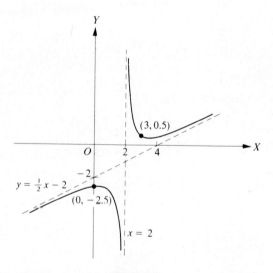

FIGURE 5.7.

EXAMPLE 2 ● Draw the graph of the equation

$$2x^2y - x^3 - 8xy + 8x^2 - 20x + 8y + 14 = 0.$$

SOLUTION. On solving for y and performing a division, we obtain

$$y = \tfrac{1}{2}x - 2 + \frac{1}{(x - 2)^2}.$$

The line $x - 2 = 0$ is a vertical asymptote and the line $y = \frac{1}{2}x - 2$ is a slant asymptote. The fraction on the right side of the equation is positive for all permissible values of x. Consequently, the graph is above the slant asymptote, as shown in Fig. 5.8. ●

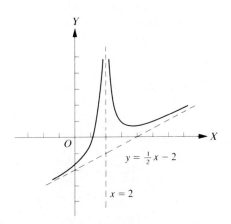

FIGURE 5.8.

EXAMPLE 3 ● Draw the graph of the equation

$$x^2 - 3xy - 13x + 12y + 39 = 0.$$

FIGURE 5.9.

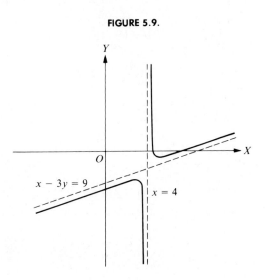

SOLUTION. On solving for y and performing a division, we get

$$y = \frac{x}{3} - 3 + \frac{1}{x - 4}.$$

Hence the line $x - 4 = 0$ is a vertical asymptote and the line $y = \frac{1}{3}x - 3$ is a slant asymptote. The term $1/(x - 4)$ on the right of the equation is positive for all values of $x > 4$ and is negative if $x < 4$. This means that the graph of the given equation is above the slant asymptote to the right of $x = 4$ and below the asymptote to the left of $x = 4$ (Fig. 5.9). ●

EXAMPLE 4 ● Analyze the equation and construct the graph of

$$x^2 + 2xy + 11x + 6y - 26 = 0.$$

SOLUTION. On solving for y and making a division, we get

$$y = -\frac{x}{2} - 4 + \frac{25}{x + 3}.$$

FIGURE 5.10.

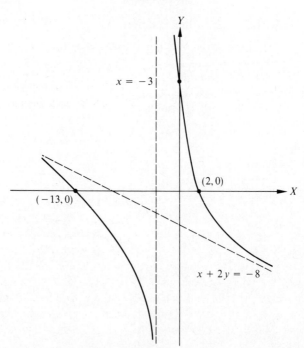

We see that $x = -3$ is a vertical asymptote of this equation, and, consequently, of the given equation. Also $y = -(x/2) - 4$ is a slant asymptote of both equations. We now picture the required graph in Fig. 5.10. ●

Graph of an Equation in Factored Form

Equations sometimes appear with one member equal to zero and the other member expressed as the product of factors in terms of x and y. When an equation is in this form, its graph can be more simply obtained by first setting each of the factors equal to zero. If the coordinates of a point make one of the factors equal to zero, they make the product equal to zero and therefore satisfy the given equation. On the other hand, the coordinates of a point that make no factor equal to zero do not satisfy the equation. Hence the graph of the given equation consists of the graphs of the equations formed by setting each of the factors of the nonzero member equal to zero.

EXAMPLE 5 ● The graph of the equation $(3x - y - 1)(y^2 - 9x) = 0$ consists of the line $3x - y - 1 = 0$ and the parabola $y^2 - 9x = 0$. ●

EXAMPLE 6 ● Discuss the graph of the equation

$$(x^2 + 1)(x - 3)(4x^2 + 9y^2 - 36) = 0.$$

SOLUTION. Equating each factor to 0, we have the equations

$$x^2 + 1 = 0, \qquad x - 3 = 0, \qquad 4x^2 + 9y^2 - 36 = 0.$$

The first equation has no real zero, and consequently has no graph. The graph of the second equation is the line parallel to the y axis and 3 units to the right of the origin. The graph of the third equation is an ellipse with center at the origin and semiaxes 3 and 2. ●

Intersections of Graphs

If the graphs of two equations in two variables have a point in common, then, from the definition of a graph, the coordinates of the point satisfy each equation separately. Hence the point of intersection gives a pair of real numbers that is a simultaneous solution of the equations. Conversely, if the two equations have a simultaneous real solution, then their graphs have the corresponding point in common. Thus simultaneous real solutions of two equations in two variables can be obtained graphically by reading the coordinates of their points of intersection. Because of the imperfections in the process, the results thus found are usually only approximate. If the graphs have no point of intersection, there is no real solution. In simple cases, the solutions, both real and imaginary, can be found by algebraic processes.

EXAMPLE 7 ● Find the points of intersection of the graphs of

$$y = x^3 \quad \text{and} \quad y = 2 - x.$$

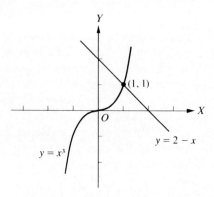

FIGURE 5.11.

SOLUTION. The graphs (Fig. 5.11) intersect in one point whose coordinates are $(1, 1)$. Eliminating y between the equations yields

$$x^3 + x - 2 = 0 \quad \text{or} \quad (x - 1)(x^2 + x + 2) = 0.$$

Hence the roots of this equation are

$$x = 1, \quad x = \frac{-1 + \sqrt{-7}}{2}, \quad x = \frac{-1 - \sqrt{-7}}{2}.$$

The corresponding values of y are obtained from the linear equation. The solutions, real and imaginary, are

$$(1, 1), \quad \left(\frac{-1 + \sqrt{-7}}{2}, \frac{5 - \sqrt{-7}}{2} \right), \quad \left(\frac{-1 - \sqrt{-7}}{2}, \frac{5 + \sqrt{-7}}{2} \right).$$

The graphical method gives only the real solution. ●

EXAMPLE 8 ● Find the points of intersection of the graphs of

$$x^2 + 4y = 0 \quad \text{and} \quad x + y = 0.$$

SOLUTION. The graphs (Fig. 5.12) intersect at the points $(0, 0)$ and $(4, -4)$. ●

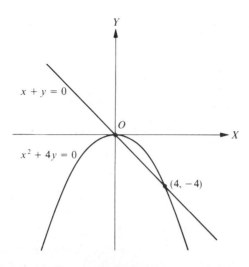

FIGURE 5.12.

Exercises

Draw the asymptotes and sketch the graph of each Exercise 1 through 14.

1. $xy + x^2 - 2 = 0$

2. $xy - x^2 + 3 = 0$

3. $x^2 + xy + x - 1 = 0$

4. $x^2 - xy + 2x - 1 = 0$

5. $x^2 + xy + x - y + 2 = 0$

6. $x^2 + xy + 3x - 2y + 1 = 0$

7. $2x^2 - 2xy + 3x + 6y + 25 = 0$

8. $2x^2 + 2xy - x - y - 2 = 0$

9. $x^2y - x^3 - 1 = 0$

10. $x^2y - x^3 + 1 = 0$

11. $x^2y - x^3 - 4xy + 12x + 4y - 14 = 0$

12. $x^2y - x^3 - 4xy + 11x + 4y - 16 = 0$

13. $x^2y - x^3 - x^2 - x + y - 2 = 0$

14. $x^2y - x^3 - x^2 - x + y = 0$

Describe the graph of each equation in Exercises 15 through 22.

15. $(x^2 + y^2)(x + 1) = 0$

16. $(x^2 + y^2 + 1)(3x + 2y) = 0$

17. $xy(x + y + 2) = 0$

18. $2x^3 + 3xy^2 = 6x$

19. $x^3y^2 + xy^4 = 4xy^2$ 20. $4x^2y - y^2 = 0$

21. $(xy + 3)(x^3 - xy^2) = 0$ 22. $(x^3 - 1)(x^3 + 1) = 0$

Construct the graph of each pair of equations in Exercises 23 through 28 and find or estimate the coordinates of any points of intersection. Check by obtaining the solutions algebraically, if possible.

23. $y = x^3$ 24. $y = x^3 - 4x$ 25. $x^2 + y^2 = 16$

    $y = x + 1$ $y = x + 4$ $y^2 = 6x$

26. $x^2 + y^2 = 13$ 27. $x^2 - 4y = 0$ 28. $x^2 + y^2 = 25$

    $2x - 3y = 0$ $x - y = 0$ $y - x = 1$

29. A hand-calculator manufacturer is willing to make y (measured in thousands) of his product each week if the product is priced at x dollars provided $y = (1/100)x^3$. A marketing firm finds that the public will buy $1000y$ of them at x dollars each if $y = 100/x$. Find the market equilibrium price and quantity that should be produced each week to maintain equilibrium. Sketch both graphs on the same coordinate axes. (See Exercise 14 after Section 2.3.)

REVIEW EXERCISES

1. Discuss the graph of the following equation with regard to intercepts, symmetry, extent of curve, and asymptotes, and sketch the graph:

$$y = \frac{x^2 - 4}{x^2 - 9}.$$

2. Draw the graph of the equation

$$y = \frac{(x - 3)(x + 1)}{(x - 1)(x + 2)}.$$

3. Sketch the graph of the equation

$$y = \sqrt{x(x^2 - 25)}.$$

4. Sketch the graph of the equation

$$y^2 = \frac{4}{x + 2}.$$

5. Draw the asymptotes and sketch the graph of the equation

$$2xy + x^2 - 6x + 4y + 14 = 0.$$

6. Describe the graph of the equation

$$(xy - 3)(x - 3y + 2)(x^2 + y^2 - 9) = 0.$$

7. Construct the graphs of the pair of equations

$$x^2 + y^2 = 25 \qquad \text{and} \qquad x - y = 1,$$

and estimate the coordinates of any points of intersection. Check by obtaining the solution algebraically.

8. Sketch a graph of $y = (x + 2)^3(x - 1)^4(x - 3)$.

6

Transcendental Functions

In the previous chapters we have studied algebraic functions and their graphs. In this chapter we will consider a class of functions that are not algebraic. They are called **transcendental functions**, and they represent a class of functions that has wide application. The student should review the definition of function before reading this chapter.

TRIGONOMETRIC FUNCTIONS

In earlier chapters of this book we measured angles in degrees because that unit is perhaps more familiar to most students. In this chapter we restrict our unit of measurement to **radians** in order to present the trigonometric functions in the way they are used in calculus and in the applied sciences. The student must know both units of measure.

Recall that an angle θ is in **standard position** if its initial side is along the horizontal axis and if it is measured *counterclockwise* for positive values of θ or *clockwise* for negative values of θ to its terminal side. Consider the circle $u^2 + v^2 = 1$ and the angle θ (measured in radians) as shown in Fig. 6.1.

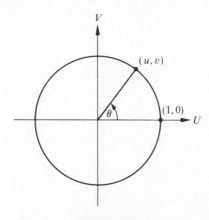

FIGURE 6.1

From trigonometry we know that the length of the arc of the circle from the point $(1,0)$ to (u, v) is θ units if θ is measured in radians. We also know that $u = \cos\theta$ and $v = \sin\theta$ and $v/u = \tan\theta$. This determines the well-known trigonometric functions **sine** and **cosine** as functions whose domain is the set of all real numbers θ and whose range is the set of all real numbers z with $-1 \leq z \leq 1$.

To plot the graph of $y = \sin x$, we may make a table (Table 6.1) by computing familiar values, by using a hand calculator, or by using a table of natural trigonometric functions (Table II in the Appendix). The graph is sketched in Fig. 6.2.

TABLE 6.1 Corresponding values of x and $\sin x$.

x	$\pi/6$	$\pi/3$	$\pi/2$	$2\pi/3$	$5\pi/6$	π
$y = \sin x$	0.5	0.87	1	0.87	0.5	0
x	$7\pi/6$	$4\pi/3$	$3\pi/2$	$5\pi/3$	$11\pi/6$	2π
$y = \sin x$	-0.5	-0.87	-1	-0.87	-0.5	0

FIGURE 6.2

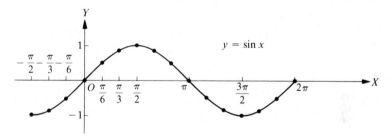

Since there are 2π radians in a complete revolution, we see that

$$\sin(\theta + 2\pi) = \sin\theta$$

for any θ. A function with such a property is said to be a **periodic function**.

DEFINITION ● *A function f with the set of all real numbers as its domain is* **periodic** *with period $p > 0$ if*

$$f(x + p) = f(x)$$

for every x in the domain and if no smaller positive number p'(< p) also has the property

$$f(x + p') = f(x)$$

for each x.

The sine function is periodic with period 2π, as is the cosine function. The period of the tangent function is π.

When the values from the table are plotted, we have one period of the sine curve (Fig. 6.2). Additional periods may be duplicated to the right and/or to the left. We can obtain graphs of $y = \cos x$ and $y = \tan x$ by proceeding in the same way as with the sine function. A portion of $y = \cos x$ is sketched in Fig. 6.3, while three periods of $y = \tan x$ are graphed in Fig. 6.4.

FIGURE 6.3

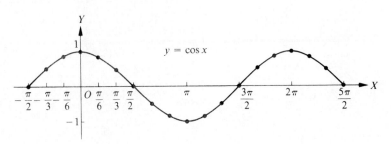

FIGURE 6.4

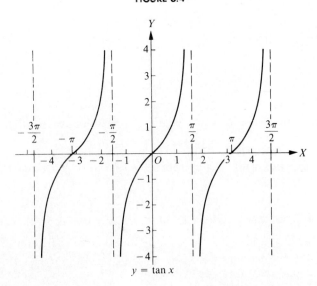

The graph of $y = a \sin x$ is obtained from the sine curve by multiplying each ordinate by a. If a is positive, this produces the graph in Fig. 6.5.

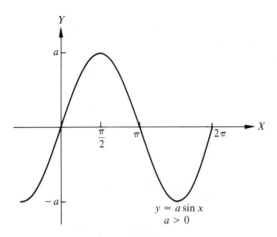

FIGURE 6.5

If $a < 0$, the graph of $y = a \sin x$ is given in Fig. 6.6.

FIGURE 6.6

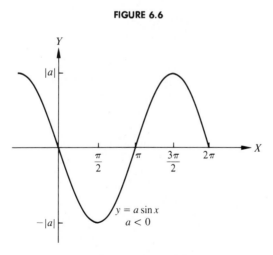

We say that the amplitude of $y = a \sin x$ is $|a|$. Likewise the graph of $y = \sin bx$ is obtained from the sine curve by altering its period to $2\pi/b$, for, as x goes from 0 to $2\pi/b$, bx goes from 0 to 2π, and $\sin bx$ goes through a complete period.

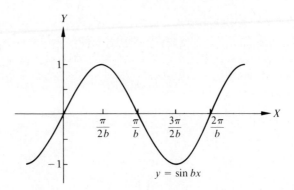

FIGURE 6.7

The graphs of the equations $y = a \sin bx$ and $y = a \sin(bx + c)$, when placed on the same coordinate axes, are alike except for position. This is true because the two values of y become equal if x is replaced by $x - (c/b)$ in the second equation. Hence a shift of c/b units along the x axis will bring the graph of the second equation into coincidence with the graph of the first equation. The shift is to the right or left depending on whether b and c have like or unlike signs.

We see, then that $y = a \sin(bx + c)$ has amplitude $|a|$, period $2\pi/b$, and phase shift c/b.

Exactly the same reasoning applies to the curve $y = a \cos(bx + c)$.

EXAMPLE 1 ● Construct the graph of the equation $y = \sin 3x$.

SOLUTION. The function defined by this equation has an amplitude of 1 and a period of $2\pi/3$. Accordingly, $\sin 3x$ and $\sin x$ have the same amplitude, but the period of $\sin 3x$ is one-third the period of $\sin x$. These facts are shown in the graphs of $y = \sin x$ and $y = \sin 3x$ (Fig. 6.8). ●

FIGURE 6.8

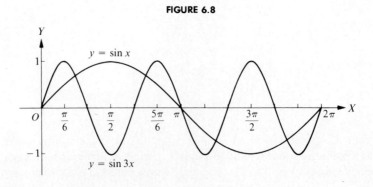

EXAMPLE 2 ● Draw the graph of $y = 2 \cos \frac{1}{2}x$.

SOLUTION. The amplitude is 2. The period, 2π divided by $\frac{1}{2}$, is 4π. The graph constructed from the accompanying table of values (Table 6.2) is exhibited in Fig. 6.9. Each value of y is twice the corresponding value of y in the graph of $y = \cos \frac{1}{2}x$, which is included for comparison. ●

TABLE 6.2 Values of $y = 2 \cos \frac{1}{2}x$.

x	0	$\dfrac{\pi}{2}$	π	$\dfrac{3\pi}{2}$	2π	$\dfrac{5\pi}{2}$	3π	$\dfrac{7\pi}{2}$	4π
$\frac{1}{2}x$	0	$\dfrac{\pi}{4}$	$\dfrac{\pi}{2}$	$\dfrac{3\pi}{4}$	π	$\dfrac{5\pi}{4}$	$\dfrac{3\pi}{2}$	$\dfrac{7\pi}{4}$	2π
$2 \cos \frac{1}{2}x$	2	1.4	0	-1.4	-2	-1.4	0	1.4	2

FIGURE 6.9

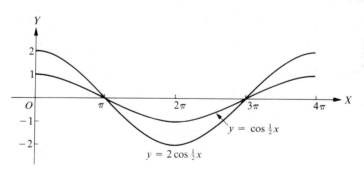

$y = \cos \frac{1}{2}x$

$y = 2 \cos \frac{1}{2}x$

EXAMPLE 3 ● Construct the graph of $y = -2 \sin \pi x$.

SOLUTION. The amplitude is $|-2| = 2$, while the period is $2\pi/\pi = 2$. The graph is shown in Fig. 6.10. ●

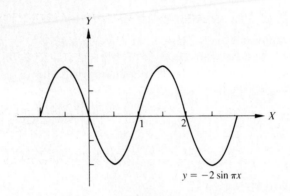

FIGURE 6.10

EXAMPLE 4 ● Draw, on the same coordinate axes, the graphs of the equations

$$y = 1.5 \sin 2x \quad \text{and} \quad y = 1.5 \sin\left(2x + \tfrac{1}{3}\pi\right).$$

SOLUTION. The graphs of the two equations are the same except for position (Fig. 6.11). The curves would coincide if the graph of the second equation were moved $\tfrac{1}{6}\pi$ units to the right. ●

FIGURE 6.11

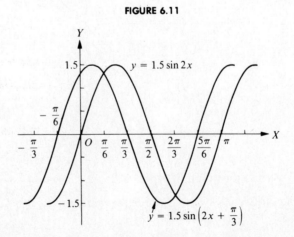

EXAMPLE 5 ● Construct the graphs of the two equations

$$y = \cos 3x \quad \text{and} \quad y = \cos\left(3x + \frac{2\pi}{3}\right).$$

SOLUTION. The graphs are drawn through two periods in Fig. 6.12. We remark that the intersection points are at $x = 2\pi/9$, $5\pi/9$, $8\pi/9$, and $11\pi/9$. The graphs would coincide if the graph of the second equation were moved $2\pi/9$ units to the right. ●

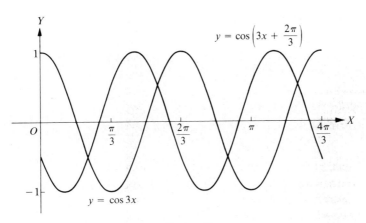

FIGURE 6.12

The trigonometric functions have been used to describe many natural phenomena. The fact that a generalized sine function is periodic with any prescribed period is of immense value in describing sound waves, electrical current, radio waves, and the swing of a pendulum. Other periodic functions may not be expressible as a generalized sine function, but may be expressible as a sum of generalized sine and cosine functions, as we shall see later.

EXAMPLE 6 ● A sound can be generated by a musical instrument, a tuning fork, an electronic organ, the hum of a motor, or an electric transformer. If the sound has the pitch of middle C, then it causes a vibration in air pressure on the ear with a frequency of about 260 Hz. (A hertz is a frequency of one cycle per second.) Suppose the power of the sound wave is 3.6×10^{10} watts per square centimeter; that is to say, the wave exerts this much pressure on the ear at its strongest point. Determine the equation of the sound wave.

SOLUTION. The amplitude of the generalized sine curve is 3.6×10^{10}, while the period is $1/260$. Its equation is then

$$y = 3.6 \times 10^{10} \sin 520\pi t.$$

This represents the air pressure t seconds after the sound begins. ●

Exercises

Find the period and the amplitude of the following equations.

1. $y = \sin 4x$

2. $y = 3 \sin \frac{1}{2}x$

3. $y = -\sin \frac{1}{2}x$

4. $y = -2 \cos 6x$

5. $y = \frac{1}{2} \cos \pi x$

6. $y = -\frac{1}{3} \cos \pi x$

7. $y = -\cos 2\pi x$

8. $y = -2 \sin 8\pi x$

9. $y = 2 \sin \pi x \cos \pi x$

Construct one period of the graph of each equation.

10. $y = \sin 4x$

11. $y = 3 \sin \frac{1}{2}x$

12. $y = -\sin \frac{1}{2}x$

13. $y = -2 \cos 6x$

14. $y = -\cos 2\pi x$

15. $y = -2 \sin 8\pi x$

16. $y = 2 \sin \pi x \cos \pi x$

Sketch one period of the graph of each pair of equations on the same coordinate axes.

17. $y = \cos x$, $y = 2 \cos x$

18. $y = -\sin x$, $y = -2 \sin x$

19. $y = \cos x$, $y = -\cos x$

20. $y = \sin x$, $y = \sin \frac{1}{2}x$

21. $y = \cos x$, $y = 2 \cos 2x$

22. $y = \sin x$, $y = \sin(x - \pi/2)$

23. $y = 2 \cos x$, $y = 2 \cos(2x - \pi/6)$

24. $y = 2 \sin x$, $y = -2 \sin 2x$

25. $y = \tan x$, $y = 2 \tan 2x$

26. During an emergency alert test, a radio station broadcasts a signal with a frequency of 60 Hz. If the intensity of the sound (measured in decibels) is 90 dB, graph the equation that represents the intensity of the sound t seconds after it begins.

27. A jet aircraft on landing makes a sound with the equation $y = 150 \sin 1000\pi t$, measured in decibels, while t is measured in seconds. What is the frequency of the sound? What is the amplitude?

6.2
THE EXPONENTIAL FUNCTION

The exponential function, to be introduced in this section, is extremely important in the study of various physical, biological, and abstract phenomena.

DEFINITION ● *If $a > 0$, $a \neq 1$, then the* **exponential function** *with base a is defined by*

$$y = f(x) = a^x,$$

where x is a real number.

We shall see that, for $a > 1$, as x increases through positive values, y grows very rapidly, and as x is taken further from the origin in the negative direction, y becomes closer to zero.

EXAMPLE 1 ● Sketch the graphs of the functions defined by the equations $y = 2^x$ and $y = 3^x$. Use the same coordinate axes.

SOLUTION. We use integral values of x in forming the table of corresponding values of x and the function values:

x	-2	-1	0	1	2
2^x	$\frac{1}{4}$	$\frac{1}{2}$	1	2	4
3^x	$\frac{1}{9}$	$\frac{1}{3}$	1	3	9

The graphs are constructed in Fig. 6.13. ●

FIGURE 6.13

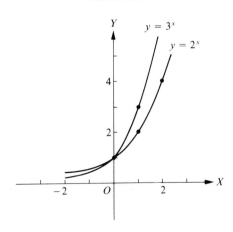

Thus the curves have the negative x axis as an asymptote. If we graphed $y = 10^x$ and exponential curves to other bases, we would conclude that, as the base is chosen larger, the curve becomes steeper for $x > 0$. An important base is $e = 2.71828\ldots$, and the graph of $y = e^x$ lies between those of $y = 2^x$ and $y = 3^x$ in Fig. 6.13.

The generalized exponential curve

$$y = ca^{bx}, \qquad a > 0,$$

can be obtained by multiplying each ordinate of $y = (a^b)^x = a^{bx}$ by c.

EXAMPLE 2 ● Draw the graph of the function defined by $y = 2^{-x}$ or, equivalently, $y = (\frac{1}{2})^x$.

SOLUTION. We form the table to aid in drawing the graph in Fig. 6.14. ●

x	-2	-1	0	1	2
y	4	2	1	$\frac{1}{2}$	$\frac{1}{4}$

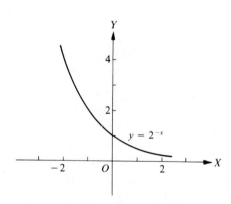

FIGURE 6.14

EXAMPLE 3 ● Sketch the graph of $y = f(x) = (1/3)e^{2x}$.

SOLUTION. We use a hand calculator, or Table III in the Appendix, to compute several values of the function. We find that $f(-1) = (1/3)e^{-2} = 0.05$, $f(0) = 0.33$, while $f(1) = (1/3)e^2 = 2.5$, and $f(2) = (1/3)e^4 = 18.2$. A sketch of the graph is given in Fig. 6.15. Note that, in Fig. 6.15, vertical and horizontal scales have been drawn unequal. This is done in order to facilitate the display of the function. ●

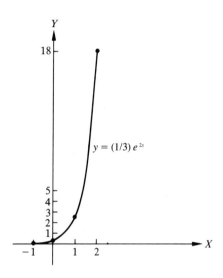

FIGURE 6.15

EXAMPLE 4 ● The number y of bacteria, measured in millions, in a culture t hours after initial introduction is given by

$$y = f(t) = 20e^{t/3}.$$

How many bacteria are present at the start? How many after one hour? How many are there after two hours? Graph the function $y = f(t)$ for $0 \leq t \leq 4$.

SOLUTION. The number of bacteria at the start is $y = f(0) = 20e^0 = 20$ million. We also find

$$f(1) = 20e^{1/3} = 20(1.396) = 27.9,$$
$$f(2) = 20(1.95) = 39,$$
$$f(3) = 20(e) = 54.4,$$
$$f(4) = 20(3.79) = 75.9.$$

So there were 20 million bacteria at the start, 27.9 million after one hour, and 39 million after two hours. The population would double in slightly over two hours. The function is graphed in Fig. 6.16. ●

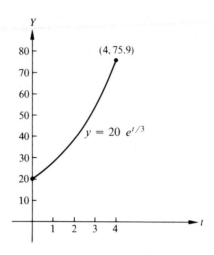

FIGURE 6.16

Exercises

Sketch the graph of the function defined by the equation in each of the following exercises, using calculators or Table III in the Appendix when necessary.

1. $y = e^x$

2. $y = e^{-x}$

3. $y = 4^x$

4. $y = -e^x$

5. $y = -e^{-x}$

6. $y = 2e^x$

7. $y = 5e^{2x}$

8. $y = e^{x^2} - 1$

9. $y = 2^{-x^2}$

10. When P dollars are invested at an annual interest rate of r per cent compounded **continuously** for t years, the amount A present after t years is given by

$$A = Pe^{rt/100}.$$

If $1000 is invested at 10% interest compounded continuously, how much is in the account after two years? How much after four years? Graph the curve for $0 < t \leq 4$.

11. The health department epidemiologist found that the number, y, of new cases of influenza, measured in hundreds, was related to the time t, in weeks, after the outbreak of the disease by:

$$y = f(t) = \begin{cases} t^2, & \text{if } 0 \leq t \leq 2, \\ t + 2, & \text{if } 2 < t \leq 3, \\ 100.5e^{-t}, & \text{if } 3 < t. \end{cases}$$

Graph the function $y = f(t)$. How many new cases were there in the fifth week? In the seventh week?

6.3

LOGARITHMS

In 1614 John Napier published a brochure introducing logarithms. The widespread applicability of logarithms to computational problems was quickly perceived by Henry Briggs, and the development of a very important branch of mathematics began. While the immediate applicability of logarithms 300 years ago was to aid the computations of such expressions as

$$\frac{(3.1742)\sqrt{1842}}{(91.45)^3(85.94)^{2/3}},$$

today we find it easier to use hand calculators for this task. Nonetheless, logarithms remain important in many areas of mathematics. They have valuable theoretical usefulness and they continue to be used to describe natural phenomena.

DEFINITION ● *For a positive number b, $b \neq 1$, the* **logarithm** *of a positive number N to the base b, $\log_b N$, is the exponent to be placed on b to yield N.*

Thus $\log_b N = x$ means $b^x = N$. We find immediately from the definition that $b^{\log_b N} = N$ and $\log_b b^N = N$. It follows from the definition that $\log_b$ is a function whose domain is the set of all positive real numbers. The range of $\log_b$ is the set of all real numbers.

While logarithms can be computed to any base b, there are two bases with particular importance, e and 10. When the base of the logarithm is 10, we suppress writing the subscript 10 in $\log_{10}$ and simply write log. This is the **common** logarithm. Likewise, when e is the base, we call $\log_e$ the **natural** logarithm and use ln instead of $\log_e$.

The equations

$$b^y = x \qquad \text{and} \qquad \log_b x = y$$

express the same relationship among the numbers b, y, and x. The first equation is in *exponential form* and the second is in *logarithmic form*. To help in understanding the meaning of logarithms, we list several logarithmic equations and the corresponding exponential equations side by side in Table 6.3.

TABLE 6.3

Logarithmic Form	Exponential Form
$\log_2 16 = 4$	$2^4 = 16$
$\log 100 = 2$	$10^2 = 100$
$\ln e = 1$	$e^1 = e$

The functions $f(x) = \ln x$ and $g(x) = e^x$ are inverses according to the following definition:

DEFINITION ● *Two functions f and g are* **inverses** *if the domain of f is the range of g, if the domain of g is the range of f, and if, for every x in the appropriate domains,*

$$f(g(x)) = x \quad \text{and} \quad g(f(x)) = x.$$

Now if $f(x) = \ln x$ and $g(x) = e^x$, we have

$$f(g(x)) = f(e^x) = \ln e^x = x$$

and

$$g(f(x)) = g(\ln x) = e^{\ln x} = x,$$

thus $\ln x$ and e^x are inverse functions. Likewise, $\log x$ and 10^x are inverse functions, as are $\log_b x$ and b^x, for any positive base b. Now $e^x = y$ if and only if $\ln y = x$, so the point (x, y) is on the graph of e^x if and only if the point (y, x) is

FIGURE 6.17

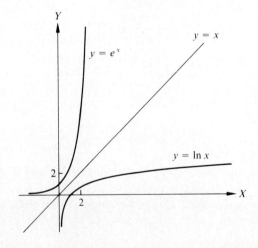

on the graph of the natural-logarithm function. Since the points (x, y) and (y, x) are equidistant from the line $y = x$, one can be obtained from the other by reflection through that line. The curves $\ln x$ and e^x are thus seen to be reflections of each other through the line $y = x$, as is shown by Fig. 6.17.

EXAMPLE 1 ● Draw, on the same coordinate axes, the graphs of the functions

$$y = \log x \qquad \text{and} \qquad y = \ln x.$$

SOLUTION. Using a calculator or Tables IV and V in the Appendix we determine several points on each curve.

x	$1/10$	$1/e$	1	e	e^2	10	e^3
$\log x$	-1	-0.43	0	0.43	0.868	1	1.30
$\ln x$	-2.30	-1	0	1	2	2.30	3

The graphs are sketched in Fig. 6.18. ●

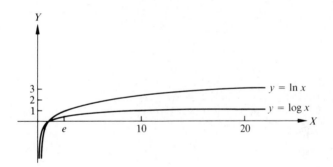

FIGURE 6.18

EXAMPLE 2 ● Sketch the graphs of the functions defined by the equations

$$y = 2^x \qquad \text{and} \qquad y = \log_2 x.$$

SOLUTION. These functions are inverses of each other. So we will draw the graph of $y = 2^x$, and then get the graph of $y = \log_2 x$ by the property of symmetry. Thus we form and prepare the following table of values of x and the corresponding values of 2^x.

x	-2	-1	0	1	2	3
2^x	$\frac{1}{4}$	$\frac{1}{2}$	1	2	4	8

These tabulated values of x and 2^x permit us to draw the graph of $y = 2^x$, as shown in Fig. 6.19. The graph of $y = \log_2 x$ is obtained by the use of symmetry. ●

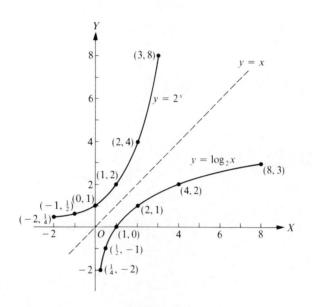

FIGURE 6.19

EXAMPLE 3 ● Find x if $\log_6 x = -3$.

SOLUTION. We use the equivalent form and get $x = 6^{-3} = \dfrac{1}{6^3} = \dfrac{1}{216}$. ●

EXAMPLE 4 ● Find a if $\log_a 49 = \frac{2}{3}$.

SOLUTION. The exponential form is $a^{2/3} = 49$. Then the $\frac{3}{2}$ power of both sides gives

$$\left(a^{2/3}\right)^{3/2} = 49^{3/2},$$
$$a = 343. ●$$

EXAMPLE 5 ● Find the value $\log_5 \frac{1}{25}$.

SOLUTION. We let $y = \log_5 \frac{1}{25}$. Then, changing to the exponential form, we get

$$5^y = \tfrac{1}{25} = 5^{-2} \quad \text{and} \quad y = -2. ●$$

EXAMPLE 6 ● If the number y, in millions, of bacteria in a culture t hours after introduction, is given by

$$y = f(t) = 20e^{t/3},$$

as in Example 4 in the last section, how long does it take for the number of bacteria to double? to triple?

SOLUTION. Since there are twenty million bacteria at $t = 0$, we seek a value of t so that

$$40 = f(t) = 20e^{t/3}.$$

Hence

$$e^{t/3} = 2$$

and

$$t = 3\ln 2$$
$$= 3(0.693)$$
$$= 2.08.$$

Also we want t so that

$$f(t) = 20e^{t/3} = 60,$$
$$e^{t/3} = 3,$$
$$t/3 = \ln 3,$$
$$t = 3\ln 3$$
$$= 3.30.$$

Thus it takes 2 hours and 5 minutes for the bacteria to double and 3 hours 18 minutes for it to triple. ●

Exercises

Give the logarithmic form of each of the following exponential equations.

1. $3^3 = 27$
2. $2^4 = 16$
3. $6^{-1} = \frac{1}{6}$
4. $8^1 = 8$
5. $15^0 = 1$
6. $27^{2/3} = 9$
7. $4^{-3/2} = \frac{1}{8}$
8. $32^{2/5} = 4$
9. $5^3 = 125$
10. $(\frac{2}{3})^{-2} = \frac{9}{4}$
11. $(\frac{3}{2})^3 = \frac{27}{8}$
12. $(\frac{1}{25})^{-3/2} = 125$

Give the exponential form of each of the following logarithmic equations.

13. $\log_3 1 = 0$
14. $\log_6 6 = 1$
15. $\log_5 25 = 2$
16. $\log_9 1 = 0$
17. $\log_2 \frac{1}{16} = -4$
18. $\log_{16} 4 = \frac{1}{2}$
19. $\log_2 \frac{1}{8} = -3$
20. $\log_4 \frac{1}{16} = -2$
21. $\log_{1/3} 27 = -3$
22. $\log_{1/4} 16 = -2$
23. $\log_{16} 64 = \frac{3}{2}$
24. $\log_{1/6} 36 = -2$

Find x, a, or y in each of the following equations.

25. $\log_5 x = 2$ 26. $\log_2 x = 4$ 27. $\log_7 x = -1$

28. $\log_3 x = -3$ 29. $\log_4 x = \frac{3}{2}$ 30. $\log_{16} x = -\frac{3}{2}$

31. $\log_{1/2} x = -3$ 32. $\log_a 4 = 2$ 33. $\log_a 4 = 1$

34. $\log_a 16 = \frac{1}{2}$ 35. $\log_a 9 = -\frac{1}{2}$ 36. $\log_a \frac{1}{4} = -\frac{1}{2}$

37. $\log_a 64 = -\frac{3}{2}$ 38. $y = \log_2 16$ 39. $y = \log_3 81$

40. $y = \log_5 \frac{1}{25}$ 41. $y = \log_4 64$ 42. $y = \log_{10} \frac{1}{1000}$

Sketch a graph of the function defined by the equation in each of the following exercises.

43. $y = \log x^2$ 44. $y = \log(x + 1)$

45. $y = \ln(x + 1)$ 46. $y = \log_2(x + 1)$

47. $y = 2 \ln x$ 48. $y = \ln 2x$

49. The Richter scale is used to measure the intensity of earthquakes. If E is the energy (in ergs) released by an earthquake, then the magnitude M on the Richter scale is related to E by

$$\log E = 1.5M + 11.4,$$

where the constants 1.5 and 11.4 are determined empirically and may be adjusted. How much energy is released by an earthquake of rank 5 on the Richter scale? Show that a quake of rank 7 is over 30 times stronger than one of rank 6.

50. The number of decibels D to measure the intensity of a sound is related to the power P of the sound by the equation

$$D = 10 \log 10^{16} P,$$

where P is measured in watts per square centimeters. Find the number of decibels in a conversation-level sound of 3.2×10^{-10} watts per square centimeter. How much power is there in a sound of 30 decibels?

6.4

ADDITION OF ORDINATES

It is often the case that we are faced with graphing a function that is expressed as a sum of two or more functions. In this section, we discuss a technique for sketching such a graph by first sketching the separate summands.

If $y = f(x) + g(x)$, we may sketch the graph by first drawing the graphs of $y_1 = f(x)$ and $y_2 = g(x)$ on the same coordinate axis. Then the **ordinates** of the curves, when added graphically, yield ordinates of the graph of $y = f(x) + g(x)$.

As we saw in Chapter 4, the presence of an xy term in a second-degree equation makes the construction of a graph more difficult. The preparation of a table of values for x and y becomes tedious in such a case. Rotation transformations can be complicated by radicals in the rotation formulas. If it is relatively easy to solve the equation for one variable, say y, in terms of x, so as to have a sum of ordinates, this can be helpful. We illustrate the technique with several examples.

EXAMPLE 1 ● Draw the graph of the equation

$$2x^2 - 2xy + y^2 + 8x - 12y + 36 = 0.$$

SOLUTION. To express y as the sum of two quantities, we treat the equation as a quadratic in y. Thus we have

$$y^2 + (-2x - 12)y + (2x^2 + 8x + 36) = 0,$$

and solving for y gives

$$y = \frac{2x + 12 \pm \sqrt{(-2x - 12)^2 - 4(2x^2 + 8x + 36)}}{2}$$

$$= x + 6 \pm \sqrt{4x - x^2}.$$

FIGURE 6.20

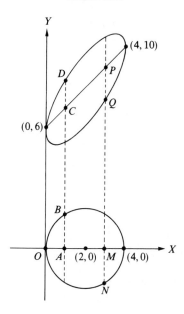

We now draw the graphs of the equations

$$y_1 = x + 6 \qquad \text{and} \qquad y_2 = \pm\sqrt{4x - x^2}.$$

The graph of the first equation is a line. By squaring and then completing the square in the x terms, we reduce the second equation to $(x - 2)^2 + y_2^2 = 4$. The graph is a circle of radius 2 and center at $(2, 0)$. The line and the circle are drawn in Fig. 6.20. The point D on the graph of the original equation is obtained by adding the ordinates AB and AC; that is, AC is extended by a length equal to AB. The addition of ordinates for this purpose must be algebraic. Thus, MN is negative, and the point Q is found by measuring downward from P so that $PQ = MN$. By plotting a sufficient number of points in this manner, the desired graph can be constructed. ●

EXAMPLE 2 ● Sketch the graph of

$$y = x + \cos x.$$

SOLUTION. Let $y_1 = x$ and $y_2 = \cos x$ and sketch each on the same coordinate axis. Now add the ordinates for the easy values ($y_1 = 0$ or $y_2 = 0$), as well as for "critical" points where $\cos x$ is a maximum or a minimum.

The heavy dots in Fig. 6.21 indicate points on the sum.

FIGURE 6.21

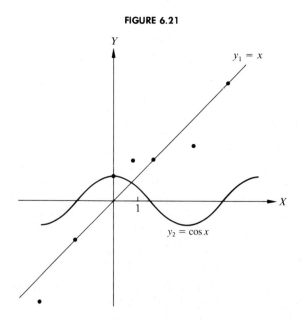

Next draw the curve through the dots, using additional points if needed, as in Fig. 6.22. ●

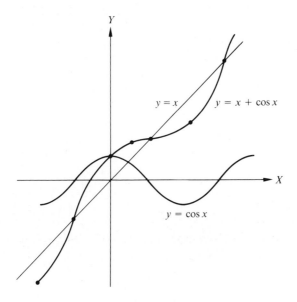

FIGURE 6.22

While studying the propagation of heat waves, J. B. J. Fourier discovered, around 1800, that any function can be expressed as a sum of "enough" generalized sine and cosine functions. This remarkable result is the theoretical backbone of many current enterprizes: electronic music such as chord organs, synthetic voice devices, almost any type of signal transmission between electronic devices such as radios, television, and computers, and the analysis of data that is presented in graphic form such as the seismographic data used in the search for oil and gas. The following example demonstrates how generalized sine and cosine curves may be added. Readers with access to a computer may find it instructive to graphically add three or more such functions.

EXAMPLE 3 ● Sketch a period of the graph of $y = f(x) = \sin x + \cos 2x$.

SOLUTION. The period of sin is 2π, while that of $\cos 2x$ is π, so the period of $f(x)$ will be 2π. The curves are sketched in Fig. 6.23. ●

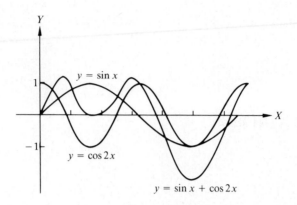

FIGURE 6.23

Exercises

Sketch the graph of each equation in Exercises 1 through 14 by addition of ordinates.

1. $y = 6 - x \pm \sqrt{4 - x^2}$ 2. $y = 1 \pm \sqrt{x}$

3. $y^2 - 2y + 2x^2 - 1 = 0$ 4. $y = 2x \pm \sqrt{5 + 6x - x^2}$

5. $y^2 + 2xy - 3x^2 + 4 = 0$ 6. $y^2 - 4xy + 3x^2 + 1 = 0$

7. $y = x + \sin x$ 8. $y = \sin x + \cos x$

9. $y = \sin x + \sin x/2$ 10. $y = x + \ln x, \quad x > 0$

11. $y = 3 + 2 \sin \pi x$ 12. $y = -1 + \ln x, \quad x > 0$

13. $y = 2 \sin x + \cos x/2$ 14. $y = \frac{1}{2} \sin \pi x + \frac{1}{3} \cos 2\pi x$

15. In psychophysics, the Weber–Fechner model is used to relate the magnitude of a stimulus to a response in a wide variety of contexts, such as hearing in response to sound. If y is the level of response to the level x of stimulus, the model states that x and y are related by the equation

$$y = a + b \ln x, \qquad a > 0.$$

Recently, evidence suggests that the equation

$$y = a + bx^c, \qquad a < 0,$$

more correctly models the situation. Sketch a graph of each of the following equations:

$$y = -0.1 + 0.2 \ln x,$$
$$y = -0.1 + 0.1x^{1/2}.$$

REVIEW EXERCISES

1. Define *sine* and *cosine* functions, *amplitude* and *period*, *exponential* and *logarithmic* functions.

2. Sketch one period of $y = 2 \sin \pi x$, stating the period and amplitude of the function.

3. Sketch, on the same coordinate axes, the graphs of
$$y = 10^x \qquad \text{and} \qquad y = \log x.$$

4. Find x if $\log_5 x = -3$.

5. Find a if $\log_a 64 = 2/3$.

6. Sketch the graph of $y = 2^{1-x}$.

7. Sketch, on the same coordinate axes,
$$y = \log_2 x \qquad \text{and} \qquad y = \log_3 x.$$

8. Sketch $y = x + 1/x$ by addition of ordinates.

7

Polar Coordinates

There are various types of coordinate systems. The rectangular system with which we have been dealing is probably the most important. In this system a point is located by its distances from two perpendicular lines. We shall introduce in this chapter a coordinate system in which the coordinates of a point in a plane are its distance from a fixed point and its direction from a fixed line. The coordinates given in this way are called **polar coordinates**. The proper choice of a coordinate system depends on the nature of the problem at hand. For some problems, either the rectangular or the polar system may be satisfactory; usually, however, one of the two is preferable; and in some situations it is advantageous to use both systems, shifting from one to the other.

7.1

THE POLAR COORDINATE SYSTEM

The reference frame in the polar-coordinate system is a half-line drawn from some point in the plane. In Fig. 7.1 a half-line is represented by OA. The point O is called the **origin** or **pole** and OA the **polar axis**. The position of any point P in the plane is definitely determined by the distance $\overline{OP}$ and the angle AOP. The segment OP, denoted by r, is referred to as the **radius vector**; the angle AOP, denoted by θ, is called the **vectorial angle**. The coordinates of P are then written as $P(r, \theta)$ or just (r, θ).

FIGURE 7.1

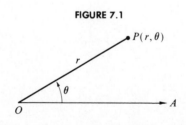

196

It is customary to regard polar coordinates as signed quantities. The vectorial angle, as in trigonometry, is defined as positive or negative depending upon whether it is measured counterclockwise or clockwise from the polar axis. The r coordinate is defined as positive if measured from the pole along the terminal side of θ and negative if measured along the terminal side extended through the pole.

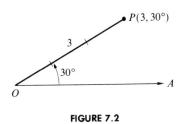

FIGURE 7.2

A given pair of polar coordinates definitely locates a point. For example, the coordinates $(3, 30°)$ determine one particular point. To plot the point, we first draw the terminal side of a 30° angle measured counterclockwise from OA (Fig. 7.2) and then lay off three units along the terminal side. While this pair of coordinates defines a particular point, there are other coordinate values that define this same point. This is evident, since the vectorial angle may have 360° added or subtracted repeatedly without changing the point represented. Additional coordinates of the point may be obtained also by using a negative value for the distance coordinate. Restricting the vectorial angle to values between $-360°$ and $360°$, we see (Figs. 7.2 through 7.5) that the following pairs of coordinates define the same point:

$$(3, 30°) \qquad (3, -330°) \qquad (-3, 210°) \qquad (-3, -150°).$$

FIGURE 7.3

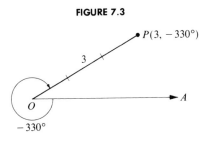

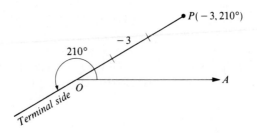

FIGURE 7.4

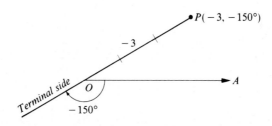

FIGURE 7.5

We next write the general formulas

$$(r, \theta) = (-r, \theta + 180°) = (-r, \theta - 180°)$$
$$= (r, \theta + 360n°), \qquad n = \text{any integer.} \qquad (7.1)$$

These formulas give different ways of representing a single point.

EXAMPLE 1 ● If we use $r = 4$, $\theta = 60°$, and $n = 1$ in the formulas, we get:

$$(4, 60°) = (-4, 60° + 180°) = (-4, 60° - 180°) = (4, 60° + 360°),$$
$$(4, 60°) = (-4, 240°) = (-4, -120°) = (4, 420°).$$

Figures 7.6 through 7.9 picture four ways of representing a single point. ●

FIGURE 7.6

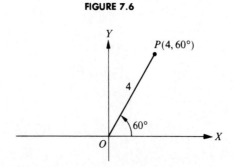

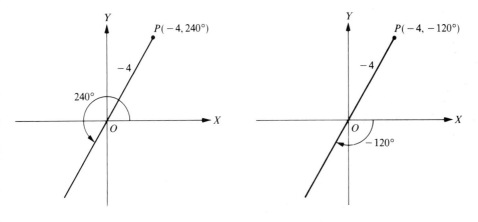

FIGURE 7.7   **FIGURE 7.8**

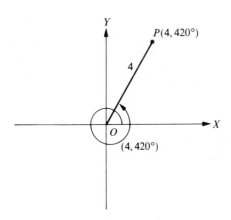

FIGURE 7.9

EXAMPLE 2 ● We now let $r = -4$, $\theta = -60°$, and $n = 1$ in formulas (7.1). Hence we obtain

$$(-4, -60°) = (4, -60° + 180°) = (4, -60° - 180°) = (-4, -60° + 360°),$$

$$(-4, -60°) = (4, 120°) = (4, -240°) = (-4, 300°). \;\bullet$$

We urge the student to ponder carefully formulas (7.1) and the two examples. Draw figures and verify in your mind the various ways of representing the same point.

Although we have used here the vectorial angles in terms of degree measure, radian measure would serve just as well. The relation between degrees and radians can be easily determined. The entire circumference of a circle subtends a central angle of 2π in terms of radians and 360 units in terms of degrees. Hence we have

$$2\pi \text{ radians} = 360° \quad \text{and} \quad \pi \text{ radians} = 180°.$$

We use the second equation to express a radian in terms of degrees and a degree in terms of radian measure. Thus

$$1 \text{ radian} = \frac{180°}{\pi} \quad \text{and} \quad 1° = \frac{\pi}{180} \text{ radian}.$$

We allow ourselves the convenience of using "angle" and "measure of an angle" interchangeably. For example, if θ stands for an angle of measure 60°, we write $\theta = 60°$ when it is really the measure of the angle that is 60°.

A point of given coordinates can be plotted by estimating the vectorial angle and the radius vector by sight. Greatly improved accuracy, however, may be obtained by the use of polar coordinate paper. This paper has equally spaced circles with their centers at the origin and equally spaced radial lines through the origin (Fig. 7.10). A point of given coordinates may be located by selecting the proper θ and r coordinates. Several points are plotted in the figure.

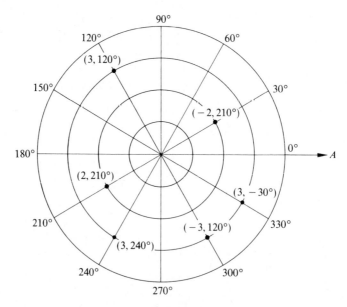

FIGURE 7.10

Exercises

Express each of the following angles in terms of radians.

1. $30°, 45°, 90°$ 2. $60°, 120°, 225°$ 3. $315°, 330°, 420°$

Change each angle from radian measure to degrees.

4. $\dfrac{\pi}{10}, \dfrac{\pi}{15}, \dfrac{3\pi}{2}$ 5. $\dfrac{3\pi}{4}, \dfrac{\pi}{9}, \dfrac{4\pi}{15}$ 6. $\dfrac{3\pi}{8}, \dfrac{11\pi}{12}, \dfrac{4\pi}{3}$

7. $\dfrac{1}{7}\pi, \dfrac{1}{11}\pi, \dfrac{1}{13}\pi$

Plot the given points on a polar coordinate system.

8. $P(3, 60°)$ 9. $P(6, -30°)$ 10. $P(2, 180°)$

11. $P(4, 15°)$ 12. $P(5, 0°)$ 13. $P(-3, 135°)$

14. $P(5, 150°)$ 15. $P(4, 0°)$ 16. $P(-5, 210°)$

17. $P(2, -90°)$ 18. $P(-2, -45°)$

19. $P(4, 210°)$ 20. $P(-4, 315°)$ 21. $P(-3, -180°)$

22. $P(-1, \pi)$ 23. $P(3, \tfrac{4}{3}\pi)$ 24. $P(2, \tfrac{1}{6}\pi)$

25. $P(-3, -\tfrac{4}{3}\pi)$ 26. $P(4, \tfrac{3}{2}\pi)$ 27. $P(2, -\tfrac{3}{2}\pi)$

28. $P(3, -\tfrac{5}{3}\pi)$ 29. $P(-3, \tfrac{7}{4}\pi)$ 30. $P(4, \tfrac{7}{4}\pi)$

31. $P(3, -\tfrac{23}{12}\pi)$ 32. $P(2, \tfrac{5}{4}\pi)$ 33. $P(5, -\tfrac{4}{3}\pi)$

Write three other pairs of polar coordinates for the point in each of Exercises 34 through 37. Restrict the vectorial angles so that they do not exceed 360° in absolute value.

34. $P(3, 60°)$ 35. $P(6, -30°)$ 36. $P(2, 180°)$

37. $P(4, 15°)$

Write three other pairs of polar coordinates for the point in each problem 38 through 41. Restrict the vectorial angles θ so that $-2\pi \le \theta \le 2\pi$.

38. $P(-1, \tfrac{5}{6}\pi)$ 39. $P(-3, \tfrac{4}{3}\pi)$ 40. $P(-4, \pi)$

41. $P(4, -\tfrac{3}{2}\pi)$

7.2

RELATIONS BETWEEN RECTANGULAR AND POLAR COORDINATES

As we have mentioned, it is often advantageous in the course of a problem to shift from one coordinate system to another. For this purpose we shall derive transformation formulas that express polar coordinates in terms of rectangular coordinates, and vice versa. In Fig. 7.11 the two systems are placed so that the origins coincide and the polar axis lies along the positive x axis. Then a point P has the coordinates (x, y) and (r, θ). From the triangle OMP, we have

$$\cos \theta = \frac{x}{r} \quad \text{and} \quad \sin \theta = \frac{y}{r},$$

and hence

$$\boxed{x = r\cos \theta,} \tag{7.2}$$

$$\boxed{y = r \sin \theta.} \tag{7.3}$$

FIGURE 7.11

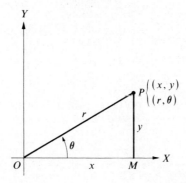

To obtain r and θ in terms of x and y, we write

$$r^2 = x^2 + y^2 \quad \text{and} \quad \tan \theta = \frac{y}{x}.$$

Then, solving for r and θ, we get

$$r = \pm\sqrt{x^2 + y^2}, \tag{7.4}$$

$$\theta = \arctan\frac{y}{x}. \tag{7.5}$$

These four equations enable us to transform the coordinates of a point, and therefore the equation of a graph, from one system to the other. The θ coordinate as given by Eq. (7.5) is not single-valued. Hence it is necessary to select a proper value for θ when applying the formula to find this coordinate of a point. See Example 2.

EXAMPLE 1 ● Find the rectangular coordinates of the point defined by the polar coordinates $(6, 120°)$.

SOLUTION. Using Eqs. (7.2) and (7.3), we have (Fig. 7.12)

$$x = r\cos\theta = 6\cos 120° = -3,$$
$$y = r\sin\theta = 6\sin 120° = 3\sqrt{3}.$$

The required coordinates are $(-3, 3\sqrt{3})$. ●

FIGURE 7.12

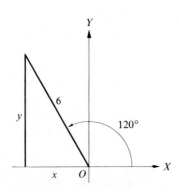

EXAMPLE 2 ● Express the rectangular coordinates $(-2, -2)$ in terms of polar coordinates.

SOLUTION. Equations (7.4) and (7.5) give

$$r = \sqrt{x^2 + y^2} = 2\sqrt{2}$$

and

$$\theta = \arctan \frac{y}{x} = \arctan 1.$$

Since the point is in the third quadrant, we select $\theta = 225°$. Hence the pair of coordinates $(2\sqrt{2}, 225°)$ is a polar representation of the given point. ●

EXAMPLE 3 ● Find the polar coordinate equation corresponding to $2x - 3y = 5$.

SOLUTION. Substituting for x and y gives

$$2(r\cos\theta) - 3(r\sin\theta) = 5$$

or

$$r(2\cos\theta - 3\sin\theta) = 5. ●$$

EXAMPLE 4 ● Transform the equation $r = 4\sin\theta$ to rectangular coordinates.

SOLUTION. Since $r = \sqrt{x^2 + y^2}$ and $\sin\theta = y/r = y/\sqrt{x^2 + y^2}$, we substitute in the given equation and obtain

$$\sqrt{x^2 + y^2} = \frac{4y}{\sqrt{x^2 + y^2}}$$

or

$$x^2 + y^2 = 4y.$$

The required equation, as well as the original equation, represents a circle. ●

EXAMPLE 5 ● Transform the polar coordinate equation

$$r = \frac{1}{\cos\theta + 3\sin\theta}$$

to the corresponding rectangular coordinate equation.

SOLUTION. We multiply both members of the equation by $\cos\theta + 3\sin\theta$ and obtain

$$r(\cos\theta + 3\sin\theta) = 1$$

or, from Eqs. (7.2) and (7.3),

$$r\left(\frac{x}{r} + \frac{3y}{r}\right) = 1.$$

Now, multiplying, we have the linear equation

$$x + 3y = 1.$$

The final equation tells us that the graph of the given equation is a straight line. ●

Exercises

1. Derive the transformation formulas (7.1) through (7.4) from a figure in which the terminal side of θ is in the (a) second quadrant, (b) third quadrant, (c) fourth quadrant.

Find the rectangular coordinates of the following polar-coordinate points.

2. $(3, 90°)$ 3. $(2\sqrt{2}, 45°)$ 4. $(5, 0°)$

5. $(0, 180°)$ 6. $(-3, 270°)$ 7. $(-2, -60°)$

8. $(4, 150°)$ 9. $(3\sqrt{2}, -135°)$ 10. $(6, 180°)$

11. $(3, 30°)$ 12. $(4, 135°)$ 13. $(5, 225°)$

Find nonnegative polar coordinates of the following rectangular-coordinate points.

14. $(0, 4)$ 15. $(4, 0)$ 16. $(0, 0)$

17. $(-2, 0)$ 18. $(0, -5)$ 19. $(\sqrt{2}, \sqrt{2})$

20. $(2\sqrt{3}, -2)$ 21. $(-4\sqrt{3}, 4)$ 22. $(-\sqrt{2}, -\sqrt{2})$

23. $(-4, -4)$ 24. $(\sqrt{3}, -1)$ 25. $(\sqrt{2}, -1)$

26. $(-4, 3)$ 27. $(5, 12)$ 28. $(-5, -12)$

Transform the following equations into the corresponding polar-coordinate equations.

29. $x = -3$ 30. $y = 4$ 31. $2x + y = 3$

32. $3x - y = 0$ 33. $y^2 = 9x$ 34. $x^2 = 4y$

35. $x^2 + y^2 = 9$ 36. $x^2 - y^2 = a^2$ 37. $x^2 - 9y = 0$

38. $x^2 - 2y^2 = 4$ 39. $x^2 + y^2 = 2x$ 40. $x^2 + y^2 = 2y$

Transform the equations to the corresponding rectangular-coordinate equations.

41. $r = 3$ 42. $\theta = 45°$ 43. $\theta = 30°$

44. $r\cos\theta = 4$ 45. $r\sin\theta = 4$ 46. $r = 6\cos\theta$

47. $r = 8 \sin \theta$

48. $r^2 \sin 2\theta = a^2$

49. $r^2 \cos 2\theta = a^2$

50. $r = \dfrac{2}{1 - \cos \theta}$

51. $r = \dfrac{2}{2 + \cos \theta}$

52. $r = \dfrac{3}{1 - 2 \cos \theta}$

53. $r = \dfrac{3}{3 \sin \theta + 4 \cos \theta}$

54. $r = \dfrac{4}{\sin \theta - 2 \cos \theta}$

55. $r = \dfrac{1}{\cos \theta + 3 \sin \theta}$

56. $r = \dfrac{2}{1 + 2 \sin \theta}$

7.3
GRAPHS OF POLAR-COORDINATE EQUATIONS

In this section we shall consider the problem of plotting the graphs of equations expressed in polar coordinates. The definition of a graph in polar coordinates and the technique of its construction are essentially the same as with an equation in rectangular coordinates.

DEFINITION 7.1 ● *The graph of an equation in polar coordinates is the set of all points whose coordinates satisfy the equation.*

The graphs of the equations most easily determined are those in which each coordinate is equal to a constant. For example, the graph of $\theta = 35°$ is the line through the origin and making an angle of $35°$ with the polar axis. If $35°$ is used for the θ coordinate at all points of the line, the corresponding r coordinates will be positive on the terminal side of $35°$ and negative on the extension of the terminal side through the origin. This is illustrated by the points $(4, 35°)$ and $(-4, 35°)$. Similarly, the graph of $r = a$, where a is any real nonzero number, is a circle with center at the origin and radius $|a|$.

EXAMPLE 1 ● Construct the graph of $r = 3 + \sin \theta$.

SOLUTION. We assign certain values to θ from $0°$ to $360°$ and prepare a table of corresponding values of θ and r (see Table 7.1). The fractional values of r are

TABLE 7.1

θ	0°	30°	45°	60°	90°	120°	150°	180°
r	3	3.5	3.7	3.9	4	3.9	3.5	3
θ	210°	225°	240°	270°	300°	315°	330°	360°
r	2.5	2.3	2.1	2	2.1	2.3	2.5	3

rounded off to one decimal place. By plotting these points and drawing a curve through them, we obtain the graph of Fig. 7.13. ●

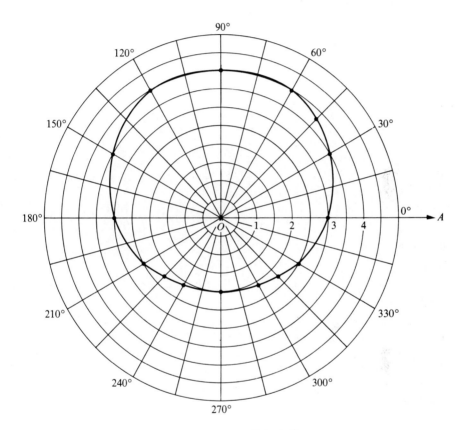

FIGURE 7.13. $r = 3 + \sin\theta$.

EXAMPLE 2 ● Draw the graph of $r = 4\cos\theta$.

SOLUTION. We first prepare a table of corresponding values of r and θ.

θ	0°	30°	45°	60°	75°	90°	120°	135°	150°	180°
r	4	3.5	2.8	2.0	1.0	0	−2.0	−2.8	−3.5	−4

This table yields the graph in Fig. 7.14. We did not extend the table to include values of θ in the interval 180° to 360°, since values of θ in this range would merely repeat the graph already obtained. For example, the point $(-3.5, 210°)$ is on the graph, but this point is also defined by the coordinates $(3.5, 30°)$.

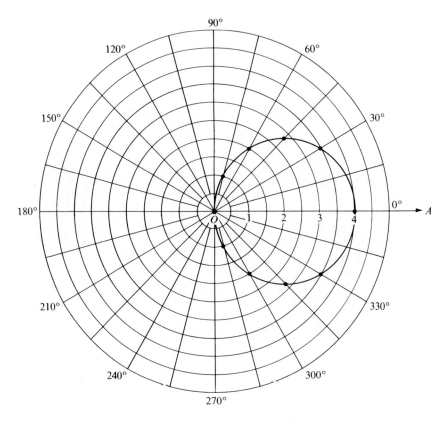

FIGURE 7.14. $r = 4\cos\theta$.

The graph appears to be a circle. This surmise is verified by transforming the equation to rectangular coordinates. The transformed equation is

$$(x - 2)^2 + y^2 = 4. \bullet$$

EXAMPLE 3 ● Draw the graph of the equation

$$r = \frac{1}{1 + \sin\theta}.$$

SOLUTION. The table of corresponding values of r and θ enables us to draw the curve in Fig. 7.15.

θ	0°	90°	180°	210°	225°	315°	330°
r	1	0.50	1	2	3.41	3.41	2

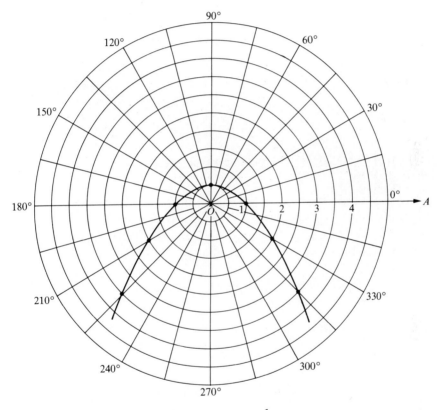

FIGURE 7.15. $r = \dfrac{1}{1 + \sin\theta}$

Exercises

Draw the graphs of the following equations. In equations involving $\sin\theta$ and $\cos\theta$, points plotted at 30° intervals will suffice.

1. $r = 4$ 2. $r = -4$ 3. $\theta = 120°$

4. $\theta = -120°$ 5. $\theta = 0°$ 6. $r = 2 - \cos\theta$

7. $r = 2 + \sin\theta$ 8. $r = 4 + \sin\theta$ 9. $r = 4 + \cos\theta$

10. $r = 4 - \sin\theta$ 11. $r = 4 - \cos\theta$ 12. $r = 3\cos\theta$

13. $r = -3\cos\theta$ 14. $r = 2\sin\theta$ 15. $r = -2\sin\theta$

16. $r = \dfrac{1}{2 + \sin\theta}$ 17. $r = \dfrac{1}{2 - \sin\theta}$ 18. $r = \dfrac{2}{2 - \sin\theta}$

19. $r = \dfrac{3}{3 + \cos\theta}$ 20. $r = \dfrac{3}{3 - \cos\theta}$ 21. $r = \dfrac{4}{2 + \sin\theta}$

22. $r = \dfrac{4}{3 + 2\sin\theta}$ 23. $r = \dfrac{-2}{3 + 2\cos\theta}$ 24. $r = \dfrac{-3}{4 - 3\sin\theta}$

7.4

AIDS IN GRAPHING POLAR-COORDINATE EQUATIONS

A careful examination of an equation in polar coordinates will often reveal shortcuts to the construction of its graph. It is better, for economy of time, to wrest useful information from an equation and thus keep at a minimum the point-by-point plotting in sketching a graph. We shall discuss and illustrate a few simple devices that facilitate the tracing of polar curves.

Variation of r with θ

Many equations are sufficiently simple so that the way in which r varies as θ increases is evident. Usually the variation of θ from 0° to 360° yields the complete graph. However, we shall find exceptions to this rule. By observing the equation and letting θ increase through the necessary values, the graph can be visualized. A rough sketch can then be made with a few pencil strokes. We illustrate with an example.

EXAMPLE 1 ● Sketch the graph of the equation

$$r = 3(1 + \sin\theta).$$

SOLUTION. If θ starts at 0° and increases in 90°-steps to 360°, it is a simple matter to see how r varies in each 90° interval (see Table 7.2). This variation is represented in the diagram. The graph (Fig. 7.16) is a heart-shaped curve called the **cardioid**. ●

TABLE 7.2

As θ increases from	$\sin \theta$ varies from	r varies from
0° to 90°	0 to 1	3 to 6
90° to 180°	1 to 0	6 to 3
180° to 270°	0 to -1	3 to 0
270° to 360°	-1 to 0	0 to 3

FIGURE 7.16. $r = 3(1 + \sin \theta)$

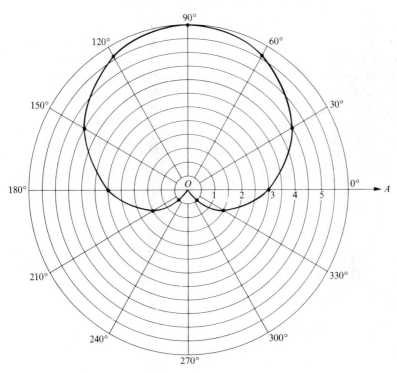

Tangent Lines at the Origin

If r shrinks to zero as θ approaches and takes a fixed value θ_0, then the line $\theta = \theta_0$ is tangent to the curve at the origin. Intuitively, this statement seems correct: it can be proved. In the equation in Example 1, $r = 3(1 + \sin \theta)$, the value of r diminishes to zero as θ increases to 270°. Hence the curve is tangent to the vertical line at the origin (Fig. 7.16). •

To find the tangents to a curve at the origin, set $r = 0$ in the equation and solve for the corresponding values of θ.

Symmetry

In Section 3.2 we discussed symmetry of curves and formulated tests applicable in a rectangular-coordinate system. The following tests for symmetry in a polar-coordinate system are deducible by referring to Fig. 7.17.

*1. The graph is symmetric with respect to the **pole** if the equation is unchanged when*

A. r is replaced with −r,
or B. θ is replaced with 180° − θ.

*2. The graph is symmetric with respect to the **polar axis** if the equation is unchanged when*

A. θ is replaced by −θ,
or B. θ is replaced by 180° − θ and r is replaced by −r.

*3. The graph is symmetric with respect to the **vertical line** $\theta = 90°$ if the equation is*

FIGURE 7.17

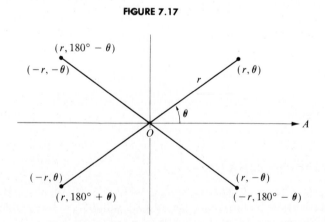

unchanged when

> A. θ is replaced with $180° - \theta$,
>
> or B. θ is replaced by $-\theta$ **and** r is replaced by $-r$.

These tests will be found to be helpful. The particular symmetry is established if *either* part, A or B, of the test is satisfied. If *both* parts A and B are not satisfied, then the curve does not have the symmetry in question. When a graph has any two of the three types of symmetry, it necessarily has the third type of symmetry. The student should verify this statement by examination of Fig. 7.17.

To use the tests for symmetry of graphs, the following trigonometric identities may be helpful.

$$\sin(-\theta) = -\sin\theta, \qquad \cos(-\theta) = \cos\theta,$$
$$\sin(180° - \theta) = \sin\theta, \qquad \cos(180° - \theta) = -\cos\theta,$$
$$\sin(180° + \theta) = -\sin\theta, \qquad \cos(180° + \theta) = -\cos\theta.$$

The following identities will be needed later in the chapter:

$$\sin^2\theta + \cos^2\theta = 1,$$
$$\sin 2\theta = 2\sin\theta\cos\theta,$$
$$\cos 2\theta = \cos^2\theta - \sin^2\theta.$$

Excluded Values

Frequently we shall meet equations in which certain values of the variables are excluded. For example, $r^2 = a^2\sin\theta$ places restrictions on both r and θ. The values of r are in the range $-a$ to a, and θ cannot have a value that makes $\sin\theta$ negative since r would then be imaginary. In particular, the angles between $180°$ and $360°$ are excluded. The graph, however, extends into the third and fourth quadrants for other values of θ because the equation satisfies the test for symmetry with respect to the origin.

Intercepts

The points at which a curve touches or cuts the horizontal and vertical lines through the pole are called **intercept** points; they are sometimes quite helpful in drawing the graph of a polar equation. The intercept points can be found by using the values $0°$, $90°$, $180°$, $270°$, or angles coterminal with these, for θ and finding the corresponding values of r. In this way, for example, we find that the intercept points of $r = 3 + \sin\theta$ are $(3, 0°)$, $(4, 90°)$, $(3, 180°)$, and $(2, 270°)$. The intercept points of $r = 2\theta$ are limitless in number. They occur at $\theta = n\pi$ and $\theta = (n + \frac{1}{2})\pi$, where n is any integer, positive, negative, or zero.

We remark, however, that a curve may pass through the origin at angles other than a quadrantal angle. The test for this situation, as already stated, is to find the values of θ when $r = 0$.

Special Types of Equations

There are several types of polar-coordinate equations whose graphs have been given special names. We consider a few of these equations.

The graphs of equations of the forms

$$r = a \sin n\theta \qquad \text{and} \qquad r = a \cos n\theta,$$

where n is a positive integer, greater than 1, are called **rose curves**. The graph of a rose curve consists of equally spaced closed loops extending from the origin. The number of loops, or leaves, depends on the integer n. If n is odd, there are n leaves; if n is even, there are $2n$ leaves.

EXAMPLE 2 ● Construct the graph of $r = \sin 2\theta$.

SOLUTION. We first apply the three tests for symmetry. If we replace r by $-r$, the equation

$$r = \sin 2\theta \quad \text{becomes} \quad -r = \sin 2\theta.$$

This does not establish symmetry with respect to the pole. But substituting $180° + \theta$ for θ yields

$$r = \sin 2(180° + \theta) = \sin(360° + 2\theta) = \sin 2\theta.$$

This result shows that the graph is symmetric with respect to the pole.

Likewise we see that the curve is symmetric with respect to the vertical line $\theta = 90°$ by part B of the test, since substitution of $-r$ and $-\theta$ for r and θ yields

$$-r = \sin 2(-\theta) = -\sin 2\theta,$$
$$r = \sin 2\theta.$$

Furthermore, we have symmetry with respect to the polar axis, since we have symmetry with respect to the pole and with respect to the line $\theta = 90°$. Note that tests A for polar symmetry and for vertical-line symmetry both fail.

Since we have all three symmetries, we need only determine the graph in the first quadrant and then use the symmetries to graph the entire curve. To determine how r varies with θ, we make Table 7.3 for θ increasing in steps of 45°. Also if $r = 0$, we find that $\theta = 0, 90°, 180°$, and $270°$. This shows that the graph is tangent to the polar axis and to the vertical line $\theta = 90°$. The completed graph, called a **four-leaved rose** because of its shape, is seen in Fig. 7.18. The numbered barbs indicate how a point would move as θ increases from $0°$ to $360°$. ●

TABLE 7.3

θ	2θ	$\sin 2\theta$, or r
$0° \rightarrow 45°$	$0° \rightarrow 90°$	$0 \rightarrow 1$
$45° \rightarrow 90°$	$90° \rightarrow 180°$	$1 \rightarrow 0$

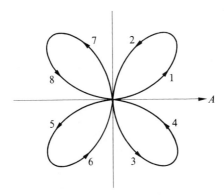

FIGURE 7.18. $r = \sin 2\theta$.

EXAMPLE 3 ● Construct the graph of the equation $r = \cos 3\theta$.

SOLUTION. If we replace θ by $-\theta$, r remains unchanged; hence the graph is symmetric with respect to the polar axis. To draw the graph we let θ increase in 30° steps from 0° to 180°. Table 7.4 indicates the way in which r changes in each of the steps. This information enables us to draw the graph in Fig. 7.19. ●

TABLE 7.4

θ	3θ	$\cos 3\theta$, or r
$0° \rightarrow 30°$	$0° \rightarrow 90°$	$1 \rightarrow 0$
$30° \rightarrow 60°$	$90° \rightarrow 180°$	$0 \rightarrow -1$
$60° \rightarrow 90°$	$180° \rightarrow 270°$	$-1 \rightarrow 0$
$90° \rightarrow 120°$	$270° \rightarrow 360°$	$0 \rightarrow 1$
$120° \rightarrow 150°$	$360° \rightarrow 450°$	$1 \rightarrow 0$
$150° \rightarrow 180°$	$450° \rightarrow 540°$	$0 \rightarrow -1$

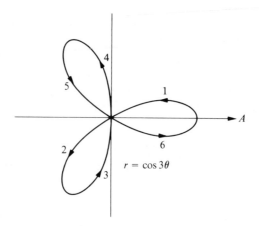

FIGURE 7.19. $r = \cos 3\theta$.

The graph of an equation of the form

$$r = b + a \sin\theta \qquad \text{or} \qquad r = b + a \cos\theta$$

is called a **limaçon**. The shape of the graph depends on the relative values of a and b. If $a = b$, the limaçon is called a cardioid from its heartlike shape, as illustrated in Fig. 7.16. If the absolute value of b is greater than the absolute value of a, the graph is a curve surrounding the origin (see Fig. 7.13). An interesting feature is introduced in the graph when the absolute value of a is greater than the absolute value of b. The graph then has an inner loop, which we illustrate in an example.

EXAMPLE 4 ● Construct the graph of the limaçon

$$r = 2 + 4 \cos\theta.$$

SOLUTION. The equation is unchanged when θ is replaced by $-\theta$, since $\cos(-\theta) = \cos\theta$. Hence the graph is symmetric with respect to the polar axis. Setting $r = 0$ gives

$$2 + 4 \cos\theta = 0,$$
$$\cos\theta = -\tfrac{1}{2},$$
$$\theta = 120°, 240°.$$

Hence the lines $\theta = 120°$ and $\theta = 240°$ are tangent to the graph at the origin. To obtain further aid in constructing the graph, we prepare Table 7.5.

TABLE 7.5

θ	$\cos \theta$	r
$0° \rightarrow 90°$	$1 \rightarrow 0$	$6 \rightarrow 2$
$90° \rightarrow 120°$	$0 \rightarrow -\frac{1}{2}$	$2 \rightarrow 0$
$120° \rightarrow 180°$	$-\frac{1}{2} \rightarrow -1$	$0 \rightarrow -2$

The graph is shown in Fig. 7.20. The lower half of the large loop and the upper half of the small loop were drawn by the use of symmetry. ●

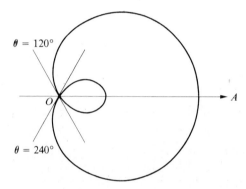

FIGURE 7.20. $r = 2 + 4 \cos \theta$.

The graphs of the polar equations

$$r^2 = a^2 \sin 2\theta \qquad \text{and} \qquad r^2 = a^2 \cos 2\theta$$

are **lemniscates**. In each of these equations, r varies from $-a$ to a, and the values of θ that make the right member negative are excluded. The excluded values in the first equation are $90° < \theta < 180°$ and $270° < \theta < 360°$. In the second equation, the excluded values are $45° < \theta < 135°$ and $225° < \theta < 315°$.

EXAMPLE 5 ● Draw the graph of the equation $r^2 = 9 \cos 2\theta$.

SOLUTION. We observe that the equation is symmetric with respect to the pole, the polar axis, and the vertical line through the pole. As θ increases from $0°$ to $45°$, the positive values of r vary from 3 to 0 and the negative values from -3 to 0. Hence this interval for θ gives rise to the half loop in the first quadrant and the half loop in the third quadrant. Either of these half loops combined with the known symmetries is sufficient for completing the graph (Fig. 7.21). Similar, except for position, is the graph of $r^2 = 9 \sin 2\theta$ (Fig. 7.22). ●

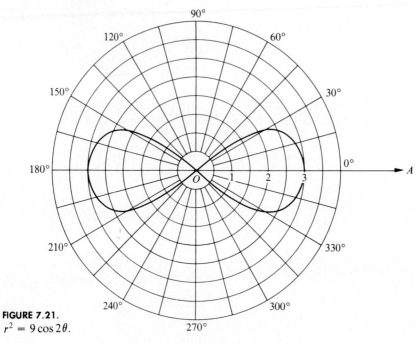

FIGURE 7.21.
$r^2 = 9 \cos 2\theta$.

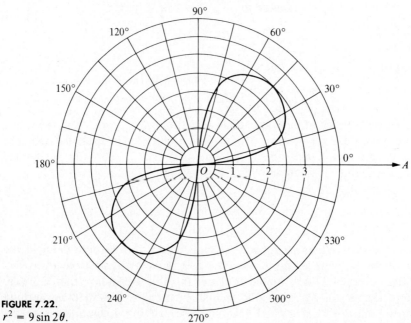

FIGURE 7.22.
$r^2 = 9 \sin 2\theta$.

Exercises

Sketch the graph of each of the following equations. First examine the equation to find properties that are helpful in tracing the graph. Where the literal constant a occurs, assign to it a convenient positive value.

1. $r = 4(1 + \sin \theta)$

2. $r = 3(1 - \sin \theta)$

3. $r = 4(1 + \cos \theta)$

4. $r = 3(1 - \cos \theta)$

5. $r = \cos 2\theta$

6. $r = \sin 4\theta$

7. $r = a \cos 3\theta$

8. $r = a \sin 3\theta$

9. $r = 2 \sin 5\theta$

10. $r = a \cos 5\theta$

11. $r = 6 - 3 \sin \theta$

12. $r = 6 + 3 \cos \theta$

13. $r = 2 + 4 \sin \theta$

14. $r = 3 + 6 \cos \theta$

15. $r = 16 \sin 2\theta$

16. $r^2 = 25 \cos 2\theta$

17. $r = 4 + 8 \cos \theta$

18. $r = 4 - 8 \sin \theta$

19. $r^2 = 16 \cos 2\theta$

20. $r^2 = \sin 2\theta$

21. $r = \sin \frac{1}{2}\theta$

22. $r = \cos \frac{1}{2}\theta$

23. $r^2 = \cos \theta$

24. $r^2 = \sin \theta$

25. $r = \dfrac{1}{1 + \cos \theta}$

26. $r = \dfrac{3}{1 - \sin \theta}$

27. $r = \dfrac{1}{1 - 2 \sin \theta}$

28. $r = \dfrac{1}{1 + 2 \cos \theta}$

7.5
POLAR EQUATIONS OF LINES AND CIRCLES

The equations of lines and circles can be obtained in polar equations by transforming the rectangular coordinate equations of these curves. The polar equations can also be derived directly. We shall derive the polar equation of a line, and also of a circle, in general positions and in certain special positions.

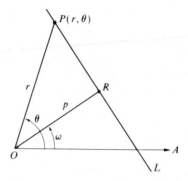

FIGURE 7.23

In Fig. 7.23 the segment OR is drawn perpendicular to the line L. We denote the length of this segment by p and the angle that it makes with the polar axis by ω. The coordinates of a variable point on the line are (r, θ). From the right triangle ORP, we have

$$\frac{p}{r} = \cos(\theta - \omega)$$

or

$$\boxed{r\cos(\theta - \omega) = p.} \qquad (7.6)$$

This equation holds for all points of the line. If P is chosen below OA, then the angle ROP is equal to $(\omega + 2\pi - \theta)$. Although this angle is not equal to $(\theta - \omega)$, we do have

$$\cos(\omega + 2\pi - \theta) = \cos(\omega - \theta) = \cos(\theta - \omega).$$

In a similar way, the equation could be derived for the line L in any other position and not passing through the origin.

Equation (7.6) is called the **polar normal form** of the equation of a straight line. If we expand $\cos(\theta - \omega)$ by the formula for the cosine of the difference of two angles, the equation becomes

$$r(\cos\theta\cos\omega + \sin\theta\sin\omega) = p. \qquad (7.7)$$

For $\omega = 0°$ this equation becomes $r\cos\theta = p$, and the corresponding line is perpendicular to the polar axis and p units to the right of the origin (we assume that p is positive). If $\omega = 180°$, Eq. (7.7) becomes $r\cos\theta = -p$. In this case the line is p units to the left of the origin. If $\omega = 90°$, we have the equation $r\sin\theta = p$, and the line is parallel to the polar axis and p units above the axis. If $\omega = 270°$, the line is p units below the axis. So for these special cases we have the equations

$$\boxed{r\cos\theta = \pm p} \tag{7.8}$$

and

$$\boxed{r\sin\theta = \pm p.} \tag{7.9}$$

The graph of $r\cos\theta = -4$ is the line perpendicular to the polar axis and 4 units to the left of the origin. The graph of $r\sin\theta = -5$ is the line parallel to the polar axis and 5 units below the axis.

The θ-coordinate is constant for points on a line passing through the origin. Hence the equation of a line through the origin with inclination α is

$$\boxed{\theta = \alpha.} \tag{7.10}$$

An equation of the form $Ax + By = C$, with $C \neq 0$ and A and B both not equal to 0, can be used to represent any line in the coordinate plane not passing through the origin. When expressed in polar form, the given equation may be reduced to

$$r = \frac{C}{A\cos\theta + B\sin\theta}.$$

If the coefficients are assigned allowable values, the graph is a particular line. Any change in the coefficients would yield another line. For example, the equation

$$r = \frac{4}{2\cos\theta - 4\sin\theta}$$

represents a particular line. Coordinates of points on the line may be determined by assigning values to θ and computing the corresponding values for r. Thus for $\theta = 0°$, the value of r is 2, and for $\theta = 90°$, the value of r is -1. So the points $(2, 0°)$ and $(-1, 90°)$ are on the line in question, and the line may be drawn.

We next write the equation of a circle of radius a and center at $R(r_1, \theta_1)$. Observing Fig. 7.24 and applying the law of cosines to the triangle ORP, we get

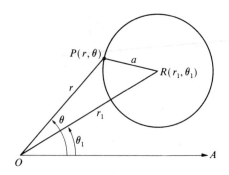

FIGURE 7.24

the equation of the circle in the form

$$r^2 + r_1^2 - 2rr_1\cos(\theta - \theta_1) = a^2. \qquad (7.11)$$

If the center is at $(a, 0°)$, then $r_1 = a$ and $\theta_1 = 0°$. The equation then reduces

$$r = 2a\cos\theta. \qquad (7.12)$$

If the center is at $(a, 90°)$, the equation becomes

$$r = 2a\sin\theta. \qquad (7.13)$$

EXAMPLE 1 ● Find the equation of the circle with center at $(5, 60°)$ and radius 3.

SOLUTION. Substituting in Eq. (7.11), we have

$$r^2 + 25 - 10r\cos(\theta - 60°) = 9,$$
$$r^2 - 10r\cos(\theta - 60°) + 16 = 0. ●$$

Exercises

1. From a figure find the equation of a line perpendicular to the polar axis (a) 4 units to the right of the pole, (b) 4 units to the left of the pole. Compare your results with formula (7.8).

2. From a figure find the equation of the line parallel to the polar axis and (a) 4 units below the axis, (b) 4 units above the axis. Compare your results with formula (7.9).

Assign convenient values to θ and find the coordinates of two points on the line represented by each of Exercises 3 through 8. Plot the points and draw the line.

3. $r = \dfrac{2}{\cos\theta + 3\sin\theta}$

4. $r = \dfrac{3}{2\cos\theta - 2\sin\theta}$

5. $r = \dfrac{-8}{4\cos\theta + 2\sin\theta}$

6. $r = \dfrac{-10}{2\cos\theta + 5\sin\theta}$

7. $r = \dfrac{3}{6\sin\theta - 3\cos\theta}$

8. $r = \dfrac{-4}{2\sin\theta + 3\cos\theta}$

Write the polar equation of each line described in Exercises 9 through 15.

9. The horizontal line through the point $(3, 90°)$.

10. The horizontal line through $(-3, 90°)$.

11. The vertical line through the point $(4, 0°)$.

12. The vertical line through the point $(4, \pi)$.

13. The vertical line through the pole.

14. The line tangent to the circle $r = 3$ at the point $\left(3, \dfrac{\pi}{3}\right)$.

15. The line tangent to the circle $r = 4$ at the point $(4, 225°)$.

Give the polar coordinates of the center and the radius of the circle defined by each of Exercises 16 through 21.

16. $r = 4\cos\theta$ 17. $r = 6\sin\theta$ 18. $r = -8\sin\theta$

19. $r = -2\cos\theta$ 20. $r = 9\cos\theta$ 21. $r = -5\sin\theta$

Find the polar equation of the circle in each Exercise 22 through 32.

22. Center at $(4, 0°)$ and radius 4.

23. Center at $(5, 0°)$ and radius 5.

24. Center at $(8, 60°)$ and radius 8.

25. Center at $(-4, 0°)$ and radius 4.

26. Center at $(5, 0°)$ and radius 4.

27. Center at $(6, 60°)$ and radius 3.

28. Center at $(5, 45°)$ and radius 4.

29. Center at $(3, 120°)$ and radius 2.

30. Center at $(8, 240°)$ and radius 7.

31. Center at $(7, 120°)$ and radius 5.

32. Center at $(9, 225°)$ and radius 8.

33. Derive formulas (7.12) and (7.13) directly from figures. You may use the condition that any angle inscribed in a semicircle is a right angle.

7.6

POLAR EQUATIONS OF CONICS

We use the focus–directrix property of conics (Section 3.3) to derive polar equations of conics. The equations can be obtained in simple forms if a focus is at the origin and the directrix is parallel or perpendicular to the polar axis. In Fig. 7.25 the directrix D is perpendicular to the polar axis and to the left of the origin. We indicate the eccentricity by e and the length of BO by p. Then for any point $P(r, \theta)$ of the conic, we have, by definition,

$$\frac{|OP|}{|EP|} = e.$$

But the numerator $|OP| = r$ and the denominator

$$|EP| = |BR| = |BO| + |OP|\cos\theta = p + r\cos\theta.$$

Hence $r/(p + r\cos\theta) = e$, and solving for r, we get

$$r = \frac{ep}{1 - e\cos\theta}. \qquad (7.14)$$

If a focus is at the pole and the directrix is p units to the right of the pole, the equation is

$$r = \frac{ep}{1 + e\cos\theta}. \qquad (7.15)$$

FIGURE 7.25

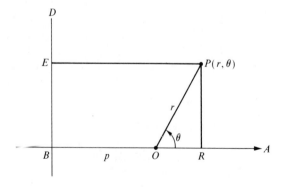

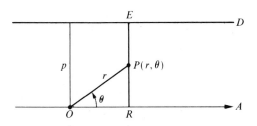

FIGURE 7.26

If a focus is at the pole and the directrix D is parallel to the polar axis and p units above the axis, then we have (Fig. 7.26)

$$\frac{|OP|}{|EP|} = e.$$

We observe that $|OP| = r$ and $|EP| = p - |PR| = p - r \sin \theta$. Hence $r/(p - r \sin \theta) = e$ and, solving for r, we get

$$r = \frac{ep}{1 + e \sin \theta}. \tag{7.16}$$

If a focus is at the pole and the directrix is p units below the polar axis, the equation is

$$r = \frac{ep}{1 - e \sin \theta}. \tag{7.17}$$

An equation in any of the forms (7.14) through (7.17) represents a parabola if $e = 1$, an ellipse if e is between 0 and 1, and a hyperbola if e is greater than 1. In any case the graph can be sketched immediately. Having observed the type of conic from the value of e, the next step is to find the points where the curve cuts the polar axis, the extension of the axis through O, and the line through the pole perpendicular to the polar axis. These are called the **intercept points**, and may be obtained by using the values 0°, 90°, 180°, and 270° for θ. Only three of these values can be used for a parabola, since one of them would make the denominator zero. The intercept points are sufficient for a rough graph. For increased accuracy a few additional points should be plotted.

EXAMPLE 1 ● Sketch the graph of the equation

$$r = \frac{8}{3 + 3 \cos \theta}.$$

SOLUTION. We divide the numerator and denominator of the fraction by 3 and have

$$r = \frac{8/3}{1 + \cos\theta}.$$

This equation is of the form (7.15) with $e = 1$. Hence the graph is a parabola with axis along the extension of the polar axis. By substituting $0°$, $90°$, and $270°$ in succession for θ, we find the intercept points to be

$$\left(\tfrac{4}{3},0°\right), \qquad \left(\tfrac{8}{3},90°\right), \qquad \left(\tfrac{8}{3},270°\right).$$

The first of these points is the vertex of the parabola and the second and third are the ends of the latus rectum. Substituting $\theta = 60°$ and $\theta = 120°$ in the given equation and noting that the graph is symmetric with respect to its axis, we find the additional points

$$\left(\tfrac{16}{9},60°\right), \qquad \left(\tfrac{16}{3},120°\right), \qquad \left(\tfrac{16}{3},240°\right), \qquad \left(\tfrac{16}{9},300°\right).$$

These points, along with the intercept points, permit us to draw the graph (Fig. 7.27). ●

FIGURE 7.27. $r = \dfrac{8}{3 + 3\cos\theta}.$

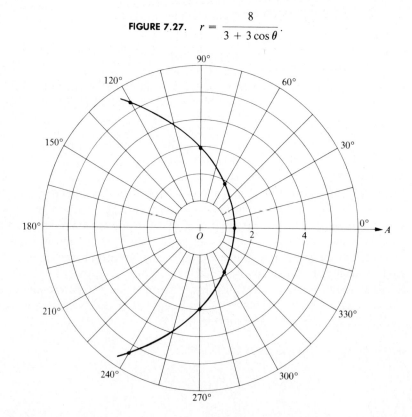

EXAMPLE 2 ● Sketch the graph of

$$r = \frac{15}{3 - 2\cos\theta}.$$

SOLUTION. The equation takes the form of Eq. (7.14) when the numerator and denominator of the right member are divided by 3. This produces

$$r = \frac{5}{1 - \left(\frac{2}{3}\right)\cos\theta}.$$

In this form we observe that $e = \frac{2}{3}$, and hence the graph is an ellipse. Substituting $0°$, $90°$, $180°$, and $270°$ in succession for θ in the original equation, we find the intercept points to be

$$(15, 0°), \qquad (5, 90°), \qquad (3, 180°), \qquad (5, 270°).$$

These points are plotted in Fig. 7.28. The points $(15, 0°)$ and $(3, 180°)$ are vertices and the other intercept points are the ends of a latus rectum. The center, midway between the vertices, is at $(6, 0°)$. Then, denoting the distance from the center to a vertex by a, the distance from the center to a focus by c, and the distance from the center to an end of the minor axis by b, we have

$$a = 9, \qquad c = 6, \qquad b = \sqrt{a^2 - c^2} = 3\sqrt{5}. ●$$

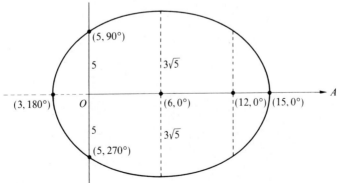

FIGURE 7.28. $r = \dfrac{15}{3 - 2\cos\theta}.$

EXAMPLE 3 ● Sketch the graph of the equation

$$r = \frac{4}{2 + 3\sin\theta}.$$

SOLUTION. Dividing the numerator and denominator of the fraction by 2, we have

$$r = \frac{2}{1 + 1.5\sin\theta}.$$

This equation is in the form (7.16) with $e = 1.5$. Hence the graph is a hyperbola. Although the graph could be roughly sketched from the intercepts, we shall plot a few additional points by use of a table of values.

θ	0°	30°	90°	150°	180°	205°	270°	335°
r	2	$\frac{8}{7}$	$\frac{4}{5}$	$\frac{8}{7}$	2	5.5	−4	5.5

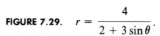

FIGURE 7.29. $r = \dfrac{4}{2 + 3\sin\theta}.$

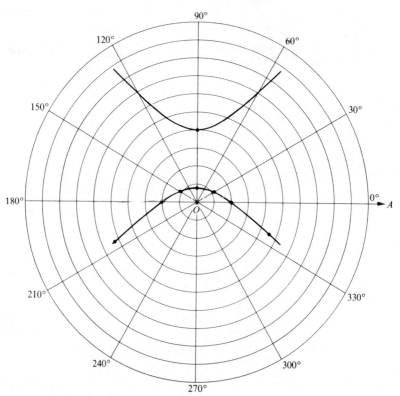

The point $(0.8, 90°)$ is the vertex of the lower branch of the hyperbola and the point $(-4, 270°)$ is the vertex of the upper branch. The center of the hyperbola, midway between the vertices, is at the point $(2.4, 90°)$. The denominator of the given fraction is equal to zero when $\sin\theta = -\frac{2}{3}$. If we denote this angle by $180° + \theta_0$, then, for all values of θ such that $180° + \theta_0 < \theta < 360° - \theta_0$, the values of r are negative and yield the points on the upper branch. We can, of course, draw the upper branch by noting that the hyperbola is symmetric with respect to its center $(2.4, 90°)$. The graph is shown in Fig. 7.29. ●

Exercises

Express each of the following equations in one of the forms (7.14) through (7.17) and sketch the graph of the conic.

1. $r = \dfrac{6}{1 + \sin\theta}$

2. $r = \dfrac{4}{1 - \cos\theta}$

3. $r = \dfrac{10}{3 - 3\sin\theta}$

4. $r = \dfrac{9}{2 + 2\cos\theta}$

5. $r = \dfrac{12}{2 + \sin\theta}$

6. $r = \dfrac{12}{2 - \cos\theta}$

7. $r = \dfrac{15}{5 - 4\sin\theta}$

8. $r = \dfrac{16}{4 + 3\cos\theta}$

9. $r = \dfrac{4}{2 + 3\cos\theta}$

10. $r = \dfrac{2}{1 - 2\cos\theta}$

11. $r = \dfrac{6}{3 - 4\sin\theta}$

12. $r = \dfrac{8}{3 + 5\cos\theta}$

7.7

TRIGONOMETRIC EQUATIONS

In this section we will solve some trigonometric equations, as a review, for you. Then we will give a set of equations for you to solve.

EXAMPLE 1 ● Find all values of θ in the interval $0 \le \theta < 2\pi$ which satisfy the equation

$$2\cos^2\theta - 5\cos\theta + 2 = 0.$$

SOLUTION. We first factor the left member and get

$$(\cos\theta - 2)(2\cos\theta - 1) = 0.$$

By setting each factor to zero, we obtain the simple equations

$$\cos\theta - 2 = 0 \quad \text{and} \quad 2\cos\theta - 1 = 0.$$

The first of these equations has no solution since the value of a cosine never exceeds 1. The second yields $\cos\theta = \frac{1}{2}$, and the solutions are

$$\theta = \frac{\pi}{3} \quad \text{and} \quad \theta = \frac{5\pi}{3}.$$

Each of these solutions may be changed by any integral of 2π. But we are limiting our solutions to the interval $0 \leq \theta < 2\pi$. We can check the results by substituting in the given equation. Thus

$$2\cos^2\frac{\pi}{3} - 5\cos\frac{\pi}{3} + 2 = 2\left(\tfrac{1}{2}\right)^2 - 5\left(\tfrac{1}{2}\right) + 2 = 0,$$

$$2\cos^2\frac{5\pi}{3} - 5\cos\frac{5\pi}{3} + 2 = 2\left(\tfrac{1}{2}\right)^2 - 5\left(\tfrac{1}{2}\right) + 2 = 0.$$

These values satisfy the given equation and hence the solution set is

$$\left\{\frac{\pi}{3}, \frac{5\pi}{3}\right\}. \; \bullet$$

EXAMPLE 2 ● Find the solution set in the interval $0 \leq \theta < 360°$ of the equation

$$3\sin^2\theta - 4\sin\theta - 2 = 0.$$

SOLUTION. We cannot find real factors of the left side, as we did in Example 1. So we will resort to the quadratic formula.* Adapting the formula to our problem, we replace x by $\sin\theta$, a by 3, b by -4, and c by -2. Hence we have

$$\sin\theta = \frac{4 \pm \sqrt{16 - 4(3)(-2)}}{6} = \frac{4 \pm \sqrt{40}}{6} = 1.72 \quad \text{or} \quad -0.387.$$

We reject 1.72 since the sine of an angle cannot exceed unity in absolute value. So we use -0.387 for the value of $\sin\theta$. The negative value shows that θ is in the third quadrant or the fourth quadrant. In Table II of the Appendix, we find that if $+\sin\theta = +0.387$, then $\theta = 23°$ (nearest degree). Hence we take, as our result, $\theta = 180° + 23° = 203°$ and $\theta = 360° - 23° = 337°$. ●

EXAMPLE 3 ● Solve the equation $1 - \sin\theta = \cos\theta$ in the interval $0 \leq \theta < 2\pi$.

SOLUTION. In order to have an equation with only one function of θ, we square both sides of the given equation and substitute for $\cos^2\theta$. Thus we get

$$1 - 2\sin\theta + \sin^2\theta = \cos^2\theta$$
$$= 1 - \sin^2\theta,$$
$$2\sin^2\theta - 2\sin\theta = 0,$$
$$2\sin\theta(\sin\theta - 1) = 0.$$

The factors of the left member give $\theta = 0, \pi/2, \pi$.

---

*The formula in terms of x is $ax^2 + bx + c = 0$, and has for its solution

$$x = \frac{-b \pm \sqrt{b^2 - 4ac}}{2a}.$$

These results are solutions of the equation found by squaring the members of the original equation.

This operation may have introduced extraneous roots. Hence we test them. Substituting in each side of the given equation, we find that

$$x = 0 \quad \text{gives } 1 - 0 = 1,$$

$$x = \frac{\pi}{2} \quad \text{gives } 1 - 1 = 0,$$

$$x = \pi \quad \text{gives } 1 - 0 \neq -1.$$

This shows that 0 and $\pi/2$ are roots but π is not a root. Hence the solution set is $\{0, \pi/2\}$. ●

EXAMPLE 4 ● Find the solution set in the interval $0 \leq \theta < 2\pi$ of the equation $2 \cos^2 2\theta + \cos 2\theta - 1 = 0$.

SOLUTION. Since the angle in this equation is 2θ instead of θ, the interval for 2θ must be $0 \leq 2\theta < 4\pi$. We first factor the left member of the given equation and have

$$(\cos 2\theta + 1)(2 \cos 2\theta - 1) = 0.$$

Equating each factor to 0, yields

$$\cos 2\theta = -1 \quad \text{and} \quad \cos 2\theta = \tfrac{1}{2}.$$

From the first factor we have

$$2\theta = \pi, \quad 3\pi, \quad 5\pi, \quad 7\pi, \quad \text{and} \quad \theta = \frac{\pi}{2}, \quad \frac{3\pi}{2}, \quad \frac{5\pi}{2}, \quad \frac{7\pi}{2}.$$

From the second factor we have

$$2\theta = \frac{\pi}{3}, \quad \frac{5\pi}{3}, \quad \frac{7\pi}{3}, \quad \frac{11\pi}{3}, \quad \text{and} \quad \theta = \frac{\pi}{6}, \quad \frac{5\pi}{6}, \quad \frac{7\pi}{6}, \quad \frac{11\pi}{6}.$$

We find the solution set to be

$$\left\{ \frac{\pi}{6}, \quad \frac{5\pi}{6}, \quad \frac{\pi}{2}, \quad \frac{7\pi}{6}, \quad \frac{11\pi}{6}, \quad \frac{3\pi}{2} \right\}. \; ●$$

Exercises

Find the solution set of each problem in the interval $0 \leq \theta < 2\pi$ or, when a table is used, in the interval $0° \leq \theta < 360°$.

1. $2 \sin 2\theta - 1 = 0$

2. $2 \cos 2\theta - 1 = 0$

3. $\sqrt{2} \sin 3\theta - 1 = 0$

4. $\sqrt{2} \cos 3\theta + 1 = 0$

5. $2 \cos 3\theta + 1 = 0$

6. $2 \sin 3\theta + 1 = 0$

7. $2 \sin^2 \theta - 5 \sin \theta + 2 = 0$

8. $8 \cos^2 \theta - 4 \cos \theta - 2 = 0$

9. $1 - \cos\theta = \sin\theta$ 10. $2\sin^2 2\theta + \sin 2\theta - 1 = 0$

11. $4\cos^2 2\theta + \cos 2\theta - 1 = 0$ 12. $2\sin^2\frac{1}{2}\theta + \sin\frac{1}{2}\theta - 1 = 0$

13. $2\cos^2\frac{1}{2}\theta + \cos\frac{1}{2}\theta - 1 = 0$ 14. $2\sin^2\frac{1}{2}\theta + 5\sin\frac{1}{2}\theta + 2 = 0$

7.8

INTERSECTIONS OF POLAR-COORDINATE GRAPHS

A simultaneous real solution of two equations in rectangular coordinates represents a point of intersection of their graphs. Conversely, the coordinates of a
point of intersection yield a simultaneous solution. In polar coordinates, however,
this converse statement does not always hold. This difference in the two systems is
a consequence of the fact that a point has more than one pair of polar
coordinates. As an illustration, consider the equations $r = -2$, $r = 1 + \sin\theta$,
and the two pairs of coordinates $(2, 90°)$, $(-2, 270°)$. The equation $r = -2$ is
satisfied by the second pair of coordinates but not by the first. The equation
$r = 1 + \sin\theta$ is satisfied by the first pair of coordinates but not by the second.
The two pairs of coordinates, however, determine the same point. Although the
two curves pass through this point, no pair of coordinates of the point satisfies
both equations. The usual process of solving two equations simultaneously does
not yield an intersection point of this kind. The graphs of the equations, of
course, show all intersections.

EXAMPLE 1 ● Solve simultaneously and sketch the graphs of

$$r = 6\sin\theta \qquad \text{and} \qquad r = 6\cos\theta.$$

SOLUTION. Equating the right members of the equations, we have

$$6\sin\theta = 6\cos\theta,$$
$$\tan\theta = 1,$$
$$\theta = 45°, 225°,$$
$$r = 3\sqrt{2}, -3\sqrt{2}.$$

The coordinates $(3\sqrt{2}, 45°)$ and $(-3\sqrt{2}, 225°)$ define the same point. The graphs
(Fig. 7.30) show this point, and show also that both curves pass through the
origin. The coordinates $(0, 0°)$ satisfy the first equation and $(0, 90°)$ satisfy the
second equation. But the origin has no pair of coordinates that satisfies both
equations. ●

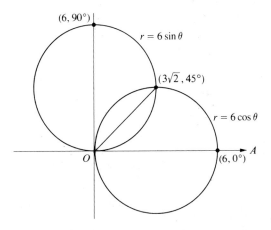

FIGURE 7.30. $r = 6 \sin \theta$; $r = 6 \cos \theta$.

EXAMPLE 2 ● Solve simultaneously and draw the graphs of

$$r = 4 \sin \theta \qquad \text{and} \qquad r = 4 \cos 2\theta.$$

SOLUTION. Eliminating r and using the trigonometric identity $\cos 2\theta = 1 - 2 \sin^2\theta$, we obtain

$$4 \sin \theta = 4(1 - 2 \sin^2\theta),$$
$$2 \sin^2\theta + \sin \theta - 1 = 0,$$
$$(2 \sin \theta - 1)(\sin \theta + 1) = 0,$$
$$\sin \theta = \tfrac{1}{2}, -1,$$
$$\theta = 30°, 150°, 270°,$$
$$r = 2, \quad 2, \quad -4.$$

The solutions are $(2, 30°)$, $(2, 150°)$, and $(-4, 270°)$. Figure 7.31 shows that the

FIGURE 7.31. $r = 4 \sin \theta$; $r = 4 \cos 2\theta$.

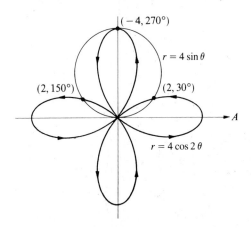

curves also cross at the origin, but the origin has no pair of coordinates that satisfies both equations. ●

EXAMPLE 3 ● Solve simultaneously and sketch the graphs of the equations

$$r\cos\theta = 2 \quad\text{and}\quad r = 2 + 4\cos\theta.$$

SOLUTION. Eliminating r between the equations and simplifying the result, we obtain

$$2\cos^2\theta + \cos\theta - 1 = 0 \quad\text{or}\quad (2\cos\theta - 1)(\cos\theta + 1) = 0.$$

We see from the last equation that $\theta = 60°$, $\theta = 300°$, and $\theta = 180°$ are solutions. The corresponding values of r are, respectively, $r = 4$, $r = 4$, $r = -2$. As may be verified, the coordinates of the points $(4, 60°)$, $(4, 300°)$, and $(-2, 180°)$ satisfy both of the given equations. The graphs are shown in Fig. 7.32. ●

FIGURE 7.32. $r\cos\theta = 2$; $r = 2 + 4\cos\theta$.

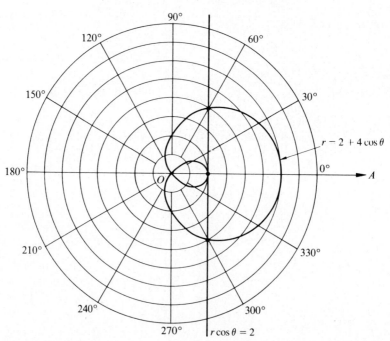

EXAMPLE 4 ● Solve simultaneously and sketch the graphs of the equations

$$r = \frac{1}{1 + \sin\theta} \quad \text{and} \quad r\sin\theta = -1.$$

SOLUTION. Eliminating r between the equations and simplifying the result, we obtain $\sin\theta = -\frac{1}{2}$, and $\theta = 210°$ and $330°$. Each of these values of θ yields $r = 2$. Hence the intersection points of the two graphs are at $(2, 210°)$ and $(2, 330°)$. (See Fig. 7.33.) ●

EXAMPLE 5 ● Solve simultaneously and sketch the graphs of the equations

$$r = 4\cos\theta \quad \text{and} \quad r\cos\theta = 1.$$

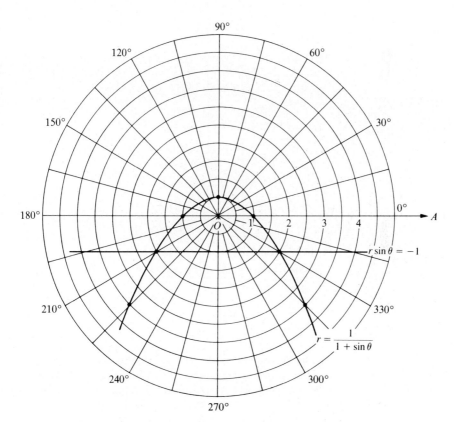

FIGURE 7.33. $r\sin\theta = -1$; $r = 1/(1 + \sin\theta)$.

SOLUTION. Eliminating r between the equations and simplifying the result, we find $\cos \theta = \pm \frac{1}{2}$ and $\theta = 60°$ and $300°$. Each of these values of θ yields $r = 2$. Hence the intersection points of the two graphs are at $(2, 60°)$ and $(2, 300°)$. The graphs are pictured in Fig. 7.34. ●

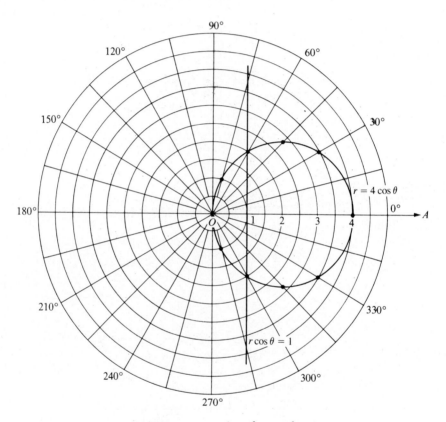

FIGURE 7.34. $r = 4 \cos \theta$; $r \cos \theta = 1$.

Exercises

In each of the following exercises, solve the pair of equations simultaneously. Sketch their graphs on the same coordinate axes. Extraneous solutions are sometimes introduced in the solving process, so we suggest that you check all your answers.

1. $r = 2 \sin \theta$
   $r = 1$

2. $r = \cos 2\theta$
   $r = 1$

3. $r = 4 \cos \theta$
   $r = 4 \sin 2\theta$

4. $r + 6\sin\theta = 0$
   $r + 6\cos\theta = 0$

5. $r = 6\sin\theta$
   $r\sin\theta = 3$

6. $r = a(1 + \cos\theta)$
   $r = 2a\cos\theta$

7. $r\sin\theta = 1$
   $r = 4\sin\theta$

8. $r^2 = 4\sin 2\theta$
   $r = 2$

9. $r = 1 + \cos\theta$
   $r = 1 + \sin\theta$

10. $r = \dfrac{2}{1 + \sin\theta}$
    $3r\sin\theta = 2$

11. $r = \dfrac{3}{4 - 3\sin\theta}$
    $r = 3\sin\theta$

12. $r = 2\cos\theta + 1$
    $r = 2\sin\theta$

13. $r^2 = a^2\cos 2\theta$
    $r = a\sqrt{2}\sin\theta$

14. $r = 4\cos\theta$
    $r = 4\sin 2\theta$

15. $r = \cos^2 2\theta$
    $r = \sin^2 2\theta$

16. $r = 2\sin\frac{1}{2}\theta$
    $r = 1$

17. $r = a(1 + \sin\theta)$
    $r = a(1 - \sin\theta)$

18. $r = 1 - \sin\theta$
    $r = \cos 2\theta$

19. $r = 4 - \cos\theta$
    $r\cos\theta = 3$

20. $r = 2\sin\theta + 1$
    $r\sin\theta = 1$

21. $r = 4\cos 2\theta$
    $r = 4\sin 2\theta$

22. $r = 4\cos\theta$
    $r = 2$

23. $r = 3\sin\theta$
    $r = 3\cos\theta$

24. $r = \sin\theta$
    $r^2 = \sin 2\theta$

25. $r^2 = a^2\cos 2\theta$
    $r = a\sqrt{2}\cos\theta$

26. $r = 2\cos\theta$
    $r^2 = 4\cos\theta$

27. $r = -\cos\theta$
    $r = 2 - 3\cos\theta$

28. $r = -3\cos\theta$
    $r = -4\sin\theta$

29. $r = 1 + \cos\theta$
    $r = 3\cos\theta$

30. $r = \sin\theta$
    $r = 1 - \cos\theta$

31. $r = 2\sin\theta$
    $r^2 = 4\sin\theta$

REVIEW EXERCISES

1. Write three other pairs of polar coordinates for the point $(5, 15°)$. Restrict the vectorial angles so that they do not exceed $360°$ in absolute value.

2. Write three other pairs of polar coordinates for the point $(3, -\frac{4}{3}\pi)$. Restrict the vectorial angles θ so that $-2\pi \le \theta \le 2\pi$.

3. Find the rectangular coordinates for the polar coordinate points $(4, 60°)$ and $(3, 225°)$.

4. Find nonnegative polar coordinates for the rectangular coordinate points $(\sqrt{2}, -\sqrt{2})$ and $(-4, -5)$.

5. Transform the equation $3x - 2y = 4$ into the corresponding polar coordinate equation.

6. Sketch the graph of the equation $r = 3 \sin \theta$.

7. Write the polar equation of the horizontal line through the point $(3, 60°)$.

8. Find the polar equation of the circle with center at $(6, 30°)$ and radius 4.

9. Sketch the graph of the conic defined by

$$r = \frac{4}{1 + \cos \theta}.$$

10. Find the solution set in the interval $0° \le \theta < 360°$ of the equation $2 \sin^2\theta + \sin \theta - 1 = 0$.

11. Solve simultaneously and draw the graphs of
$$r = 3(1 + \sin \theta) \qquad \text{and} \qquad r = 3(1 - \sin \theta).$$

8

Parametric Equations

We have obtained graphs of equations in the two variables x and y. Another way of defining a graph is to express x and y separately in terms of a third variable. Equations of this kind are very important because the mathematical treatment of many problems is facilitated by their use. Before illustrating the two-equation situation, we state the following definition.

DEFINITION 8.1 ● *The equation of a graph in two dimensions is said to be in* **parametric form** *if each coordinate of a general point $P(x, y)$ is expressed in terms of a third variable. The third variable, usually denoted by a letter, is called a* **parameter**.

The equations

$$x = t - 1 \quad \text{and} \quad y = 2t + 3,$$

for example, are parametric equations, and t is the parameter. The equations define a graph. If t is assigned a value, corresponding values are determined for x and y. The pair of values for x and y constitute the coordinates of a point of the graph. The complete graph consists of the set of all points determined in this way as t varies through all its chosen values. We can eliminate t between the equations and obtain an equation involving x and y. Thus, solving either equation for t and substituting in the other, we get

$$2x - y + 5 = 0.$$

The graph of this equation, which is also the graph of the parametric equations, is a straight line.

Often the parameter can be eliminated, as illustrated here, to obtain an equation in x and y. Sometimes, however, the process is not easy (or not even possible) because the parameter is included in a complicated way. The equations

$$x = t^5 + \log t \quad \text{and} \quad y = t^3 + \tan t$$

illustrate this statement.

It is sometimes helpful in solving a problem to change an equation in x and y to parametric form. We illustrate this process with the equation

$$x^2 + 2x + y = 4,$$

which defines a parabola. If we substitute $2t$ for x and solve the resulting equation for y, we get $y = 4 - 4t - 4t^2$. Hence the parametric equations

$$x = 2t \qquad \text{and} \qquad y = 4 - 4t - 4t^2$$

also represent the parabola. It is evident that other representations could be obtained by equating x to other expressions in t. Again, this procedure is inconvenient or perhaps impossible in equations that contain both variables in a complicated way.

The representation of equations by parametric equations is frequently used in mathematical situations. We note, however, that there is no general procedure for choosing a pair of simple parametric equations.

The parameter, as used in this chapter, plays a different role from the parameter in Sections 2.5 and 2.7. Here a curve is determined by letting the parameter vary. In the earlier use, the parameter gave rise to a set of curves (lines and circles).

8.1

PARAMETRIC EQUATIONS OF THE CONICS

We have already illustrated parametric representations of a straight line and parabola. We turn now to a consideration of the circle and the remaining conics.

To find a parametric representation of the circle of radius a and center at the origin, we select for the parameter the angle θ, as indicated in Fig. 8.1. Recalling

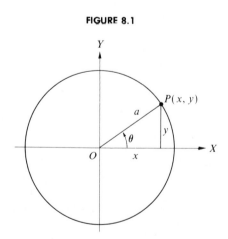

FIGURE 8.1

the definitions of the sine and cosine of an angle, we have

$$\frac{x}{a} = \cos\theta \qquad \text{and} \qquad \frac{y}{a} = \sin\theta$$

or, equivalently, the pair of equations

$$x = a\cos\theta \qquad \text{and} \qquad y = a\sin\theta.$$

If we let θ increase from $0°$ to $360°$, the point $P(x, y)$ defined by these equations starts at $(a, 0)$ and moves counterclockwise around the circle.

Next, let the center of a circle be at (h, k) with radius a (Fig. 8.2). For the directed distances $\overrightarrow{CQ}$ and $\overrightarrow{QP}$, we obtain

$$\overrightarrow{CQ} = x - h = a\cos\theta \qquad \text{and} \qquad \overrightarrow{QP} = y - k = a\sin\theta,$$

or the pair of equations

$$x = h + a\cos\theta,$$
$$y = k + a\sin\theta.$$

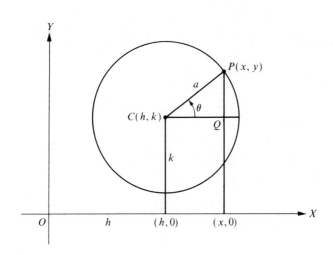

FIGURE 8.2

Hence these equations constitute a parametric representation of a circle with center at (h, k) and radius a.

To find a parametric representation of the ellipse, we use the equations

$$x = a\cos\theta \qquad \text{and} \qquad y = b\sin\theta,$$

where $a > b$.

These equations define an ellipse, as may be verified by eliminating the parameter θ. Thus, writing the equations as $x/a = \cos\theta$ and $y/b = \sin\theta$, squaring the members of each equation, and adding, we get

$$\left(\frac{x}{a}\right)^2 + \left(\frac{y}{b}\right)^2 = \cos^2\theta + \sin^2\theta$$

or

$$\frac{x^2}{a^2} + \frac{y^2}{b^2} = 1.$$

From this result we see that the parametric equations represent an ellipse with a and b as semiaxes. The geometric significance of θ can be determined by referring to Fig. 8.3. The radius of the smaller circle is b and the radius of the larger circle is a. The terminal side of θ cuts the circles at B and A. The horizontal line through B and the vertical line through A intersect at $P(x, y)$. For this point, we have

$$x = \overrightarrow{OM} = |OA|\cos\theta = a\cos\theta,$$
$$y = \overrightarrow{MP} = \overrightarrow{NB} = |OB|\sin\theta = b\sin\theta.$$

FIGURE 8.3

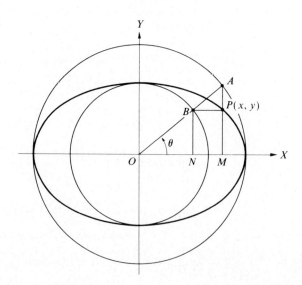

Hence $P(x, y)$ is a point of the ellipse. As θ varies, P moves along the ellipse. If θ starts at $0°$ and increases to $360°$, the point P starts at $(a, 0)$ and traverses the ellipse in a counterclockwise direction.

To obtain a parametric representation of a hyperbola, we use the equations

$$x = a \sec \theta \qquad \text{and} \qquad y = b \tan \theta,$$

where a and b are positive numbers. Since $\sec^2\theta - \tan^2\theta = 1$, we may write

$$\left(\frac{x}{a}\right)^2 - \left(\frac{y}{b}\right)^2 = \sec^2\theta - \tan^2\theta,$$

or

$$\frac{x^2}{a^2} - \frac{y^2}{b^2} = 1.$$

We note that $\sec \theta$ and $\tan \theta$ exist for all angles except those for which θ is an odd multiple of $90°$. Suppose we let θ take all values such that $0° \leq \theta < 90°$. At $0°$, $\sec \theta = 1$ and $\tan \theta = 0$. As θ increases in this specified interval, $\sec \theta$ starts at 1 and assumes all positive values greater than 1, and $\tan \theta$ starts at 0 and assumes all positive values. Hence, x starts at a and assumes all positive values greater than a, and y starts at 0 and assumes all positive values. We see, then, that this chosen interval for θ provides for the portion of the hyperbola in the first quadrant. Similarly, the values of θ such that $90° < \theta \leq 180°$ represents the portion of the hyperbola in the third quadrant. The student may continue the discussion for the remaining quadrants.

We proceed to the problem of constructing the graph defined by two parametric equations. The method is straightforward. We first assign to the parameter a set of values and compute the corresponding values of x and y. The plotted points (x, y) furnish a guide for drawing the graph. Usually only a few plotted points are necessary. This is especially true when certain properties of the graph, such as the extent, the intercepts, and the symmetry are apparent from the equations. The student with computer graphic capabilities should be able to plot graphs of parametric equations on the computer with a high degree of satisfaction.

If a pair of parametric equations is to represent a rectangular equation, it is well to determine whether the equations and the domain of the parameter are suitably chosen. That is, if the coordinates $P(x, y)$ satisfy the rectangular equation, there must be a value of the parameter so that the parametric equations yield the same coordinates. Conversely, any pair of coordinates obtained from the parametric equations must satisfy the rectangular equation. In some cases, the graph of the parametric equations and the graph of the corresponding rectangular equation do not coincide throughout. Example 2 below illustrates such a case.

EXAMPLE 1 ● Sketch the graph of the parametric equations

$$x = 2 + t \quad \text{and} \quad y = 3 - t^2.$$

SOLUTION. By inspecting the equations, we see that x may have any real value and that y may have any real value not exceeding 3. We are able to sketch the part of the graph shown in Fig. 8.4 by using the values listed in Table 8.1. The true graph extends indefinitely far into the third and fourth quadrants. ●

TABLE 8.1

t	-3	-2	-1	0	1	2	3
x	-1	0	1	2	3	4	5
y	-6	-1	2	3	2	-1	-6

FIGURE 8.4

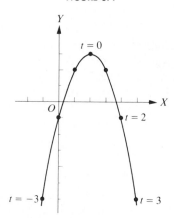

EXAMPLE 2 ● Construct the graph of the equations

$$x = \cos^2\theta \quad \text{and} \quad y = 2\sin\theta.$$

SOLUTION. Let us eliminate the parameter θ from these equations. The second equation yields $y^2/4 = \sin^2\theta$. Then, adding the corresponding members of this equation and the first given equation, we obtain

$$\frac{y^2}{4} + x = 1,$$

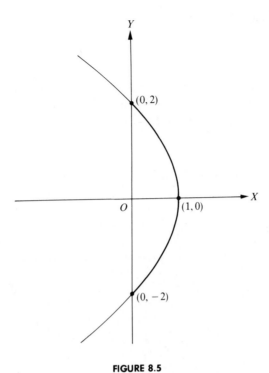

FIGURE 8.5

and consequently,

$$y^2 = -4(x - 1).$$

The graph of this equation is the parabola drawn in Fig. 8.5. We note, however, that the graph of the parametric equations does not include the part of the parabola to the left of the y axis. This results from the fact that the values of x are nonnegative. ●

It may happen that an equation in its original form is difficult to plot, and yet can be represented by more manageable equations in parametric form. We illustrate this situation in an example.

EXAMPLE 3 ● Find a parametric representation of the equation

$$y^{2/3} + x^{2/3} = a^{2/3}.$$

SOLUTION. Solving the equation for $y^{2/3}$, we get

$$y^{2/3} = a^{2/3} - x^{2/3}$$

$$= a^{2/3}\left[1 - \left(\frac{x}{a}\right)^{2/3}\right].$$

We observe that the bracketed expression may be simplified by setting $(x/a)^{2/3}$ $= \sin^2\theta$ or $x = a\sin^3\theta$. When x has this value, we find that $y = a\cos^3\theta$. Hence the given equation is represented in parametric form by the equations

$$x = a\sin^3\theta,$$
$$y = a\cos^3\theta.$$

We can visualize the graph by letting θ increase, in 90° steps, from 0° to 360°. Thus, in the first step x increases from 0 to a and y decreases from a to 0. The graph (Fig. 8.6) is called a **hypocycloid of four cusps**. It can be shown that the path traced by a given point on a circle of radius $\frac{1}{4}a$ as it rolls inside and along a circle of radius a is a hypocycloid. ●

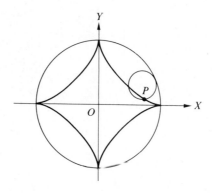

FIGURE 8.6

Exercises

In Exercises 1 through 8, sketch the graph represented by the parametric equations. Then compare the graph with that of the rectangular equation obtained by eliminating the parameter.

1. $x = 3t$; $y = 2t$

2. $x = 4 - 3t$; $y = 1 + t$

3. $x = t^2 - 2$; $y = t + 3$

4. $x = 1 + 2t$; $y = 2 - t^2$

5. $x = 2 + t$; $y = 1 + t^2$

6. $x = 3 - t$; $y = t^2 - 2$

7. $x = 1 + t^2$; $y = 1 + t$

8. $x = 2 + t^2$; $y = 3 + t$

Find the rectangular form of each pair of parametric equations. Then sketch the graph, using the simpler of the two forms.

9. $x = 5 \cos \theta$; $y = 5 \sin \theta$

10. $x = 2 + \cos \theta$; $y = 2 + \sin \theta$

11. $x = 3 \sin \theta$; $y = 4 \cos \theta$

12. $x = 3 \tan^2 \theta$; $y = 2 \sec^2 \theta$

13. $x = 2 \sin \theta$; $y = \cos 2\theta$

14. $x = \tan \theta$; $y = \sec \theta$

15. $x = 3 + 4 \sin \theta$; $y = -4 + 3 \cos \theta$

16. $x = -2 + 2 \sin \theta$; $y = 3 + 2 \cos \theta$

17. $x = \dfrac{2}{1 + t^2}$; $y = \dfrac{2t}{1 + t^2}$

18. $x = \dfrac{6t}{1 + t^2}$; $y = \dfrac{6t^2}{1 + t^2}$

Eliminate the parameter from each pair of equations. Draw the graph of the resulting equation and tell what part of the graph is covered by the parametric equations.

19. $x = 3 \cos^2 \theta$; $y = 2 \sin^2 \theta$

20. $x = \sec^2 \theta$; $y = \tan^2 \theta$

21. $x = 1 + 3 \cos^2 \theta$; $y = 1 + 3 \sin^2 \theta$

22. $x = 2 \cos \theta$; $y = \sin^2 \theta$

23. $x = 4 \sin \theta$; $y = 2 \cos^2 \theta$

24. $x = t$; $y = 6t^{-1}$

Using the accompanying equation, express each rectangular equation in parametric form.

25. $2x + xy - 1 = 0$; $y = t + 2$

26. $x^3 + y^3 + 3xy = 0$; $y = tx$

27. $x^{1/2} + y^{1/2} = a^{1/2}$; $x = a \sin^4 \theta$

28. $x^2(y + 3) = y^3$; $x = ty$

29. $x^2 y^2 = x^2 - y^2$; $y = \sin \theta$

30. $y^3 - 2y^2 = x^2$; $y = t^2 + 2$

Let t take values in steps of $\pi/2$ from 0 to 4π and find the corresponding values of x and y. Plot the points $P(x, y)$ thus determined, and draw a smooth curve through the points by going from point to point in order of increasing t.

31. $x = t \sin t$; $y = \cos t$

32. $x = t \cos t$; $y = \sin t$

Find parametric representations of each rectangular equation.

33. $xy = 1$

34. $y^2 - x^2 = 3$

35. $y^{2/3} - x^{2/3} = a^{2/3}$

36. $bx^2 + a^2y = a^2b$ 37. $y = x + 1$ 38. $x^2/a^2 + y^2/b^2 = 1$

8.2

APPLICATIONS OF PARAMETRIC EQUATIONS*

The equations of certain curves can be determined more readily by the use of a parameter than otherwise. In fact, this is one of the principal uses of parametric equations. In the remainder of this chapter, parametric equations of curves are required. These curves have interesting properties and also have important practical and theoretical applications.

We consider first the path of a projectile in air. Suppose that a body is given an initial upward velocity of v_0 feet per second in a direction that makes an angle α with the horizontal. If the resistance of the air is small and can be neglected without great error, the object will move subject to the vertical force of gravity. This means that there is no horizontal force to change the speed in the horizontal direction. Observing Fig. 8.7 with the origin of coordinates at the point where the projectile is fired, we see that the velocity in the x-direction is $v_0\cos \alpha$. Then the distance traveled horizontally at the end of t seconds is $(v_0\cos \alpha)t$ feet. Now the projectile is started with a vertical component of velocity of $v_0\sin \alpha$ feet per second. This velocity would cause the projectile to rise upward to a height of $(v_0\sin \alpha)t$ feet in t seconds. But the effect of the pull of gravity lessens this distance. According to a formula of physics the amount to be subtracted is $\frac{1}{2}gt^2$, where g is a constant and approximately equal to 32. Hence the parametric

FIGURE 8.7

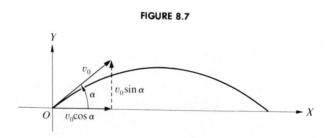

---

*If desired, the remainder of this chapter, or any part of it, may be omitted.

equations of the path are

$$x = (v_0\cos\alpha)t, \qquad y = (v_0\sin\alpha)t - \tfrac{1}{2}gt^2. \tag{8.1}$$

If we solve the first equation for t and substitute the result in the second, we obtain the equation of the path in the rectangular form

$$y = (\tan\alpha)x - \frac{gx^2}{2v_0^2\cos^2\alpha}. \tag{8.2}$$

This equation, which is of the second degree in x and the first degree in y, represents a parabola.

EXAMPLE 1 ● A stone is thrown with a velocity of 160 ft/sec in a direction 45° above the horizontal. Find how far away the stone strikes the ground and its greatest height.

SOLUTION. We substitute $v_0 = 160$, $\alpha = 45°$, and $g = 32$ in Eq. (8.1). This gives the parametric equations

$$x = 80t\sqrt{2}, \qquad y = 80t\sqrt{2} - 16t^2.$$

The stone reaches the ground when $y = 0$. We substitute this value for y in the second equation and find $t = 5\sqrt{2}$ sec as the time of flight. The value of x at this time is $x = 80\sqrt{2}(5\sqrt{2}) = 800$ ft. We know that the stone moves along a parabola that opens downward and that a parabola is symmetric with respect to its axis. Hence the greatest height is the value of y when t is half the flight time. Substituting $t = 5\sqrt{2}/2$ in the second equation, we find $y = 200$ ft. The stone strikes the ground 800 ft from the starting point and reaches a maximum height of 200 ft.

Alternatively, we can obtain the desired results by using the rectangular equation of the path. Thus substituting for v_0, α, and y in Eq. (8.2), we have

$$y = x - \frac{x^2}{800}.$$

This equation, reduced to standard form, becomes

$$(x - 400)^2 = -800(y - 200).$$

The vertex, at $(400, 200)$, is the highest point. Letting $y = 0$, we find $x = 800$. Hence the stone strikes the ground at the point $(800, 0)$. ●

The path traced by a given point on the circumference of a circle that rolls along a line is called a **cycloid**. To derive the equation of the cycloid, we select the line as the x axis and take the origin at a position where the tracing point is in contact with the x axis.

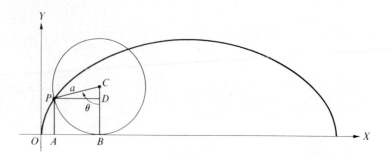

FIGURE 8.8

In Fig. 8.8 the radius of the rolling circle is a, and P is the tracing point. In the position drawn, the circle has rolled so that CP makes an angle θ (radians) with the vertical. Since the circle rolls without slipping, the line segment OB and the arc PB are of equal length. Hence

$$\overrightarrow{OB} = \text{arc}\,\overrightarrow{PB} = a\theta.$$

Observing the right triangle PDC, we may write

$$x = \overrightarrow{OA} = \overrightarrow{OB} - \overrightarrow{PD} = a\theta - a\sin\theta,$$
$$y = \overrightarrow{AP} = \overrightarrow{BC} - \overrightarrow{DC} = a - a\cos\theta.$$

The equations of the cycloid in parametric form are

$$\boxed{x = a(\theta - \sin\theta), \qquad y = a(1 - \cos\theta).}$$

The result of eliminating θ from these equations is the complicated equation

$$x = a\arccos\frac{a - y}{a} \pm \sqrt{2ay - y^2}.$$

If a circle rolls beneath a line, a point of the circle would generate an inverted cycloid (Fig. 8.9). This curve has an interesting and important physical property. A body sliding without friction would move from A to B, two points on a downward part of the curve, in a shorter time than would be required along any other path connecting the two points. A proof of this property is too difficult to be included in this text.

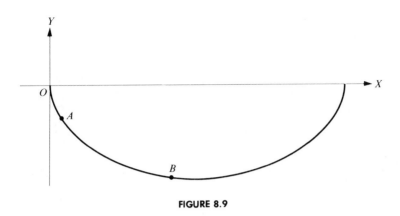

FIGURE 8.9

Exercises

In Exercises 1 through 4, write the parametric equations of the path of the object, using $g = 32$. Write also the rectangular equation of the path and give the requested information.

1. A ball is thrown with an initial velocity of 96 ft/sec and at an angle of 45° above the horizontal. How high does the ball ascend and how far away does the ball strike the ground? Assume the ground to be level.

2. A projectile is fired with an initial velocity of 192 ft/sec and at an angle of 30° above the horizontal. Find the coordinates of its position at the end of (a) 1 sec, (b) 3 sec, (c) 5 sec. At what times is the projectile 96 ft above the ground?

3. A projectile is fired horizontally ($\alpha = 0°$) from a building 96 ft high. If the initial velocity is v_0 ft/sec, find how far downward and how far horizontally the projectile travels in 2 sec.

4. A pitcher throws a baseball horizontally with an initial velocity of 108 ft/sec. If the point of release is 6 ft above the ground,

    (a) at what height does the ball reach home plate, 60.5 ft from the pitcher's box?

    (b) What would be the answer to this question if the initial velocity is 132 ft/sec? Assume that the air resistance is negligible.

5. A circle of radius a rolls along a line. A point on a radius, b units from the center, describes a path. Paralleling the derivation in Section 8.2, show that

the path is represented by the equations

$$x = a\theta - b\sin\theta, \qquad y = a - b\cos\theta.$$

The curve is called a **curtate cycloid** if $b < a$ and a **prolate cycloid** if $b > a$.

6. Sketch the curve of the equations in Exercise 5, taking $a = 4$ and $b = 3$. Sketch the curve if $a = 4$ and $b = 6$.

7. A circle of radius 4 rolls along a line and makes a revolution in 2 sec. A point, starting downward on a vertical radius, moves from the center to the circumference along the radius at a rate of 2.5 ft/sec. Find the equations of the path of the point.

8. The end of a thread kept in the plane of a circle describes a path called the **involute** of the circle, as it is unwound tautly from the circle. Use Fig. 8.10 to show that the parametric equations of the involute are

$$x = a(\cos\theta + \theta\sin\theta), \qquad y = a(\sin\theta - \theta\cos\theta).$$

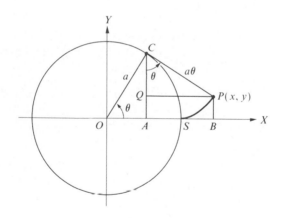

FIGURE 8.10

9. In Fig. 8.11 a circle of radius a is tangent to the two parallel lines OX and AC. The line OC cuts the circle at B, and $P(x, y)$ is the intersection of a horizontal line through B and a vertical line through C. Show that the equations of the graph of P, as C moves along the upper tangent, are

$$x = 2a\cot\theta, \qquad y = 2a\sin^2\theta.$$

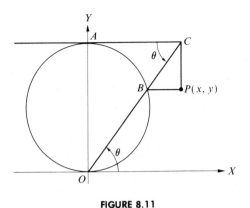

FIGURE 8.11

This curve is called the **witch of Agnesi**. Show that its rectangular equation is

$$y = \frac{8a^3}{x^2 + 4a^2}.$$

10. In Fig. 8.12, $OP = AB$. Show that the equations of the path traced by P, as A moves around the circle, are

$$x = 2a \sin^2\theta, \qquad y = 2a \sin^2\theta \tan \theta.$$

The curve is called the **cissoid of Diocles**. The rectangular equation is

$$y^2 = \frac{x^3}{2a - x}.$$

FIGURE 8.12

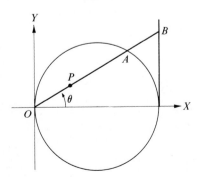

REVIEW EXERCISES

1. Sketch the graph represented by the parametric equations $x = 2t$; $y = 3t$.

2. Find the rectangular form of the pair of parametric equations $x = 4\sin\theta$ and $y = 3\cos\theta$. Then sketch the graph represented, using the simpler of the two forms.

3. Eliminate the parameter from the parametric equations $x = 2\sin^2\theta$ and $y = 3\cos^2\theta$. Draw the graph of the resulting equation and tell what part of the graph is covered by the parametric equations.

4. An object is dropped from a height of 64 ft above the ground with a constant wind of 11 miles per hour (16 ft/sec) from the north. The location t seconds after being dropped is given by

$$y(t) = -16t^2 + 64,$$
$$x(t) = 16t.$$

Graph the path of the object. When and where does it hit the ground?

5. A stone is thrown with a velocity of 80 ft/sec in a direction 45° above the horizontal. Find how far away the stone strikes the ground and its greatest height.

6. Using the accompanying equation, express the rectangular equation in parametric form:

$$2x + xy - 1 = 0; \qquad y = t + 2.$$

9

Space Coordinates and Surfaces

As we will see, much of our knowledge of plane analytic geometry will serve as a basis and as a model for solid analytic geometry. Indeed, many of the formulae which we will develop for space coordinates are "extensions" of the corresponding formulae in two dimensions. In our study thus far, we have dealt with equations in two variables and have pictured equations in a plane coordinate system. When we introduce a third variable, a plane will not suffice for the illustration of an equation. For this purpose our coordinate system is extended to three dimensions.

9.1

SPACE COORDINATES

Let OX, OY, and OZ be three mutually perpendicular lines (Fig. 9.1). These lines constitute the x axis, the y axis, and the z axis of a **three-dimensional rectangular coordinate system**. In this drawing, and others which we shall make, the y axis and the z axis are in the plane of the page. The x axis is to be visualized as perpendicular to the page. The z axis may be regarded as vertical and the others as horizontal. The axes, in pairs, determine three mutually perpendicular planes called **coordinate planes**. The planes are designated the XOY plane, the XOZ plane, and the YOZ plane or, more simply, the xy plane, the xz plane, and the yz plane. The coordinate planes divide space into eight regions, called **octants**.

We next establish a number scale on each axis with the point O as the origin. The position of a point P in this coordinate system is determined by its distances from the coordinate planes. The distance of P from the yz plane is called the x

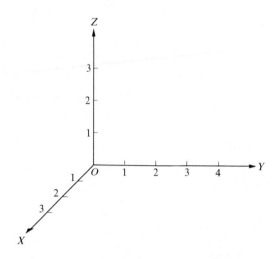

FIGURE 9.1

coordinate, the distance from the xz plane the y **coordinate**, and the distance from the xy plane the z **coordinate**. The coordinates of a point are written in the form (x, y, z), in this order, x first, y second, and z third. To plot the point $(1.5, -1, 2)$, for example, we go 1.5 units from the origin along the positive x axis, then 1 unit to the left parallel to the y axis, and finally 2 units upward parallel to the z axis. The signs of the coordinates determine the octant in which a point lies. Points whose coordinates are all positive are said to belong to the **first octant**; the other octants are not customarily assigned numbers. If a point is on a coordinate axis, two of its coordinates are zero.

In plotting points and drawing figures, we shall make unit distances on the y and z axes equal. A unit distance on the x axis will be represented by an actual length of about 0.7 of a unit. The x axis will be drawn at an angle of 135° with the y axis. This position of the x axis and the foreshortening in the x direction aid in visualizing space figures. Look at the cube and the plotted points in Fig. 9.2.

In our first application of the three-dimensional coordinate system, we consider the distance between two points of known coordinates. We took up this question for the two-dimensional case in Section 1.1. Exactly the same plan will serve in our new system except for the additional axis. Thus, $P(x_1, y_1, z_1)$ and $Q(x_2, y_1, z_1)$ denote the endpoints of the line segment PQ parallel to the x axis. The distance from P to Q is $x_2 - x_1$. This distance is positive if $x_2 > x_1$ and negative if $x_2 < x_1$. In either case, using the absolute-value symbols, $|PQ| = |x_2 - x_1|$. A similar situation applies to line segments parallel to the y axis and

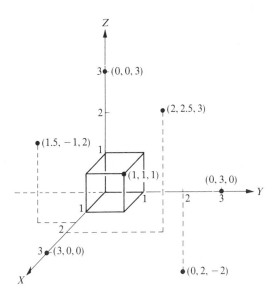

FIGURE 9.2

the z axis. With this understanding, we are ready to derive a formula for the distance between two points of known coordinates.

THEOREM 9.1 ● *Let* $P_1(x_1, y_1, z_1)$ *and* $P_2(x_2, y_2, z_2)$ *be the coordinates of two points in a three-dimensional coordinate system. Then the distance between* P_1 *and* P_2 *is given by*

$$|P_1P_2| = \sqrt{(x_2 - x_1)^2 + (y_2 - y_1)^2 + (z_2 - z_1)^2}. \qquad (9.1)$$

PROOF. In Fig. 9.3 each edge of the rectangular parallelepiped (box-shaped figure) is parallel to a coordinate axis and each face is parallel to a coordinate plane. We indicate the coordinates of P_1, P_2, Q, and R, respectively, by

$$P_1(x_1, y_1, z_1), \qquad P_2(x_2, y_2, z_2), \qquad Q(x_2, y_1, z_1), \qquad R(x_2, y_2, z_1).$$

We observe that P_1QR is a right triangle with P_1R the hypotenuse, and P_1RP_2 is a right triangle with P_1P_2 the hypotenuse. Furthermore,

$$|P_1Q| = |x_2 - x_1|, \qquad |QR| = |y_2 - y_1|, \qquad |RP_2| = |z_2 - z_1|.$$

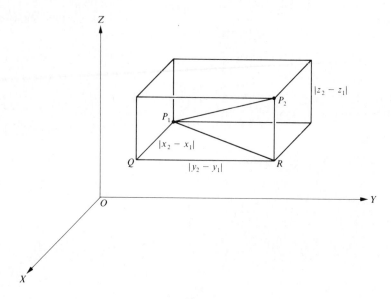

FIGURE 9.3

Hence,

$$|P_1P_2|^2 = |P_1R|^2 + |RP_2|^2$$
$$= |P_1Q|^2 + |QR|^2 + |RP_2|^2$$
$$= (x_2 - x_1)^2 + (y_2 - y_1)^2 + (z_2 - z_1)^2$$

or

$$|P_1P_2| = \sqrt{(x_2 - x_1)^2 + (y_2 - y_1)^2 + (z_2 - z_1)^2}.$$

We have written the coordinates of four of the vertices of the parallelepiped. The student may write coordinates of the other four vertices.

EXAMPLE 1 ● Find the distance between the points $P_1(-4, 4, 1)$ and $P_2(-3, 5, -4)$.

SOLUTION. Substituting in formula (9.1), we get

$$|P_1P_2| = \sqrt{(-3 + 4)^2 + (5 - 4)^2 + (-4 - 1)^2}$$
$$= \sqrt{1 + 1 + 25} = 3\sqrt{3}. ●$$

The division of a line segment discussed in Section 1.7 can be extended readily to the three-dimensional case.

THEOREM 9.2 ● *The coordinates $P(x, y, z)$ of the midpoint of the line segment joining $A(x_1, y_1, z_1)$ and $B(x_2, y_2, z_2)$ are given by the equations*

$$x = \frac{x_1 + x_2}{2}, \qquad y = \frac{y_1 + y_2}{2}, \qquad z = \frac{z_1 + z_2}{2}. \qquad (9.2)$$

This theorem may be generalized by letting $P(x, y, z)$ be any division point of the line through A and B. If the ratio of $\overline{AP}$ to $\overline{AB}$ is a number r, then

$$x = x_1 + r(x_2 - x_1), \qquad y = y_1 + r(y_2 - y_1), \qquad z = z_1 + r(z_2 - z_1). \qquad (9.3)$$

EXAMPLE 2 ● Find the coordinates of the midpoint of the line segment joining $A(3, -2, 4)$ and $B(-6, 5, 8)$.

SOLUTION. Substituting in formulas (9.2), we get

$$x = \frac{3 - 6}{2} = -\frac{3}{2}, \qquad y = \frac{-2 + 5}{2} = \frac{3}{2}, \qquad z = \frac{4 + 8}{2} = 6.$$

Hence the coordinates of the midpoint of the line segment are $P(-\frac{3}{2}, \frac{3}{2}, 6)$. ●

EXAMPLE 3 ● Find the coordinates of $P(x, y, z)$ which is one-third of the way from $A(1, 3, 5)$ to $B(5, 7, 9)$.

SOLUTION. We use formulas (9.3) with $r = \frac{1}{3}$. Thus

$$x = 1 + \tfrac{1}{3}(5 - 1) = \tfrac{13}{3},$$
$$y = 3 + \tfrac{1}{3}(7 - 3) = \tfrac{13}{3},$$
$$z = 5 + \tfrac{1}{3}(9 - 5) = \tfrac{19}{3}.$$

The desired coordinates are $(\frac{13}{3}, \frac{13}{3}, \frac{19}{3})$. This is the trisection point of the line segment that is nearer the point A. ●

EXAMPLE 4 ● Given $A = (1, 4, 7)$ and $B = (5, -1, 11)$, find the point P so that the ratio of $\overrightarrow{AP}$ to $\overrightarrow{PB}$ is equal to 4 to 7.

SOLUTION. To use formulas (9.3), we need the ratio of $\overrightarrow{AP}$ to $\overrightarrow{AB}$, which is 4 to 11. Therefore we use formulas (9.3) with $r = \frac{4}{11}$. Thus we find

$$x = 1 + \tfrac{4}{11}(5 - 1) = \tfrac{27}{11},$$
$$y = 4 + \tfrac{4}{11}(4 + 1) = \tfrac{64}{11},$$
$$z = 7 + \tfrac{4}{11}(11 - 7) = \tfrac{93}{11}.$$

The desired coordinates are $(\frac{27}{11}, \frac{64}{11}, \frac{93}{11})$. ●

The Graph of an Equation

The graph of an equation in the three-dimensional system is defined in exactly the same way as in the two-dimensional system.

DEFINITION 9.1 ● *The* **graph** *of an equation consists of the set of all points, and only those points, whose coordinates satisfy the equation.*

In studying graphs we shall use the idea of symmetry analogous to that discussed in Section 3.2. For this purpose we state the following:

1. *If an equation is unchanged when x is replaced by* $-x$, *then the graph of the equation is symmetric with respect to the yz plane.*
2. *If an equation is unchanged when y is replaced by* $-y$, *then the graph of the equation is symmetric with respect to the xz plane.*
3. *If an equation is unchanged when z is replaced by* $-z$, *then the graph of the equation is symmetric with respect to the xy plane.*

In the two-dimensional system we found lines and curves as the graphs of equations. In three dimensions the graph of an equation is called a **surface**. There are equations whose graphs, in three dimensions, are space curves (curves not lying in a plane). We are excluding space curves from consideration. We have observed, of course, that some two-dimensional equations have no graphs, and that others consist of one or more isolated points. Similarly, there are exceptional cases in a three-dimensional system. However, we shall be interested in equations whose graphs exist and are surfaces.

We shall begin our study of graphs by considering equations in one and two variables. As a further restriction, we shall use equations of only the first and second degrees. The graphs of equations of this class are comparatively easy to determine.

EXAMPLE 5 ● To find the graph of the equation

$$y = 4,$$

we observe that the equation is satisfied only by giving y the value 4. Since the equation does not contain x or z, no restrictions are placed on these variables; hence the graph consists of all points that have the y coordinate equal to 4. Clearly the graph is the plane parallel to the xz plane and 4 units to the right. ●

EXAMPLE 6 ● Graph the equation

$$2x + 3z = 6.$$

SOLUTION. In the xz plane this equation represents a line. Consider now a plane through this line and parallel to the y axis (Fig. 9.4). Corresponding to any point $P(x, y, z)$ on this plane there is a point on the line with the same x and z coordinates. Hence the coordinates of P satisfy the given equation. We conclude, therefore, that the plane is the graph of the equation. ●

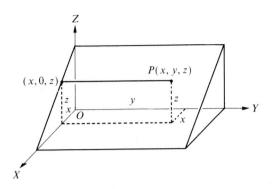

FIGURE 9.4

The two examples illustrate the correctness of the following theorem.

THEOREM 9.3 ● *The graph of a first-degree equation in one or two variables is a plane parallel to the axis of each missing variable.*

EXAMPLE 7 ● Graph the equation

$$x^2 + (y - 2)^2 = 4.$$

SOLUTION. In the xy plane the graph of this equation is a circle of radius 2 with the center on the positive y axis 2 units from the origin (Fig. 9.5). Let $(x, y, 0)$ be the coordinates of any point of the circle. Then the point (x, y, z), where z is any

FIGURE 9.5

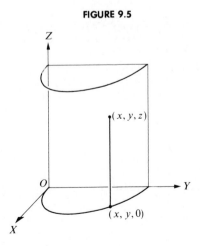

real number, satisfies the equation. Thus we see that the graph of the given equation is a surface generated by a line that moves so that it keeps parallel to the z axis and intersects the circle. The surface therefore is a right circular cylinder that is symmetric with respect to the *yz* plane. In the figure, only the part of the surface in the first octant is indicated. ●

A surface generated by a line that moves so that it keeps parallel to a fixed line and intersects a fixed curve in a plane is called a **cylindrical surface** or **cylinder**. The curve is called the **directrix**, and the generating line in any position is called an **element** of the cylinder. In accordance with this definition, a plane is a special case of a cylinder with a straight line as the directrix. Hence the graph of each of the three equations that we have considered is a cylinder.

It is easy to generalize the preceding discussion to apply to equations in two variables, even without restriction to the degree, and establish the following theorem.

THEOREM 9.4 ● *The graph of an equation in two variables is a cylinder whose elements are parallel to the axis of the missing variable.*

The General Linear Equation

In rectangular coordinates of two dimensions we found that a linear equation, in either one or two variables, represents a line. In our three-dimensional system we might, by analogy, surmise that linear equations in one, two, or three variables represent surfaces of the same type. The surmise is correct. At this point, however, we merely state the fact as a theorem and reserve the proof for the next chapter.

THEOREM 9.5 ● *The graph, in three dimensions, of the equation*

$$Ax + By + Cz + D = 0,$$

where the constants A, B, and C are not all zero, is a plane.

The curves in which a plane, or other surface, intersects the coordinate planes are called **traces**. The curves in which a surface intersects planes drawn parallel to the coordinate planes are called cross sections, or simply **sections**. Traces and sections may be used to advantage in sketching the surface.

EXAMPLE 8 ● Sketch the surface

$$y = x^2.$$

SOLUTION. Since z is missing from the equation, the surface is a cylinder parallel to the z axis. The trace in the *xy* plane, obtained by setting z = 0, is the parabola

$y = x^2$. The trace in the xz plane (set $y = 0$) is $x = 0$ or the z axis. The trace in the yz plane (set $x = 0$) is $y = 0$ or again the z axis. Additional help in visualizing the surface is obtained by setting $z = c$ to obtain sections parallel to the xy plane. These sections are again the parabola $y = x^2$, since z does not appear in the equation. Sections parallel to the xz plane ($y = c$, $c > 0$) are the lines $x = \pm \sqrt{c}$. The surface is sketched in Fig. 9.6. ●

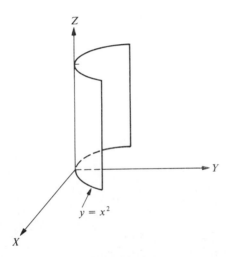

FIGURE 9.6

EXAMPLE 9 ● Sketch the plane

$$3x + 4y + 6z = 12.$$

SOLUTION. The x intercept of the plane, obtained by setting $y = 0$ and $z = 0$, is 4. Hence the plane passes through the point $(4, 0, 0)$. Similarly, the plane passes through the points $(0, 3, 0)$ and $(0, 0, 2)$. The traces of the given plane are the lines passing through these points and forming the triangle drawn in Fig. 9.7. The triangle is an excellent aid in visualizing the position of the given plane. ●

A plane parallel to a coordinate axis, and not passing through the origin, is easily sketched. For example, the trace of the plane $2y + 3z - 6 = 0$ in the yz plane passes through the points $(0, 3, 0)$ and $(0, 0, 2)$. The trace in the xy plane passes through $(0, 3, 0)$ and is parallel to the x axis, and the trace in the yz plane passes through $(0, 0, 2)$ and is parallel to the x axis.

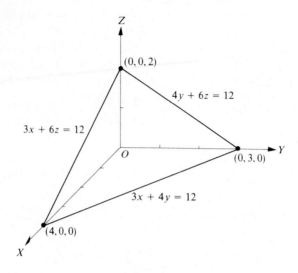

FIGURE 9.7

Exercises

Plot the points in each Exercise 1 through 8.

1. $(3, 0, 0)$ 2. $(0, 0, 3)$ 3. $(0, 3, 0)$ 4. $(3, 5, 0)$

5. $(-3, 4, 2)$ 6. $(4, -3, 0)$ 7. $(-3, -5, -2)$ 8. $(4, 5, -6)$

Find the distance between the points A and B in Exercises 9 through 12.

9. $A(-2, 0, -3)$, $B(7, -6, -1)$ 10. $A(-5, 2, 4)$, $B(4, 0, -2)$

11. $A(1, -6, -2)$, $B(5, 4, 1)$ 12. $A(1, -2, 3)$, $B(3, 4, -5)$

Find the coordinates of the midpoint of the line segment joining A and B in Exercises 13 through 16.

13. $A(-3, -4, -5)$, $B(5, 6, 7)$ 14. $A(0, -5, 8)$, $B(8, -3, 12)$

15. $A(1, 2, -3)$, $B(7, -10, 9)$ 16. $A(1, -8, 2)$, $B(11, 9, 5)$

17. Find the coordinates of the point that is two-thirds of the way from $A(-7, -5, 3)$ to $B(-4, 13, 6)$.

18. Find the coordinates of the point one-fifth of the way from $A(-7, 2, 1)$ to $B(3, 7, 11)$.

19. Given $A(1, -1, 2)$ and $B(3, 3, -1)$. Find the coordinates of a point P on AB if $AP : PB = 3 : 7$.

20. Given $A(-5, 8, -3)$ and $B(6, 5, -5)$. Find the coordinates of a point P on AB extended through B so that $|\overrightarrow{AP}| = 2|\overrightarrow{AB}|$.

21. Draw a cube that has the origin and the point $(6, 6, 6)$ as opposite vertices. Write the coordinates of the other vertices if the cube has a face in each coordinate plane.

22. Draw the rectangular parallelepiped that has three of its faces in the coordinate planes and the line joining $(0, 0, 0)$ and $(5, 6, 8)$ as the ends of a diagonal. Write the coordinates of the remaining vertices.

Describe the surface represented by each of the following equations.

23. $y = 0$ 24. $z = 0$ 25. $x = 0$ 26. $x = 6$

27. $z = -4$ 28. $y = -5$ 29. $x = -3$ 30. $z = 1$

Describe the surface represented by each of the following equations. Make a sketch for each one.

31. $2x + 3z = 6$ 32. $2y + 3z = 12$ 33. $2x + 5y = 10$

34. $x^2 + y^2 = 9$ 35. $y^2 + z^2 = 4$ 36. $4x^2 + y^2 = 36$

37. $(y - 2)^2 + z^2 = 4$ 38. $(x - 3)^2 + z^2 = 1$

39. $y^2 = 4z$ 40. $x^2 = z$

41. $x^2 = y - 1$ 42. $3x + 2y + 2z = 12$

43. $3x + 2y + z = 6$ 44. $3x + 3y + z = 9$

9.2
SURFACE OF REVOLUTION AND QUADRIC SURFACES

When a plane curve is revolved about a fixed line in the plane of the curve, a **surface of revolution** is said to be generated. The fixed line is called the **axis** of the surface. The path of each point of the curve is a circle with its center on the axis of the surface. Surfaces of revolution are used extensively in the applications of mathematics.

EXAMPLE 1 ● Find the equation of the surface generated by revolving the ellipse

$$\frac{x^2}{a^2} + \frac{y^2}{b^2} = 1$$

about the x axis.

SOLUTION. The surface is symmetric with respect to each coordinate plane. And, if $a > b$, the surface is shaped somewhat like a football. Let $P(x, y, z)$ be any point on the surface and let a plane perpendicular to the x axis pass through P (Fig. 9.8). The intersection of the plane and the surface is a circle. The center of the circle is at $C(x, 0, 0)$ and an intersection of the circle and the given ellipse is at $A(x, y', 0)$. The segments CP and CA, being radii of a circle, have the same length. Hence $|CP|^2 = |CA|^2$. But

$$|CP|^2 = y^2 + z^2 \quad \text{and} \quad |CA|^2 = y'^2 = \frac{b^2}{a^2}(a^2 - x^2)$$

and, therefore

$$y^2 + z^2 = \frac{b^2}{a^2}(a^2 - x^2)$$

or

$$\frac{x^2}{a^2} + \frac{y^2}{b^2} + \frac{z^2}{b^2} = 1.$$

This is the equation of the surface of revolution. Note that we would obtain this equation also by replacing y^2 by $y^2 + z^2$ in the given equation of the ellipse. ●

FIGURE 9.8

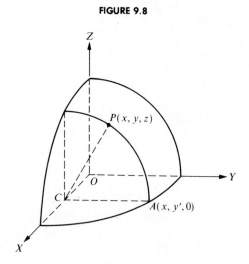

EXAMPLE 2 ● A straight line, making a constant acute angle θ with the z axis, is rotated about the z axis. Find the equation of the surface thus generated.

SOLUTION. Let $P(x, y, z)$ denote any point on the surface and let a plane perpendicular to the z axis pass through P (Fig. 9.9). The plane intersects the surface in a circle with center at $C(0, 0, z)$. The triangle OCP is a right triangle with C the vertex of the right angle. Consequently $|CP|/|OC| = \tan \theta$ and

$$|CP|^2 = |OC|^2 \tan^2 \theta.$$

But $|CP|^2 = x^2 + y^2$ and $|OC|^2 = z^2$, and therefore

$$x^2 + y^2 = z^2 \tan^2 \theta$$

or

$$x^2 + y^2 - k^2 z^2 = 0,$$

where $k = \tan \theta$. This is the equation of a right circular cone. ●

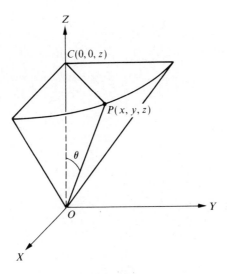

FIGURE 9.9

A sphere can be generated by revolving a circle (or just a semicircle) about a diameter. However, we shall not use the revolving method in finding the equation of a sphere.

EXAMPLE 3 Find the equation of the set of points that are at a distance a from the point $C(h, k, l)$.

SOLUTION. Using the distance formula (Section 9.1), we write immediately

$$\boxed{(x - h)^2 + (y - k)^2 + (z - l)^2 = a^2.}$$ (9.4)

This is called the center–radius form of the equation of a spherical surface. ●

By performing the indicated squares in Eq. (9.4), we get an equation in the general form

$$\boxed{x^2 + y^2 + z^2 + Dx + Ey + Fz - H = 0.}$$ (9.5)

Conversely, we can reduce an equation of the form (9.5) to one of the form (9.4). We illustrate with the equation

$$x^2 + y^2 + z^2 - 2x + 6y + 8z + 17 = 0.$$

Completing the squares in the x, y, and z terms gives

$$(x^2 - 2x + 1) + (y^2 + 6y + 9) + (z^2 + 8z + 16) = -17 + 1 + 9 + 16$$

or

$$(x - 1)^2 + (y + 3)^2 + (z + 4)^2 = 9.$$

This equation, in center–radius form, represents a sphere with center at $(1, -3, -4)$ and radius 3.

Second-Degree Equations

The graph of a second-degree equation is called a **quadric surface**. It is not easy to determine the characteristics and location of a quadric surface corresponding to a general equation. To simplify the situation, we shall consider only second-degree terms that are squares of variables and not the product of two variables.

The main device in examining the graph of an equation consists of observing its traces in the coordinate planes and additionally, if necessary, sections made by planes parallel to the coordinate planes. The idea of symmetry, of course, may be used to advantage. The traces and sections may be drawn "loosely." In fact, it is customary to distort these curves somewhat when it will help in visualizing the surface. To illustrate the method, we examine some examples.

EXAMPLE 4 ● Sketch the surface whose equation is $x^2 + y^2 = 4z$.

SOLUTION. Replacing x by $-x$ and y by $-y$ does not alter the equation; hence the surface is symmetric to the yz and xz planes. Negative values must not be assigned to z and, consequently, no part of the surface is below the xy plane. Since $x = 0$, $y = 0$, and $z = 0$ satisfies the equation, the origin falls on the surface. The trace in the yz plane ($x = 0$) is the parabola $y^2 = 4z$.

The trace in the xz plane ($y = 0$) is the parabola

$$x^2 = 4z,$$

and the trace in the xy plane ($z = 0$) is

$$x^2 + y^2 = 0,$$

the origin. When these traces alone are drawn it may prove difficult to visualize the shape of the surface. If, however, we now obtain some sections, it may tend to "tie the ends together" for us. Let us substitute $z = 1$ to obtain a section parallel to the xy plane. The plane $z = 1$ intersects the surface in the circle.

$$x^2 + y^2 = 4.$$

Circles of greater radii are obtained as the intersecting plane is taken farther and farther from the xy plane. We should now have sufficient information to form a mental picture of the surface, which somewhat resembles a cereal bowl and, as we shall see (in the subsection entitled "Quadric Surfaces"), is called a paraboloid. (See Fig. 9.10.) As a matter of interest, though, we observe further that sections parallel to the xz and yz planes are parabolas. When $y = 2$, for example, we find that the other coordinates must satisfy the equation

$$x^2 = 4(z - 1).$$

This parabola, shown in the figure, has its vertex at $(0, 2, 1)$. ●

FIGURE 9.10

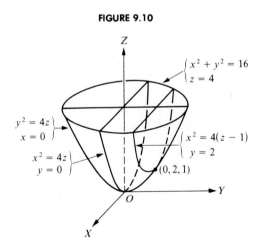

EXAMPLE 5 ● Sketch the surface whose equation is

$$4x^2 + y^2 + 4z^2 = 16.$$

SOLUTION. Replacing x by $-x$, y by $-y$, or z by $-z$ does not alter the equation, and hence the surface is symmetric to each of the coordinate planes. We next obtain and sketch the traces. The trace in the yz plane is the ellipse

$$\frac{y^2}{16} + \frac{z^2}{4} = 1.$$

The trace in the xz plane is the circle

$$x^2 + z^2 = 4.$$

The trace in the xy plane is the ellipse

$$\frac{x^2}{4} + \frac{y^2}{16} = 1.$$

When these traces are drawn, a surface like that of a football is likely to be suggested. (See Fig. 9.11.) If there is still difficulty in visualizing the surface, sections in planes parallel to the coordinate planes should be found. For example, when $y = \pm 2$, the other coordinates must satisfy the equation

$$x^2 + z^2 = 3,$$

which represents a circle of radius $\sqrt{3}$. If there is still difficulty, sections parallel to the xy plane may be obtained. If $z = \pm 1$, the other coordinates must satisfy the

FIGURE 9.11

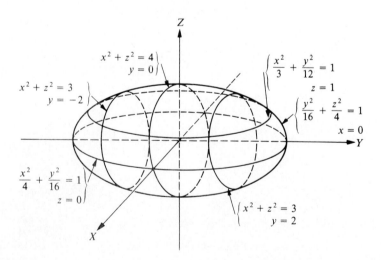

equation

$$\frac{x^2}{3} + \frac{y^2}{12} = 1,$$

which represents an ellipse. Other sections may be found, but use your imagination freely; the figure may become cluttered if you draw too many auxiliary curves. This surface is called an **ellipsoid**. ●

Quadric Surfaces

We shall now discuss a number of second-degree, or quadratic, equations that are said to be in standard forms. The study of these equations and their graphs, though presently of only geometric interest, furnishes information and experience that will prove helpful in other mathematical situations, particularly in the calculus. Throughout this section we assume that a, b, and c stand for positive constants.

THE ELLIPSOID.   The surface represented by the equation

$$\frac{x^2}{a^2} + \frac{y^2}{b^2} + \frac{z^2}{c^2} = 1$$

is called an **ellipsoid** (see Fig. 9.12). We see at once that the surface is symmetric with respect to each coordinate plane. By setting one of the variables at a time

FIGURE 9.12

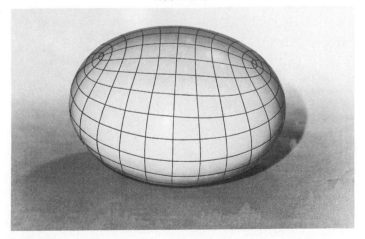

equal to zero, we find the trace equations to be

$$\left.\begin{matrix}\dfrac{x^2}{a^2} + \dfrac{y^2}{b^2} = 1\\[2mm] z = 0\end{matrix}\right\}, \qquad \left.\begin{matrix}\dfrac{x^2}{a^2} + \dfrac{z^2}{c^2} = 1\\[2mm] y = 0\end{matrix}\right\}, \qquad \left.\begin{matrix}\dfrac{y^2}{b^2} + \dfrac{z^2}{c^2} = 1\\[2mm] x = 0\end{matrix}\right\}.$$

The traces are all ellipses. Next we assign to y a definite value $y = y_0$ such that $0 < y_0 < b$, and write the given equation in the form

$$\frac{x^2}{a^2} + \frac{z^2}{c^2} = 1 - \frac{y_0^2}{b^2}.$$

This equation shows that sections made by planes parallel to the xz plane are ellipses. Furthermore, the elliptic sections decrease in size as the intersecting plane moves farther from the xz plane. When the moving plane reaches a distance b from the xz plane, the equation of the section becomes simply

$$\frac{x^2}{a^2} + \frac{z^2}{c^2} = 0,$$

and the section, therefore, is the point $(0, b, 0)$. We see then that each of the planes $y = b$ and $y = -b$ contains one point of the ellipsoid, and all other points of the surface lie between these planes. Similarly, elliptic sections are obtained for values of x between $-a$ and a, and for z between $-c$ and c. So, by the method of sections, we get a clear mental picture of the ellipsoid (Fig. 9.13).

FIGURE 9.13

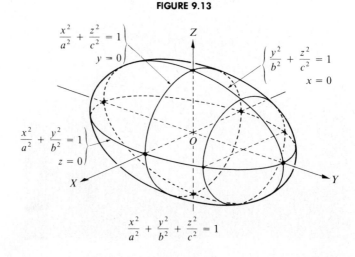

$$\left.\begin{matrix}\dfrac{x^2}{a^2} + \dfrac{z^2}{c^2} = 1\\[2mm] y = 0\end{matrix}\right\}$$

$$\left.\begin{matrix}\dfrac{y^2}{b^2} + \dfrac{z^2}{c^2} = 1\\[2mm] x = 0\end{matrix}\right.$$

$$\left.\begin{matrix}\dfrac{x^2}{a^2} + \dfrac{y^2}{b^2} = 1\\[2mm] z = 0\end{matrix}\right\}$$

$$\frac{x^2}{a^2} + \frac{y^2}{b^2} + \frac{z^2}{c^2} = 1$$

In our discussion, we have assumed that a, b, and c are any positive constants. If, however, two of the quantities are equal, the sections parallel to one of the coordinate planes are circles. Taking $a = c$, for example, and choosing an appropriate value y_0 for y, we have the equation

$$x^2 + z^2 = \frac{a^2}{b^2}\left(b^2 - y_0^2\right).$$

Thus we see that planes parallel to the xz plane cut the surface in circles. The ellipsoid in this instance could be generated by revolving the xy trace or the yz trace about the y axis. Finally, if $a = b = c$, the ellipsoid is a sphere.

THE HYPERBOLOID OF ONE SHEET. The surface represented by the equation

$$\boxed{\frac{x^2}{a^2} + \frac{y^2}{b^2} - \frac{z^2}{c^2} = 1}$$

is called a **hyperboloid of one sheet** (Fig. 9.14). The surface is symmetric with

FIGURE 9.14

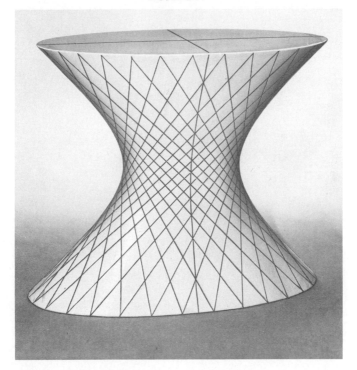

respect to each of the coordinate planes. Setting $z = 0$, we get the equation

$$\frac{x^2}{a^2} + \frac{y^2}{b^2} = 1.$$

Hence the xy trace is an ellipse. If we replace z in the given equation by a fixed value z_0, we obtain

$$\frac{x^2}{a^2} + \frac{y^2}{b^2} = 1 + \frac{z_0^2}{c^2}.$$

This equation shows that sections parallel to the xy plane are ellipses and that the sections increase in size as the intersecting plane $z = z_0$ recedes from the origin. If $a = b$, the sections are circles, and the surface is a surface of revolution.

The traces in the xz and yz planes, respectively, are the hyperbolas

$$\left.\begin{array}{c} \dfrac{x^2}{a^2} - \dfrac{z^2}{c^2} = 1 \\ y = 0 \end{array}\right\} \quad \text{and} \quad \left.\begin{array}{c} \dfrac{y^2}{b^2} - \dfrac{z^2}{c^2} = 1 \\ x = 0 \end{array}\right\}.$$

The sections parallel to the xz and yz planes are likewise hyperbolas.

FIGURE 9.15

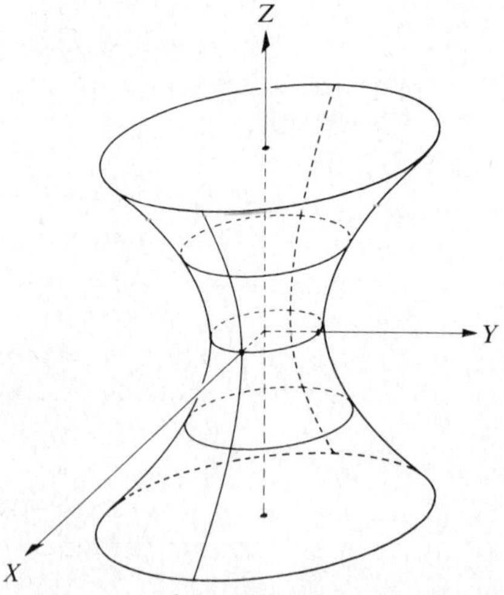

$$\frac{x^2}{a^2} + \frac{y^2}{b^2} - \frac{z^2}{c^2} = 1$$

Likewise, each of the equations

$$\frac{x^2}{a^2} - \frac{y^2}{b^2} + \frac{z^2}{c^2} = 1 \quad \text{and} \quad -\frac{x^2}{a^2} + \frac{y^2}{b^2} + \frac{z^2}{c^2} = 1$$

represents a hyperboloid of one sheet. The first encloses the y axis and the second the x axis.

It is interesting to note that the cooling towers at nuclear power generating plants are hyperboloids of one sheet. Such a surface has exceptional strength capabilities.

THE HYPERBOLOID OF TWO SHEETS.   The surface represented by the equation

$$\frac{x^2}{a^2} - \frac{y^2}{b^2} - \frac{z^2}{c^2} = 1$$

is called a **hyperboloid of two sheets** (see Fig. 9.16). The surface possesses

FIGURE 9.16

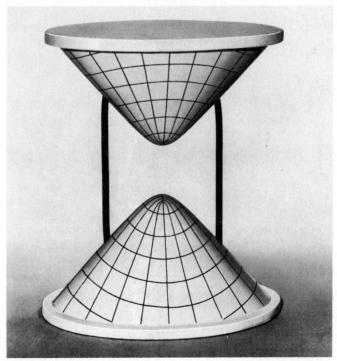

symmetry with respect to each coordinate plane. By setting each variable in turn equal to zero, we get the equations

$$\left.\begin{array}{c} \dfrac{x^2}{a^2} - \dfrac{y^2}{b^2} = 1 \\ z = 0 \end{array}\right\}, \qquad \left.\begin{array}{c} \dfrac{x^2}{a^2} - \dfrac{z^2}{c^2} = 1 \\ y = 0 \end{array}\right\}, \qquad \left.\begin{array}{c} -\dfrac{y^2}{b^2} - \dfrac{z^2}{c^2} = 1 \\ x = 0 \end{array}\right\}.$$

These equations reveal that the xy and xz traces are hyperbolas, and that there is no trace in the yz plane. The sections made by the plane $x = x_0$ are given by the equation

$$\frac{y^2}{b^2} + \frac{z^2}{c^2} = \frac{x_0^2}{a^2} - 1.$$

This equation represents a point or an ellipse depending upon whether the absolute value of x_0 is equal to or greater than a. Hence the graph of the given equation consists of two separate parts. Sections parallel to the xz and xy planes are hyperbolas. If $b = c$, the sections parallel to the yz plane are circles, and, in this case, the hyperboloid of two sheets is a surface of revolution.

FIGURE 9.17

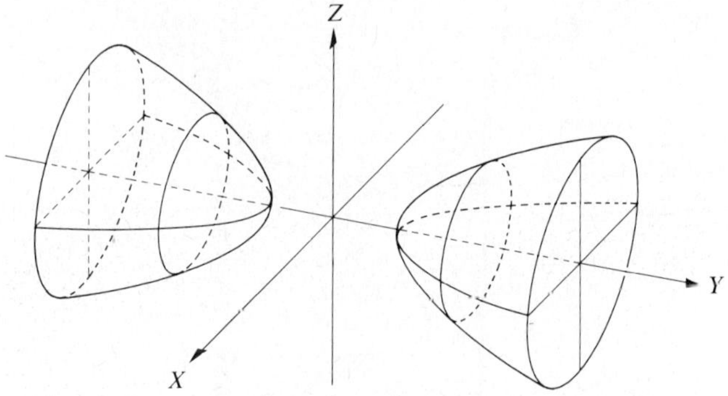

$$-\frac{x^2}{a^2} + \frac{y^2}{b^2} - \frac{z^2}{c^2} = 1$$

Hyperboloids of two sheets are also represented by the equations

$$-\frac{x^2}{a^2} - \frac{y^2}{b^2} + \frac{z^2}{c^2} = 1 \qquad \text{and} \qquad -\frac{x^2}{a^2} + \frac{y^2}{b^2} - \frac{z^2}{c^2} = 1.$$

The surface corresponding to this last equation is pictured in Fig. 9.17.

THE ELLIPTIC PARABOLOID.   The surface represented by the equation

$$\frac{x^2}{a^2} + \frac{y^2}{b^2} = cz$$

is called an **elliptic paraboloid** (Fig. 9.18). The xy trace, obtained by setting $z = 0$,

FIGURE 9.18

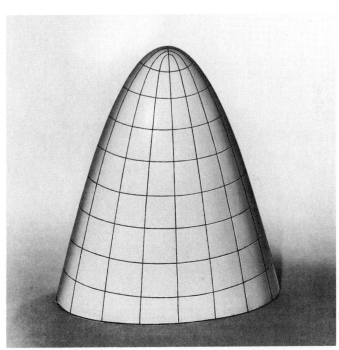

is the origin. Since we are assuming that a, b, and c are positive constants, the surface, except for the origin, is above the xy plane. A plane parallel to the xy plane and cutting the surface makes an elliptic section, which increases in size as the plane recedes from the origin. From this information, the surface can be readily visualized (see Fig. 9.19). If $a = b$, the sections parallel to the xy plane are circles. In this case the surface is obtainable by revolving either the xz trace or the yz trace about the z axis.

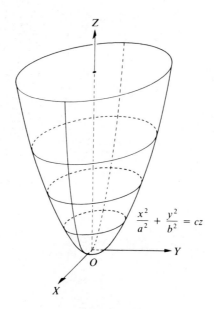

$$\frac{x^2}{a^2} + \frac{y^2}{b^2} = cz$$

FIGURE 9.19

Elliptic paraboloids are also represented by the equations

$$\frac{x^2}{a^2} + \frac{z^2}{c^2} = by \qquad \text{and} \qquad \frac{y^2}{b^2} + \frac{z^2}{c^2} = ax.$$

THE HYPERBOLIC PARABOLOID.   The surface represented by the equation

$$\boxed{\frac{x^2}{a^2} - \frac{y^2}{b^2} = -cz}$$

is called the **hyperbolic paraboloid** (Fig. 9.20). The surface is symmetric with respect to the yz and the xz planes. The xy trace is given by the pair of simultaneous equations

$$\left.\begin{array}{c} \dfrac{x^2}{a^2} - \dfrac{y^2}{b^2} = 0 \\[2mm] z = 0 \end{array}\right\} \qquad \text{or} \qquad \left.\begin{array}{c} \left(\dfrac{x}{a} + \dfrac{y}{b}\right)\left(\dfrac{x}{a} - \dfrac{y}{b}\right) = 0 \\[2mm] z = 0 \end{array}\right\}.$$

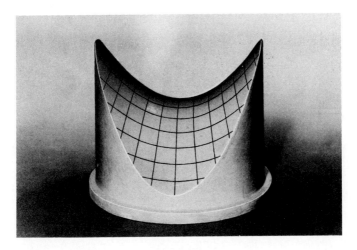

FIGURE 9.20

Hence the trace is a pair of lines intersecting at the origin. We represent the section of a plane parallel to the xy axis by the equation

$$\frac{x^2}{a^2} - \frac{y^2}{b^2} = -cz_0.$$

This is a hyperbola with transverse axis parallel to the y axis when z_0 is positive and parallel to the x axis when z_0 is negative. Sections parallel to the xz plane and the yz plane are parabolas. The preceding analysis suggests that the hyperbolic paraboloid is a saddle-shaped surface. Further aid in visualizing the surface may be gained from Fig. 9.21.

FIGURE 9.21

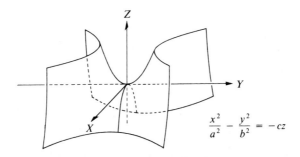

$$\frac{x^2}{a^2} - \frac{y^2}{b^2} = -cz$$

THE ELLIPTIC CONE.   The surface represented by the equation

$$\frac{x^2}{a^2} + \frac{y^2}{b^2} = \frac{z^2}{c^2}$$

is called an **elliptic cone** (Fig. 9.22). Setting x, y, and z in turn equal to zero, we

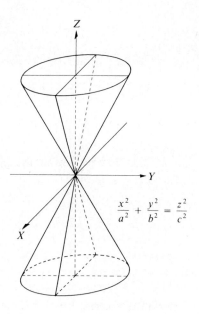

$$\frac{x^2}{a^2} + \frac{y^2}{b^2} = \frac{z^2}{c^2}$$

FIGURE 9.22

find the equations for the traces to be

$$\left.\begin{array}{c} \dfrac{x^2}{a^2} + \dfrac{y^2}{b^2} = 0 \\[2mm] z = 0 \end{array}\right\}, \qquad \left.\begin{array}{c} \dfrac{x^2}{a^2} = \dfrac{z^2}{c^2} \\[2mm] y = 0 \end{array}\right\}, \qquad \left.\begin{array}{c} \dfrac{y^2}{b^2} = \dfrac{z^2}{c^2} \\[2mm] x = 0 \end{array}\right\}.$$

The equations reveal that the xy trace is the origin, and that each of the other traces is a pair of lines intersecting at the origin. Sections parallel to the xy plane are ellipses, and those parallel to the other coordinate planes are hyperbolas. If

$a = b$, the cone is a right circular cone. Elliptic cones are also represented by the equations

$$\frac{x^2}{a^2} + \frac{z^2}{c^2} = \frac{y^2}{b^2} \qquad \text{and} \qquad \frac{y^2}{b^2} + \frac{z^2}{c^2} = \frac{x^2}{a^2}.$$

Translation of the axes to the new origin (h, k, l) may be accomplished by use of the equations

$$x' = x - h, \qquad y' = y - k, \qquad z' = z - l.$$

Thus the equation

$$(x - 1)^2 + (y - 2)^2 = 4(z - 3)$$

represents the same paraboloid as Example 4 of this section, except that the vertex is now located at the point $(1, 2, 3)$. To graph the equation, one has but to translate to the point $(1, 2, 3)$, construct a set of x', y', and z' axes, and graph the equation

$$x'^2 + y'^2 = 4z'.$$

Further, if the equation were presented in the form

$$x^2 + y^2 - 2x - 4y - 4z + 17 = 0,$$

very little additional trouble would be encountered, since completing the squares and translating again brings us to the same equation. It is worth noting that all of the equations mentioned in this section might be encountered in these forms.

Exercises

The following equations represent spheres. Find the coordinates of the center of the sphere and the radius and sketch each one.

1. $x^2 + y^2 + z^2 - 4x = 0$ 2. $x^2 + y^2 + z^2 - 2y - 6z = 0$

3. $x^2 + y^2 + z^2 - 6x + 4y - 8z - 7 = 0$

4. $x^2 + y^2 + z^2 + 4x + 6y + 8z = 0$

5. Find the equation of the surface generated by revolving the ellipse

$$\frac{x^2}{a^2} + \frac{y^2}{b^2} = 1$$

   about the y axis.

6. Find the equation of the surface generated by revolving the parabola $y^2 = 4x$ about the x axis.

Identify and sketch each quadric surface in Exercises 7–16.

7. $\dfrac{x^2}{4} + \dfrac{y^2}{16} + \dfrac{z^2}{9} = 1$

8. $\dfrac{x^2}{4} + \dfrac{y^2}{9} + \dfrac{z^2}{9} = 1$

9. $\dfrac{x^2}{16} + \dfrac{y^2}{4} - z^2 = 1$

10. $16x^2 - 16y^2 + 9z^2 = 144$

11. $\dfrac{y^2}{4} - \dfrac{x^2}{9} - \dfrac{z^2}{16} = 1$

12. $z^2 - 4x^2 - 9y^2 = 36$

13. $\dfrac{x^2}{4} + \dfrac{y^2}{9} = z$

14. $x^2 + y^2 + z + 4 = 0$

15. $9x^2 + 9y^2 = z^2$

16. $4x^2 + 9y^2 = z^2$

Sketch each quadric surface in Exercises 17–20 after completing the squares and translating the axes.

17. $x^2 + 2y^2 - 2x - 8y - 3z + 9 = 0$

18. $9x^2 + 16y^2 - 36z^2 - 18x + 72z = 171$

19. $4x^2 + 4y^2 + z^2 - 16x - 8y - 6z + 25 = 0$

20. $x^2 + y^2 - z^2 - 2x + 2y + 2 = 0$

21. The surface of a TV-satellite dish is obtained by revolving a parabola about its axis (see Fig. 9.23). The receiver is 24 inches from the vertex, as in Exercise 42, Section 3.2. Find the equation of the surface of the parabolic dish.

FIGURE 9.23

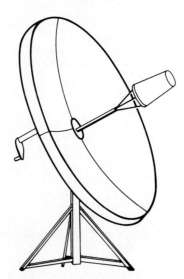

REVIEW EXERCISES

In each of Exercises 1 through 4 find the distance from A to B, the midpoint of AB, the point P that is $\frac{1}{5}$ of the way from A to B, and a point P on AB extended through B so that AB is $\frac{1}{4}$ of BP.

1. $A(1, 2, 3)$, $B(11, 12, 5)$
2. $A(0, -2, 3)$, $B(5, 3, -7)$
3. $A(1, -1, -1)$, $B(-4, 9, 4)$
4. $A(-2, -3, -4)$, $B(3, 7, 11)$

Draw the planes whose equations are:

5. $y = 4$
6. $z = -3$
7. $x = 5$
8. $x + 2y = 4$
9. $3x + 4y + z = 12$

10. What is the general equation of a plane if it passes through the origin?

11. What is the equation of (a) the xy plane? (b) the yz plane? (c) the xz plane?

12. What surfaces are represented by (a) $xy = 0$? (b) $xz = 0$? (c) $yz = 0$?

13. Given the plane whose equation is $Ax + By + Cz + D = 0$. Find the intercepts on each of the axes and the trace in each coordinate plane.

Describe the surface represented by each of the following equations. Make a sketch for each one.

14. $x - y = 0$
15. $4x + 3z = 12$
16. $y^2 = 9$
17. $x^2 - 6x + 9 = 0$
18. $x^2 + y^2 = 9$
19. $(x - 3)^2 + z^2 = 4$
20. $y = 4x^2$
21. $y = \sin x$
22. $x^2 + y^2 + z^2 = 25$

In Exercises 23 through 28 identify and sketch each quadric surface.

23. $\dfrac{x^2}{9} + \dfrac{y^2}{4} + \dfrac{z^2}{16} = 1$
24. $\dfrac{x^2}{9} + \dfrac{y^2}{4} - \dfrac{z^2}{9} = 1$
25. $\dfrac{x^2}{16} + \dfrac{y^2}{9} = z$
26. $\dfrac{x^2}{9} + \dfrac{y^2}{4} = \dfrac{z^2}{16}$
27. $\dfrac{y^2}{16} - \dfrac{x^2}{9} - \dfrac{z^2}{4} = 1$
28. $x^2 + y^2 = 4z^2$

10

Vectors, Planes, and Lines

Many physical quantities possess the properties of **magnitude** and **direction**. A quantity of this kind is called a **vector quantity**. A force, for example, is characterized by its magnitude and direction of action. The force would not be completely specified by one of these properties without the other. The velocity of a moving body is determined by its speed (magnitude) and direction of motion. Acceleration and displacement are other examples of vector quantities.

Vectors are of great importance in physics and engineering. They are also used to much advantage in pure mathematics. The study of solid analytic geometry, in particular, is facilitated by the application of the vector concept. Our immediate objective in the introduction of vectors, however, is their use in dealing with planes and lines in space.

It is essential to note, however, that the treatment of vectors presented here is primitive in comparison with the many sophisticated formulations in use by mathematicians and engineers today. Vectors, at one level of abstraction or another, have become one of the most fruitful ideas in modern mathematics. Much of applied mathematics depends on a firm grasp of the notion of vector.

To obtain a geometric representation of a vector quantity, we employ a **directed line segment** (Section 1.1), whose length and direction represent the magnitude and direction, respectively, of the vector quantity. In order to get a working basis for investigating problems involving vector quantities, we shall introduce certain definitions and establish a number of useful theorems.

DEFINITION 10.1 ● *A directed line segment, when used to represent a vector quantity, is called a **vector**.*

We shall denote a vector by a boldface letter, or by giving its starting point and its ending point. Thus the vector (arrow) in Fig. 10.1 is drawn from O to P and we let $\overline{OP}$ or **A** indicate the vector. The point O is called the **foot** of the vector and the point P is called the **head**. The vectors **B** and $-\mathbf{B}$ in the figure have the same length as **A**. The vectors **A** and **B** have the same direction, but $-\mathbf{B}$

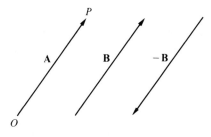

FIGURE 10.1

is oppositely directed. The three vectors are related in accordance with the following definition.

DEFINITION 10.2 ● *Two vectors* **A** *and* **B** *are* ***equal*** (**A** = **B**) *if they have the same length and direction. The* ***negative*** *of a vector* **B**, *denoted by* −**B**, *is a vector having the same length as* **B** *but pointing in the opposite direction.*

In view of this definition, we note that vectors of the same length and different directions are not equal, and vectors of the same direction and different lengths are not equal.

10.1

OPERATIONS ON VECTORS

The operations of addition, subtraction, and multiplication of vectors are defined differently from the corresponding operations on real numbers. The new operations are designed so as to establish an appropriate theory for studying forces, velocities, and other physical concepts.

DEFINITION 10.3 ● *Let* **A** *and* **B** *denote vectors, and draw from the head of* **A** *a vector equal to* **B**. *Then the* ***sum*** (**A** + **B**) *of* **A** *and* **B** *is the vector extending from the foot of* **A** *to the head of* **B**.

This definition is illustrated in Fig. 10.2. The sum of the vectors is called the **resultant** and each of the vectors forming the sum is called a **component**. The

FIGURE 10.2

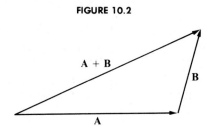

triangle formed by the three vectors, **A**, **B**, and **A** + **B**, is called a **vector triangle**. This method of adding vectors is used in physics, where it is shown, for example, that two forces applied at a point of a body have exactly the same effect as a single force equal to their resultant.

THEOREM 10.1 ● *Vector addition is commutative.*

PROOF. We need to show that if **A** and **B** are any two vectors, then **A** + **B** = **B** + **A**. For this purpose we draw **A** and **B** from the point O (Fig. 10.3), and then complete a parallelogram with the vectors forming adjacent sides. Since the opposite sides of a parallelogram are equal and parallel, we see that the foot of **B** in the lower part of the figure is at the head of **A**. Hence the sum **A** + **B** is along the diagonal extending from O; and from the upper part of the figure, we see that **B** + **A** is along the same diagonal. We conclude then that **A** + **B** = **B** + **A**, which means the vectors are commutative with respect to addition.

FIGURE 10.3

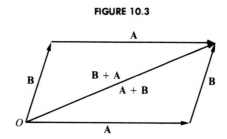

THEOREM 10.2 ● *Vectors obey the associative law of addition.*

PROOF. Given three vectors, **A**, **B**, and **C**, we are to prove that

$$(\mathbf{A} + \mathbf{B}) + \mathbf{C} = \mathbf{A} + (\mathbf{B} + \mathbf{C}).$$

From Fig. 10.4 we observe that $\overrightarrow{OE} = \mathbf{A} + \mathbf{B}$ and that **C** added to this sum yields

$$\overrightarrow{OF} = (\mathbf{A} + \mathbf{B}) + \mathbf{C}.$$

Similarly, $\overrightarrow{DF} = \mathbf{B} + \mathbf{C}$ and adding this sum to **A** gives

$$\overrightarrow{OF} = \mathbf{A} + (\mathbf{B} + \mathbf{C}).$$

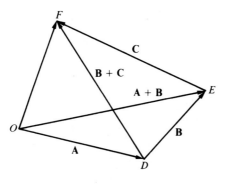

FIGURE 10.4

Hence the sums $(\mathbf{A} + \mathbf{B}) + \mathbf{C}$ and $\mathbf{A} + (\mathbf{B} + \mathbf{C})$ have the same resultant, and consequently vector addition is associative.

DEFINITION 10.4 ● *A vector* **B** *subtracted from a vector* **A** *is equal to the sum of* **A** *and the negative of* **B**. *That is,*

$$\mathbf{A} - \mathbf{B} = \mathbf{A} + (-\mathbf{B}).$$

Referring to the parallelogram $OPQR$ in Fig. 10.5, we observe that the vector from O to P is equal to $\mathbf{A} - \mathbf{B}$, and the vector from R to Q is also equal to $\mathbf{A} - \mathbf{B}$. So, alternatively, if $\mathbf{A}$ and $\mathbf{B}$ are drawn from a common point, the vector from the head of $\mathbf{B}$ to the head of $\mathbf{A}$ is equal to $\mathbf{A} - \mathbf{B}$.

FIGURE 10.5

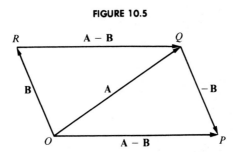

When numbers and vectors are involved in a problem or discussion, the numbers are sometimes called **scalars** to distinguish them from vectors.

DEFINITION 10.5 ● *The **product** of a scalar m and a vector* **A**, *expressed by* m**A**, *is a vector m times as long as* **A**, *and has the direction of* **A** *if m is positive and the opposite direction if m is negative.*

This definition is illustrated in Fig. 10.6, where the scalar m has the values -2 and $\frac{3}{4}$.

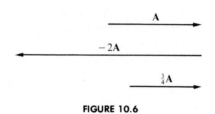

FIGURE 10.6

THEOREM 10.3 ● *If m and n are scalars and* **A** *and* **B** *are vectors, then*

$$(m + n)\mathbf{A} = m\mathbf{A} + n\mathbf{A} \tag{10.1}$$

and

$$m(\mathbf{A} + \mathbf{B}) = m\mathbf{A} + m\mathbf{B}. \tag{10.2}$$

PROOF. We leave the proof of Eq. (10.1) to the student. To establish Eq. (10.2), we note that **A**, **B**, and **A** + **B** form the sides of a triangle (Fig. 10.7). If each of these vectors is multiplied by a nonzero scalar m, the products $m\mathbf{A}$, $m\mathbf{B}$, and $m(\mathbf{A} + \mathbf{B})$ can be placed so as to form a triangle in which $m(\mathbf{A} + \mathbf{B}) = m\mathbf{A} + m\mathbf{B}$ (Fig. 10.8). Hence, as expressed by Eqs. (10.1) and (10.2), scalars and vectors obey the distributive law of multiplication.

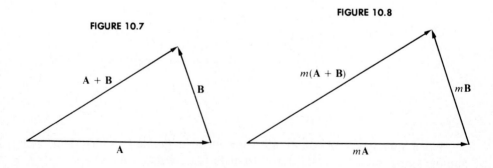

FIGURE 10.7

FIGURE 10.8

Let us seek an interpretation of the difference of a vector and itself and the product of zero and a vector. That is, if **A** is a vector, what meaning should be given to **A** − **A** and (0)**A**? For these quantities to be vectors, Definitions (10.4) and (10.5) require the length in each case to be equal to zero. To handle a situation of this kind, it is customary to enlarge the concept of a vector to include one of zero length, which is called the **zero vector**.

It is sometimes desirable to find two vectors whose sum is equal to a given vector. The given vector is then said to be **resolved** into two components. The components may be along any two directions in a plane containing the given vector. A graphical construction of the components may be obtained by forming a vector triangle of which the given vector is a side. The components are then along the other sides.

Vectors in a Rectangular-Coordinate Plane

In our consideration of vectors thus far we have not used a coordinate system. Many operations on vectors, however, can be carried out advantageously by the aid of a coordinate system. To begin our study of vectors in a coordinate plane, we introduce two special vectors each of unit length. One of the vectors, denoted by **i**, has the direction of the positive x axis; the other vector, denoted by **j**, has the direction of the positive y axis. Each of these vectors, as well as any vector of unit length, is said to be a **unit vector**.

Since vectors of the same length and same direction are equal (Definition 10.2), each of the vectors **i** and **j** may extend from any chosen point of the coordinate plane, but it is usually convenient to place them so that they extend from the origin (Fig. 10.9).

FIGURE 10.9

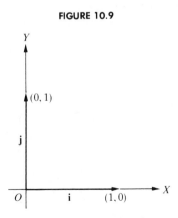

The product $m\mathbf{i}$ is a vector of length m units and has the direction of $\mathbf{i}$ if m is positive and the opposite direction if m is negative (Definition 10.5). A similar statement applies to the product $m\mathbf{j}$. Using these facts, we shall point out that any vector can be expressed in terms of the unit vectors $\mathbf{i}$ and $\mathbf{j}$. Let $\mathbf{V}$ be a vector from the origin to the point (a, b), as shown in Fig. 10.10. Clearly $a\mathbf{i}$ and $b\mathbf{j}$ are the horizontal and vertical components of the given vector, and therefore $\mathbf{V} = a\mathbf{i} + b\mathbf{j}$. The vector $a\mathbf{i}$ is called the x **component** of $\mathbf{V}$ and $b\mathbf{j}$ the y **component**.

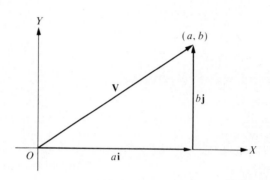

FIGURE 10.10

Continuing with the vector $\mathbf{V}$, we make the following observations. Since the length of the x component of $\mathbf{V}$ is a and the length of the y component is b, we may employ the Pythagorean theorem to find the length or magnitude of $\mathbf{V}$. Thus, denoting the length by $|\mathbf{V}|$, we have

$$|\mathbf{V}| = \sqrt{a^2 + b^2}.$$

If $\mathbf{V}$ is divided by $|\mathbf{V}|$, the result is a unit vector with the same direction as that of $\mathbf{V}$. The length of $\mathbf{V} = 3\mathbf{i} - 4\mathbf{j}$, for example, is

$$|\mathbf{V}| = \sqrt{9 + 16} = 5,$$

and

$$\frac{\mathbf{V}}{|\mathbf{V}|} = \frac{3\mathbf{i} - 4\mathbf{j}}{5} = \frac{3}{5}\mathbf{i} - \frac{4}{5}\mathbf{j}$$

is a unit vector having the same direction as $3\mathbf{i} - 4\mathbf{j}$.

THEOREM 10.4 ● *If the vectors* V_1 *and* V_2, *in terms of their x components and y components, are*

$$V_1 = a_1 i + b_1 j \qquad and \qquad V_2 = a_2 i + b_2 j,$$

then

$$V_1 + V_2 = (a_1 + a_2)i + (b_1 + b_2)j \qquad (10.3)$$

and

$$V_1 - V_2 = (a_1 - a_2)i + (b_1 - b_2)j. \qquad (10.4)$$

PROOF. If the foot of V_1 is at the origin, the head will be at the point (a_1, b_1). Then if the foot of V_2 is at the head of V_1, the head will be at the point $(a_1 + a_2, b_1 + b_2)$, and the vector from the origin to this point is expressed by

$$(a_1 + a_2)i + (b_1 + b_2)j,$$

which, by Definition 10.3, is equal to $V_1 + V_2$.

Although formula (10.3) follows readily from the definition of the sum of two vectors, it is worth noting that the formula can be established by use of Theorems 10.1, 10.2, and 10.3. Thus

$$\begin{aligned} V_1 + V_2 &= (a_1 i + b_1 j) + (a_2 i + b_2 j) \\ &= a_1 i + a_2 i + b_1 j + b_2 j \\ &= (a_1 + a_2)i + (b_1 + b_2)j. \end{aligned}$$

We leave the proof of formula (10.4) to the student.

It should be mentioned here that if $P_1(x_1, y_1)$ and $P_2(x_2, y_2)$ are two points, -then

$$\overrightarrow{P_1 P_2} = (x_2 - x_1)i + (y_2 - y_1)j.$$

This is easy to see since, if O is the origin, then

$$\overrightarrow{OP_1} = x_1 i + y_1 j, \qquad \overrightarrow{OP_2} = x_2 i + y_2 j, \qquad and \qquad \overrightarrow{P_1 P_2} = \overrightarrow{OP_2} - \overrightarrow{OP_1}.$$

EXAMPLE 1 ● Vectors are drawn from the origin to the points $P(3, -2)$ and $Q(1, 5)$. Indicating these vectors by $\overrightarrow{OP} = A$ and $\overrightarrow{OQ} = B$, find $A + B$ and $A - B$.

SOLUTION. The given vectors, in terms of their x components and y components, are

$$A = 3i - 2j \qquad and \qquad B = i + 5j.$$

Then, by the preceding theorem,

$$\begin{aligned} A + B &= (3 + 1)i + (-2 + 5)j \\ &= 4i + 3j \end{aligned}$$

and

$$\begin{aligned}\mathbf{A} - \mathbf{B} &= (3 - 1)\mathbf{i} + (-2 - 5)\mathbf{j} \\ &= 2\mathbf{i} - 7\mathbf{j}.\end{aligned}$$

These two results are pictured in Fig. 10.11. The vector $\mathbf{A} - \mathbf{B}$, as we pointed out by use of Fig. 10.5, is equal to the vector extending from the head of $\mathbf{B}$ to the head of $\mathbf{A}$. The foot of $\mathbf{A} - \mathbf{B}$, in the figure, is not at the origin. But a vector equal to $\mathbf{A} - \mathbf{B}$ with its foot at the origin would have $(2, -7)$ as the coordinates of its head. ●

FIGURE 10.11

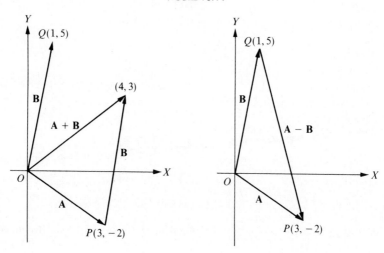

EXAMPLE 2 ● Find the coordinates of the midpoint of the line segment joining the points $P(-2, 4)$ and $Q(8, 2)$.

SOLUTION. We first find the vector from the origin to the midpoint of the line segment. This vector is equal to the vector from the origin to P plus half the vector from P to Q (Fig. 10.12). Indicating the vectors from the origin to P and Q by $\mathbf{A}$ and $\mathbf{B}$, respectively, we have

$$\begin{aligned}\mathbf{A} &= -2\mathbf{i} + 4\mathbf{j}, \\ \mathbf{B} &= 8\mathbf{i} + 2\mathbf{j}, \\ \mathbf{B} - \mathbf{A} &= 10\mathbf{i} - 2\mathbf{j}.\end{aligned}$$

The desired vector $\mathbf{V}$ then is

$$\begin{aligned}\mathbf{V} &= \mathbf{A} + \tfrac{1}{2}(\mathbf{B} - \mathbf{A}) \\ &= (-2\mathbf{i} + 4\mathbf{j}) + \tfrac{1}{2}(10\mathbf{i} - 2\mathbf{j}) \\ &= 3\mathbf{i} + 3\mathbf{j}.\end{aligned}$$

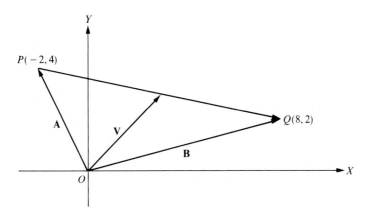

FIGURE 10.12

This result shows that the head of **V** is at the point $(3, 3)$, and these are the coordinates of the midpoint of PQ. ●

EXAMPLE 3 ● Find the vectors from the origin to the trisection points of the line segment joining the points $P(1, 3)$ and $Q(4, -3)$. Give the coordinates of the trisection points.

SOLUTION. One of the required vectors is equal to the vector from the origin to P plus one-third of the vector from P to Q (Fig. 10.13). The other required vector is

FIGURE 10.13

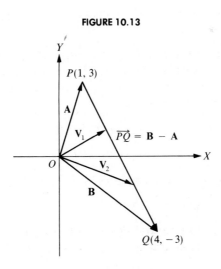

equal to the vector from the origin to P plus two-thirds of the vector from P to Q. Denoting the vectors from the origin to P and Q by $\mathbf{A}$ and $\mathbf{B}$, respectively, we write

$$\mathbf{A} = \mathbf{i} + 3\mathbf{j},$$
$$\mathbf{B} = 4\mathbf{i} - 3\mathbf{j},$$
$$\mathbf{B} - \mathbf{A} = 3\mathbf{i} - 6\mathbf{j}.$$

Hence one of the required vectors $\mathbf{V}_1$ is

$$\begin{aligned}
\mathbf{V}_1 &= \mathbf{A} + \tfrac{1}{3}(\mathbf{B} - \mathbf{A}) \\
&= (\mathbf{i} + 3\mathbf{j}) + \tfrac{1}{3}(3\mathbf{i} - 6\mathbf{j}) \\
&= 2\mathbf{i} + \mathbf{j}.
\end{aligned}$$

The other vector $\mathbf{V}_2$ is

$$\begin{aligned}
\mathbf{V}_2 &= (\mathbf{i} + 3\mathbf{j}) + \tfrac{2}{3}(3\mathbf{i} - 6\mathbf{j}) \\
&= 3\mathbf{i} - \mathbf{j}.
\end{aligned}$$

The vectors $2\mathbf{i} + \mathbf{j}$ and $3\mathbf{i} - \mathbf{j}$ tell us that the coordinates of the trisection points are $(2, 1)$ and $(3, -1)$. ●

Examples 2 and 3 duplicate, in vector terminology, the results of Section 1.3. They illustrate a more general technique. To find the equation of all vectors $\mathbf{V}$ from the origin to the line through the endpoints of vectors $\mathbf{A} = a_1\mathbf{i} + a_2\mathbf{j}$ and $\mathbf{B} = b_1\mathbf{i} + b_2\mathbf{j}$, we note that $\mathbf{V} = \mathbf{A} + t(\mathbf{B} - \mathbf{A})$, for some real number t. Then

$$\begin{aligned}
\mathbf{V} &= \mathbf{A} + t(\mathbf{B} - \mathbf{A}) \\
&= a_1\mathbf{i} + a_2\mathbf{j} + t(b_1 - a_1)\mathbf{i} + t(b_2 - a_2)\mathbf{j} \\
&= ((1 - t)a_1 + tb_1)\mathbf{i} + ((1 - t)a_2 + tb_2)\mathbf{j}.
\end{aligned}$$

This determines the vector $\mathbf{V}$ in terms of the parameter t. Note that, when $t = 0$, $\mathbf{V} = \mathbf{A}$, and when $t = 1$, $\mathbf{V} = \mathbf{B}$. For $0 < t < 1$, the end of $\mathbf{V}$ lies along $\mathbf{B} - \mathbf{A}$.

EXAMPLE 4 ● Find the parametric equation for the vectors from the origin to the line joining $P(1, 3)$ and $Q(4, -3)$.

SOLUTION. The points are the same as those of Example 3, so Fig. 10.13 continues to suffice. From the discussion we know that the vector $\mathbf{V}$ from the origin to

$\overrightarrow{PQ} = \mathbf{B} - \mathbf{A}$ is represented by

$$\begin{aligned} \mathbf{V} &= \mathbf{A} + t(\mathbf{B} - \mathbf{A}) \\ &= (\mathbf{i} + 3\mathbf{j}) + t(3\mathbf{i} - 6\mathbf{j}) \\ &= (3t + 1)\mathbf{i} + (3 - 6t)\mathbf{j}, \end{aligned}$$

where t is an arbitrary real number. ●

EXAMPLE 5 ● A force is represented in magnitude (pounds) and direction by the vector $6\mathbf{i} + 8\mathbf{j}$. A second force is represented by $10\mathbf{i} + 24\mathbf{j}$. Find a vector representation for a single force that is exactly equivalent to these two forces acting simultaneously. What are the magnitudes of the three forces? Give the direction of the resultant by stating the angle (to nearest degree) it makes with the positive x axis. What angle (to nearest degree) does the resultant make with the direction of the first force?

SOLUTION. Let $\mathbf{V}_1$, $\mathbf{V}_2$, and $\mathbf{V}$ be the vectors representing the first, second, and resultant forces, respectively (Fig. 10.14). Then

$$\begin{aligned} \mathbf{V} &= \mathbf{V}_1 + \mathbf{V}_2, \\ \mathbf{V} &= (6\mathbf{i} + 8\mathbf{j}) + (10\mathbf{i} + 24\mathbf{j}) \\ &= 16\mathbf{i} + 32\mathbf{j}. \end{aligned}$$

FIGURE 10.14

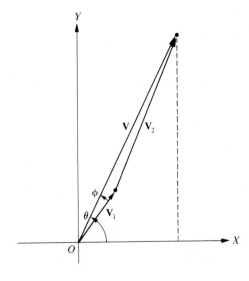

This is the vector representation for the one force replacing the other two. Now,

$$|\mathbf{V}_1| = \sqrt{36 + 64} = \sqrt{100} = 10 \text{ pounds,}$$
$$|\mathbf{V}_2| = \sqrt{100 + 576} = \sqrt{676} = 26 \text{ pounds,}$$

and

$$|\mathbf{V}| = \sqrt{256 + 1024} = \sqrt{1280} = 16\sqrt{5} \text{ pounds.}$$

If θ is the angle that the resultant makes with the positive x axis (Fig. 10.14), then

$$\tan\theta = \frac{32}{16} = 2,$$
$$\theta = \text{Arctan } 2 = 63°.$$

If ϕ is the angle that the resultant makes with the direction of the first force (Fig. 10.14), then, by the law of cosines,

$$\cos\phi = \frac{10^2 + (16\sqrt{5})^2 - (26)^2}{2(10)(16\sqrt{5})}$$
$$= 0.9839,$$
$$\phi = \text{Arccos } 0.9839 = 10°. \bullet$$

Exercises

In each of Exercises 1 through 8, let $\mathbf{P}$ be the vector drawn from the origin to the first point and $\mathbf{Q}$ the vector drawn from the origin to the second point. Find in component form $\mathbf{P}$, $\mathbf{Q}$, $\overrightarrow{PQ}$, $\mathbf{P} + \mathbf{Q}$, and $\mathbf{P} - \mathbf{Q}$. Find the length or magnitude of $\mathbf{P}$ in each case. Draw all vectors.

1. $P(3, 2)$, $Q(5, -4)$ 2. $P(4, 0)$, $Q(0, 5)$

3. $P(-4, 5)$, $Q(-2, -3)$ 4. $P(5, 6)$, $Q(-3, -3)$

5. $P(0, 3)$, $Q(-4, -2)$ 6. $P(-1, 7)$, $Q(2, -3)$

7. $P(-1, -3)$, $Q(3, 4)$ 8. $P(3, -12)$, $Q(4, -1)$

Determine a unit vector having the direction of the vector in each of Exercises 9 through 12.

9. $4\mathbf{i} - 3\mathbf{j}$ 10. $5\mathbf{i} + 12\mathbf{j}$ 11. $\mathbf{i} + 5\mathbf{j}$ 12. $-\mathbf{i} - 7\mathbf{j}$

Determine a vector having the direction of the given vector and with the given length in Exercises 13 through 16.

13. $3\mathbf{i} + 4\mathbf{j}$, length 8   14. $-5\mathbf{i} + 12\mathbf{j}$, length 15

15. $\mathbf{i} - \mathbf{j}$, length 5   16. $2\mathbf{i} - 3\mathbf{j}$, length 10

Use the method of Example 2, to find the coordinates of the midpoint of the line segment joining P and Q in each of Exercises 17 through 20.

17. $P(2,3)$, $Q(6,-5)$ 18. $P(3,8)$, $Q(-7,2)$

19. $P(-3,-2)$, $Q(7,4)$ 20. $P(a_1,b_1)$, $Q(a_2,b_2)$

Use the method of Example 3, to find the coordinates of the trisection points of the line segment joining P and Q in each of Exercises 21 through 24.

21. $P(-4,-4)$, $Q(5,2)$ 22. $P(-8,5)$, $Q(1,-7)$

23. $P(3,3)$, $Q(-6,-3)$ 24. $P(a_1,b_1)$, $Q(a_2,b_2)$

Use the method of Example 4 to find the parametric representation of a vector from the origin to the line through P and Q in each of Exercises 25 through 28.

25. $P(-4,-4)$, $Q(5,2)$ 26. $P(3,8)$, $Q(-7,2)$

27. $P(0,1)$, $Q(-3,-2)$ 28. $P(0,a)$, $Q(b,0)$

29. A force is represented by $2\mathbf{i} + 8\mathbf{j}$ and a second force is represented by $3\mathbf{i} + 4\mathbf{j}$. Find the vector representing the resultant force and the angle (to nearest degree) it makes with the direction of the second force.

30. The resultant of two forces is given by $9\mathbf{i} + 15\mathbf{j}$. If one of the forces is given by $4\mathbf{i} + 3\mathbf{j}$, find a vector for the second force and the angle (to nearest degree) between the positive directions of the two component forces.

10.2

VECTORS IN SPACE

In the three-dimensional rectangular-coordinate system, the unit vectors from the origin to the points $(1,0,0)$, $(0,1,0)$, and $(0,0,1)$ are denoted, respectively, by $\mathbf{i}$, $\mathbf{j}$, and $\mathbf{k}$. Any vector can be expressed in terms of these unit vectors. Thus the vector from the origin to the point $P(a,b,c)$ is given by

$$\overrightarrow{OP} = \mathbf{A} = a\mathbf{i} + b\mathbf{j} + c\mathbf{k}.$$

The vectors $a\mathbf{i}$, $b\mathbf{j}$, and $c\mathbf{k}$ are the x, y, and z components of the vector $\mathbf{A}$. The length of $\mathbf{A}$ may be obtained by using the lengths of the sides of the vector right triangles OCP and ODC (Fig. 10.15). From the Pythagorean relation, we find

$$|\overrightarrow{OP}|^2 = |\overrightarrow{OC}|^2 + |\overrightarrow{CP}|^2$$
$$= |\overrightarrow{OD}|^2 + |\overrightarrow{DC}|^2 + |\overrightarrow{CP}|^2$$
$$= a^2 + b^2 + c^2.$$

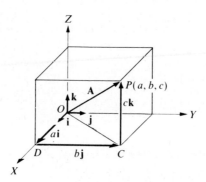

FIGURE 10.15

THEOREM 10.5 ● *If the vectors V_1 and V_2 in terms of their x, y, and z components are*

$$V_1 = a_1\mathbf{i} + b_1\mathbf{j} + c_1\mathbf{k} \quad and \quad V_2 = a_2\mathbf{i} + b_2\mathbf{j} + c_2\mathbf{k},$$

then

$$V_1 + V_2 = (a_1 + a_2)\mathbf{i} + (b_1 + b_2)\mathbf{j} + (c_1 + c_2)\mathbf{k} \tag{10.5}$$

and

$$V_1 - V_2 = (a_1 - a_2)\mathbf{i} + (b_1 - b_2)\mathbf{j} + (c_1 - c_2)\mathbf{k}. \tag{10.6}$$

PROOF. If the foot of V_1 is at the origin, the head will be at the point (a_1, b_1, c_1). Then if the foot of V_2 is at the head of V_1, the head of V_2 will be at the point $(a_1 + a_2, b_1 + b_2, c_1 + c_2)$; and, by Definition 10.3, the vector from the origin to this point is equal to the sum of the two given vectors. Hence

$$V_1 + V_2 = (a_1 + a_2)\mathbf{i} + (b_1 + b_2)\mathbf{j} + (c_1 + c_2)\mathbf{k}.$$

This establishes formula (10.5). We leave the proof of formula (10.6) to the student.

We remark that formulas (10.5) and (10.6) can also be obtained by use of Theorems 10.1, 10.2, and 10.3.

As was the case in the plane, and for the same reasons, we mention that if $P_1(x_1, y_1, z_1)$ and $P_2(x_2, y_2, z_2)$ are two points in space, then

$$\overline{P_1 P_2} = (x_2 - x_1)\mathbf{i} + (y_2 - y_1)\mathbf{j} + (z_2 - z_1)\mathbf{k}.$$

EXAMPLE 1 ● Vectors extend from the origin to the points $P(3, -2, 4)$ and $Q(1, 5, -1)$. Indicating these vectors by $\overrightarrow{OP} = \mathbf{A}$ and $\overrightarrow{OQ} = \mathbf{B}$, find $\mathbf{A} + \mathbf{B}$ and $\mathbf{A} - \mathbf{B}$.

SOLUTION. The given vectors in terms of their x, y, and z components are

$$\mathbf{A} = 3\mathbf{i} - 2\mathbf{j} + 4\mathbf{k} \qquad \text{and} \qquad \mathbf{B} = \mathbf{i} + 5\mathbf{j} - \mathbf{k}.$$

Then, by the preceding theorem,

$$\mathbf{A} + \mathbf{B} = (3+1)\mathbf{i} + (-2+5)\mathbf{j} + (4-1)\mathbf{k} = 4\mathbf{i} + 3\mathbf{j} + 3\mathbf{k},$$
$$\mathbf{A} - \mathbf{B} = (3-1)\mathbf{i} + (-2-5)\mathbf{j} + (4+1)\mathbf{k} = 2\mathbf{i} - 7\mathbf{j} + 5\mathbf{k}. \;\bullet$$

EXAMPLE 2 ● Find the coordinates of the midpoint of the line segment joining the points $P(2, 4, -1)$ and $Q(3, 0, 5)$.

SOLUTION. We first find the vector from the origin to the midpoint of the line segment. This vector is equal to the vector from the origin plus half the vector from P to Q. Indicating the vectors from the origin to P and Q by $\mathbf{A}$ and $\mathbf{B}$, respectively, we have

$$\mathbf{A} = 2\mathbf{i} + 4\mathbf{j} - \mathbf{k},$$
$$\mathbf{B} = 3\mathbf{i} + 0\mathbf{j} + 5\mathbf{k},$$
$$\mathbf{B} - \mathbf{A} = \mathbf{i} - 4\mathbf{j} + 6\mathbf{k}.$$

Since $\mathbf{B} - \mathbf{A}$ is the vector from the head of $\mathbf{A}$ to the head of $\mathbf{B}$, the desired vector $\mathbf{V}$ is

$$\mathbf{V} = \mathbf{A} + \tfrac{1}{2}(\mathbf{B} - \mathbf{A})$$
$$= (2\mathbf{i} + 4\mathbf{j} - \mathbf{k}) + \tfrac{1}{2}(\mathbf{i} - 4\mathbf{j} + 6\mathbf{k})$$
$$= \tfrac{5}{2}\mathbf{i} + 2\mathbf{j} + 2\mathbf{k}.$$

From this result, we see that the head of $\mathbf{V}$ is at the point $(\tfrac{5}{2}, 2, 2)$, and these are the coordinates of the midpoint of PQ. ●

EXAMPLE 3 ● Find the coordinates of the point that is $\tfrac{3}{4}$ of the way from $P(2, 5, 6)$ to $Q(6, -7, -2)$.

SOLUTION. Noting that the vector from P to Q is $\overrightarrow{OQ} - \overrightarrow{OP}$, we desire to find the head of the vector $\overrightarrow{OP} + \tfrac{3}{4}(\overrightarrow{OQ} - \overrightarrow{OQ} - \overrightarrow{OP})$. Thus we have

$$\overrightarrow{OP} = 2\mathbf{i} + 5\mathbf{j} + 6\mathbf{k},$$
$$\overrightarrow{OQ} = 6\mathbf{i} - 7\mathbf{j} - 2\mathbf{k},$$
$$\overrightarrow{OQ} - \overrightarrow{OP} = 4\mathbf{i} - 12\mathbf{j} - 8\mathbf{k}.$$

Then the vector $\mathbf{V}$ from the origin to the point in question is

$$\mathbf{V} = 2\mathbf{i} + 5\mathbf{j} + 6\mathbf{k} + \tfrac{3}{4}(4\mathbf{i} - 12\mathbf{j} - 8\mathbf{k})$$
$$= 5\mathbf{i} - 4\mathbf{j} + 0\mathbf{k}.$$

Hence $(5, -4, 0)$ are the coordinates of the point $\tfrac{3}{4}$ of the way from P to Q. ●

Exercises

In each of Exercises 1 through 8, let **P** be the vector drawn from the origin to the first point and **Q** the vector drawn from the origin to the second point. Find, in component form, **P**, **Q**, $\overrightarrow{PQ}$, **P** + **Q**, and **P** − **Q**. Find the length or magnitude of **P** in each case.

1. $P(1, 2, 2)$, $Q(1, -4, -2)$ 2. $P(3, 0, 4)$, $Q(0, 7, 8)$

3. $P(3, -2, 4)$, $Q(-1, -2, -3)$ 4. $P(7, 8, 5)$, $Q(3, -3, 2)$

5. $P(-4, -4, -4)$, $Q(1, 3, 7)$ 6. $P(-2, 1, -2)$, $Q(5, 1, 0)$

7. $P(0, 8, -6)$, $Q(4, -3, 6)$ 8. $P(4, -2, -4)$, $Q(1, 9, 12)$

Determine a unit vector having the direction of the vector in each of Exercises 9 through 12.

9. $2\mathbf{i} - 2\mathbf{j} + \mathbf{k}$ 10. $3\mathbf{i} - 4\mathbf{j} + 4\mathbf{k}$

11. $3\mathbf{j} - 4\mathbf{k}$ 12. $\mathbf{i} - 5\mathbf{j} - 2\mathbf{k}$

Determine a vector having the direction of the given vector and with the given length in Exercises 13 through 16.

13. $\mathbf{i} + 2\mathbf{j} - 2\mathbf{k}$,   length 12   14. $\mathbf{i} + 3\mathbf{j} - 4\mathbf{k}$,   length 3

15. $5\mathbf{i} + 6\mathbf{j} + 7\mathbf{k}$,   length 5   16. $2\mathbf{i} - 3\mathbf{j} - 4\mathbf{k}$,   length 10

Find the vectors from the origin to the midpoint of the line segment joining the points P and Q in each of Exercises 17 through 20. What are the coordinates of the heads of these vectors?

17. $P(3, -4, 5)$, $Q(5, -6, 3)$ 18. $P(-7, -4, -5)$, $Q(-3, 2, -11)$

19. $P(8, 7, -6)$, $Q(3, 4, 5)$ 20. $P(3, 5, 4)$, $Q(4, -2, 7)$

Find the vectors from the origin to the trisection points of the line segment joining P and Q in Exercises 21 and 22. Give the coordinates of the heads of these vectors.

21. $P(5, 1, -1)$, $Q(11, 10, 5)$ 22. $P(-2, 3, -4)$, $Q(-8, 15, 14)$

Find the parametric representation of a vector from the origin to the line through P and Q in each of Exercises 23 and 24.

23. $P(3, 5, 6)$, $Q(7, 1, 4)$ 24. $P(a, b, c)$, $Q(d, e, f)$

25. The line segment from $P(-2, 3, 5)$ to $Q(1, 2, -2)$ is extended through each end by an amount equal to its own length. Use vectors to find the coordinates of the new ends.

10.3

THE SCALAR PRODUCT OF TWO VECTORS

So far we have not defined a product of two vectors. Actually two kinds of vector products are important in physics, engineering, and other fields. We shall introduce the simpler of the two products.

DEFINITION 10.6 ● *The **scalar product** of two vectors* **A** *and* **B***, denoted by* **A** · **B***, is the product of their lengths times the cosine of the angle* θ *between them. That is,*

$$\mathbf{A} \cdot \mathbf{B} = |\mathbf{A}| |\mathbf{B}| \cos \theta.$$

The name scalar is used because the product is a scalar quantity. The product is also called the **dot product**. It makes no difference whether the angle θ is positive or negative, since $\cos \theta = \cos(-\theta)$. However, we shall restrict θ to the interval $0° \leq \theta \leq 180°$. The angle is equal to $0°$ if **A** and **B** point in the same direction, and it is equal to $180°$ if they point oppositely.

Since $\cos 90° = 0$ and $\cos 0° = 1$, it is evident that the scalar product of two perpendicular vectors is zero, and the scalar product of two vectors in the same direction is the product of their lengths. The dot product of a vector on itself is the square of the length of the vector. That is

$$\mathbf{A} \cdot \mathbf{A} = |\mathbf{A}|^2.$$

In Fig. 10.16 the point M is the foot of the perpendicular to the vector **A** drawn from the tip of **B**. The vector O to M is called the **vector projection** of **B** on **A**. The vector projection and **A** point in the same direction, since θ is an acute angle. If θ exceeds $90°$, then **A** and the vector from O to M point oppositely. The **scalar projection** of **B** on **A** is defined as $|\mathbf{B}| \cos \theta$; the sign of the scalar projection depends on $\cos \theta$. Using the idea of scalar projection of one vector on another, we

FIGURE 10.16

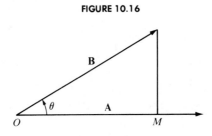

can interpret the dot product geometrically as

$$\mathbf{A} \cdot \mathbf{B} = |\mathbf{A}||\mathbf{B}| \cos \theta$$
$$= (\text{length of } \mathbf{A}) \text{ times (the scalar projection of } \mathbf{B} \text{ on } \mathbf{A}).$$

We could also say that the dot product of **A** and **B** is the length of **B** times the scalar projection of **A** on **B**.

Perhaps the simplest example of the scalar product is obtained illustrating mechanical work. Let a force **F** be applied to an object O (Fig. 10.17). If the force moves the object a distance represented in magnitude and direction by a vector **S**, then the force **F** is said to do work. The measure of the work is defined to be the product of the distance moved and the component of the **F** in the direction of **S**. That is,

$$\text{Work} = |\mathbf{S}||\mathbf{F}| \cos \theta$$

where θ is the angle between **F** and **S**. This may be equivalently expressed as:

$$\text{Work} = \mathbf{F} \cdot \mathbf{S}.$$

FIGURE 10.17

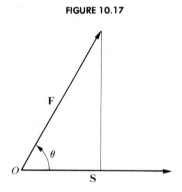

EXAMPLE 1 ● In Fig. 10.17, let the length of $\mathbf{F} = 10$ ft, length of $\mathbf{S} = 6$ ft., and $\theta = 60°$. Then the work done by **F** is given by

$$\text{Work} = \tfrac{1}{2}(10)(6) = 30 \text{ ft-lbs.}$$

By definition, $\mathbf{A} \cdot \mathbf{B}$ and $\mathbf{B} \cdot \mathbf{A}$ have exactly the same scalar factors and therefore

$$\mathbf{A} \cdot \mathbf{B} = \mathbf{B} \cdot \mathbf{A}. \quad ● \qquad\qquad (10.7)$$

This equation expresses the commutative property for the scalar multiplication of vectors. We next establish the distributive law.

THEOREM 10.6 ● *The scalar product of vectors is distributive. That is,*

$$\mathbf{A} \cdot (\mathbf{B} + \mathbf{C}) = \mathbf{A} \cdot \mathbf{B} + \mathbf{A} \cdot \mathbf{C}. \qquad\qquad (10.8)$$

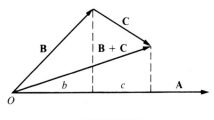

FIGURE 10.18

PROOF. If we let b and c stand for the scalar projections of **B** and **C** on **A**, we see (Fig. 10.18) that the sum of the scalar projections of **B** and **C** on **A** is the same as the scalar projection of (**B** + **C**) on **A**. Hence

$$|A|(b + c) = |A|b + |A|c$$

and

$$A \cdot (B + C) = A \cdot B + A \cdot C. \tag{10.9}$$

From Eqs. (10.7) and (10.8) we can deduce that the scalar product of sums of vectors may be carried out as in multiplying two algebraic expressions, each of more than one term. Thus, for example,

$$(A + B) \cdot (C + D) = A \cdot (C + D) + B \cdot (C + D)$$
$$= A \cdot C + A \cdot D + B \cdot C + B \cdot D.$$

If m and n are scalars, then

$$(mA) \cdot (nB) = mn(A \cdot B). \tag{10.10}$$

The equation is true if either m or n is equal to zero. But if m and n are both positive or both negative, then mn is positive, and consequently

$$(mA) \cdot (nB) = |mA||nB|\cos\theta$$
$$= mn(A \cdot B).$$

If m and n have opposite signs, the angle between **A** and **B** and the angle between m**A** and n**B** differ by 180°. Then observing Fig. 10.19 and noting that mn is negative, we may write

$$(mA) \cdot (nB) = |mA||nB|\cos(180° - \theta)$$
$$= -|mA||nB|\cos\theta$$
$$= mn(A \cdot B).$$

Hence Eq. (10.10) is true for all scalars m and n and all vectors **A** and **B**.

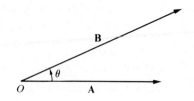

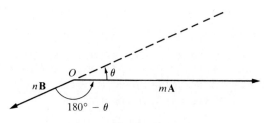

FIGURE 10.19

THEOREM 10.7 ● *If the vectors* **A** *and* **B** *are expressed in terms of the unit vectors* **i**, **j**, *and* **k** *by*

$$\mathbf{A} = a_1\mathbf{i} + b_1\mathbf{j} + c_1\mathbf{k},$$
$$\mathbf{B} = a_2\mathbf{i} + b_2\mathbf{j} + c_2\mathbf{k},$$

then

$$\boxed{\mathbf{A} \cdot \mathbf{B} = a_1a_2 + b_1b_2 + c_1c_2.} \tag{10.11}$$

PROOF. To obtain the dot product of **A** and **B**, we determine first the dot products of the unit vectors **i**, **j**, and **k**. Thus we have

$$\mathbf{i} \cdot \mathbf{i} = \mathbf{j} \cdot \mathbf{j} = \mathbf{k} \cdot \mathbf{k} = 1,$$
$$\mathbf{i} \cdot \mathbf{j} = \mathbf{j} \cdot \mathbf{k} = \mathbf{k} \cdot \mathbf{i} = 0.$$

Then we find

$$\mathbf{A} \cdot \mathbf{B} = (a_1\mathbf{i} + b_1\mathbf{j} + c_1\mathbf{k}) \cdot (a_2\mathbf{i} + b_2\mathbf{j} + c_2\mathbf{k})$$
$$= a_1\mathbf{i} \cdot (a_2\mathbf{i} + b_2\mathbf{j} + c_2\mathbf{k}) + b_1\mathbf{j} \cdot (a_2\mathbf{i} + b_2\mathbf{j} + c_2\mathbf{k})$$
$$\quad + c_1\mathbf{k} \cdot (a_2\mathbf{i} + b_2\mathbf{j} + c_2\mathbf{k})$$
$$= a_1a_2\mathbf{i} \cdot \mathbf{i} + 0 + 0 + 0 + b_1b_2\mathbf{j} \cdot \mathbf{j} + 0 + 0 + 0 + c_1c_2\mathbf{k} \cdot \mathbf{k}.$$

Hence

$$\mathbf{A} \cdot \mathbf{B} = a_1a_2 + b_1b_2 + c_1c_2.$$

This equation shows that the dot product is obtained by the simple process of adding the products of the corresponding coefficients of **i**, **j**, and **k**.

Before proceeding to some examples we wish to emphasize that if **A** and **B** are nonzero vectors, then **A** is perpendicular to **B** if and only if $\mathbf{A} \cdot \mathbf{B} = 0$; and **A** is parallel to **B** if and only if $\mathbf{A} \cdot \mathbf{B} = \pm|\mathbf{A}||\mathbf{B}|$.

EXAMPLE 2 ● Determine whether the vectors

$$\mathbf{A} = 3\mathbf{i} + 4\mathbf{j} - 8\mathbf{k},$$
$$\mathbf{B} = 4\mathbf{i} - 7\mathbf{j} - 2\mathbf{k}$$

are perpendicular.

SOLUTION. The scalar product is

$$\mathbf{A} \cdot \mathbf{B} = (3)(4) + (4)(-7) + (-8)(-2) = 0.$$

Since this product is zero, the vectors are perpendicular. ●

EXAMPLE 3 ● Prove that the vectors

$$\mathbf{A} = 2\mathbf{i} - 3\mathbf{j} - 4\mathbf{k},$$
$$\mathbf{B} = -6\mathbf{i} + 9\mathbf{j} + 12\mathbf{k}$$

are parallel.

SOLUTION. The scalar product is

$$\mathbf{A} \cdot \mathbf{B} = -12 - 27 - 48 = -87.$$

Now,

$$|\mathbf{A}| = \sqrt{4 + 9 + 16} = \sqrt{29}\,,$$
$$|\mathbf{B}| = \sqrt{36 + 81 + 144} = \sqrt{261} = 3\sqrt{29}\,,$$

and

$$|\mathbf{A}||\mathbf{B}| = 3(29) = 87.$$

Since in this case $\mathbf{A} \cdot \mathbf{B} = -|\mathbf{A}||\mathbf{B}|$, the vectors are parallel. ●

EXAMPLE 4 ● Vectors are drawn from the origin to the points $P(6, -3, 2)$ and $Q(-2, 1, 2)$. Find the angle *POQ*.

SOLUTION. Indicating $\overrightarrow{OP}$ by **A** and $\overrightarrow{OQ}$ by **B**, we write

$$\mathbf{A} = 6\mathbf{i} - 3\mathbf{j} + 2\mathbf{k},$$
$$\mathbf{B} = -2\mathbf{i} + \mathbf{j} + 2\mathbf{k}.$$

To find the angle, we substitute in both members of the equation

$$\mathbf{A} \cdot \mathbf{B} = |\mathbf{A}||\mathbf{B}|\cos\theta.$$

The product in the left member is $\mathbf{A} \cdot \mathbf{B} = -12 - 3 + 4 = 11$. The lengths of **A** and **B** are $|\mathbf{A}| = \sqrt{36 + 9 + 4} = 7$, $|\mathbf{B}| = \sqrt{4 + 1 + 4} = 3$. Hence

$$\cos\theta = \frac{\mathbf{A} \cdot \mathbf{B}}{|\mathbf{A}||\mathbf{B}|} = \frac{-11}{21}$$
$$= -0.524$$
$$\theta = 122° \quad \text{(to the nearest degree).} \ ●$$

EXAMPLE 5 ● Find the scalar projection and the vector projection of

$$\mathbf{B} = 2\mathbf{i} - 3\mathbf{j} - \mathbf{k} \qquad \text{on} \qquad \mathbf{A} = 3\mathbf{i} - 6\mathbf{j} + 2\mathbf{k}.$$

SOLUTION. The scalar projection of **B** on **A** is $|\mathbf{B}|\cos\theta$, where θ is the angle between the vectors. Using the equation

$$\mathbf{A} \cdot \mathbf{B} = |\mathbf{A}||\mathbf{B}|\cos\theta,$$

we have

$$
\begin{aligned}
|\mathbf{B}|\cos\theta &= \frac{\mathbf{A} \cdot \mathbf{B}}{|\mathbf{A}|} \\
&= \frac{6 + 18 - 2}{\sqrt{9 + 36 + 4}} \\
&= \frac{22}{7}.
\end{aligned}
$$

The scalar projection of **B** on **A** is $\frac{22}{7}$. Since the scalar projection is positive, the vector projection is in the direction of **A**. The vector projection is, therefore, the product of the scalar projection and a unit vector in the direction of **A**. This unit vector is **A** divided by its length. Hence the vector projection of **B** on **A** is

$$\frac{22}{7} \cdot \frac{3\mathbf{i} - 6\mathbf{j} + 2\mathbf{k}}{7} = \frac{22}{49}(3\mathbf{i} - 6\mathbf{j} + 2\mathbf{k}). \ ●$$

Exercises

In Exercises 1 through 6, find $\mathbf{A} \cdot \mathbf{B}$ and the cosine of the angle between the vectors.

1. $\mathbf{A} = 8\mathbf{i} + 8\mathbf{j} - 4\mathbf{k}$,
   $\mathbf{B} = \mathbf{i} - 2\mathbf{j} - 3\mathbf{k}$

2. $\mathbf{A} = \mathbf{i} - 2\mathbf{j} + 2\mathbf{k}$,
   $\mathbf{B} = \mathbf{i} + \mathbf{j} + \mathbf{k}$

3. $\mathbf{A} = \mathbf{i} + 10\mathbf{j} + 2\mathbf{k}$,
   $\mathbf{B} = 3\mathbf{i} - 4\mathbf{k}$

4. $\mathbf{A} = \mathbf{i} - \mathbf{j} - \mathbf{k}$,
   $\mathbf{B} = 4\mathbf{i} - 8\mathbf{j} + \mathbf{k}$

5. $\mathbf{A} = 2\mathbf{i} - 2\mathbf{j} - \mathbf{k}$,
   $\mathbf{B} = 16\mathbf{i} + 8\mathbf{j} + 2\mathbf{k}$

6. $\mathbf{A} = 2\mathbf{i} - \mathbf{j} + 3\mathbf{k}$,
   $\mathbf{B} = \mathbf{i} + \mathbf{j} - \mathbf{k}$

Exercises 7 and 8: Find the scalar projection and the vector projection of **B** on **A**.

7. $\mathbf{A} = 4\mathbf{i} - 4\mathbf{j} + 2\mathbf{k}$
   $\mathbf{B} = \mathbf{i} + \mathbf{j} + 2\mathbf{k}$

8. $\mathbf{A} = 8\mathbf{i} + 4\mathbf{j} - \mathbf{k}$
   $\mathbf{B} = 2\mathbf{i} - 3\mathbf{j} + \mathbf{k}$

Exercises 9 and 10: Find the scalar projection and the vector projection of **A** on **B**.

9. $\mathbf{A} = \mathbf{i} - \mathbf{j} - \mathbf{k}$
   $\mathbf{B} = 5\mathbf{i} - 12\mathbf{k}$

10. $\mathbf{A} = 2\mathbf{i} + \mathbf{j} + \mathbf{k}$
    $\mathbf{B} = 2\mathbf{i} - 2\mathbf{j} - \mathbf{k}$

In Exercises 11 through 14, determine which of the pairs of vectors are perpendicular and which are parallel.

11. $\mathbf{A} = \mathbf{i} + 3\mathbf{j} - 5\mathbf{k}$,
    $\mathbf{B} = 4\mathbf{i} + 2\mathbf{j} + 2\mathbf{k}$

12. $\mathbf{A} = 2\mathbf{i} - 3\mathbf{j} + 4\mathbf{k}$,
    $\mathbf{B} = -\mathbf{i} + \frac{4}{2}\mathbf{j} - 2\mathbf{k}$

13. $\mathbf{A} = 6\mathbf{i} + 9\mathbf{j} - 15\mathbf{k}$,
    $\mathbf{B} = 2\mathbf{i} + 3\mathbf{j} - 5\mathbf{k}$

14. $\mathbf{A} = 3\mathbf{i} - 2\mathbf{j} + 7\mathbf{k}$,
    $\mathbf{B} = \mathbf{i} - 2\mathbf{j} - \mathbf{k}$

15. Find the angle that a diagonal of a cube makes with one of its edges.

16. From a vertex of a cube, a diagonal of a face and a diagonal of the cube are drawn. Find the angle thus formed.

The points in Exercises 17 and 18 are vertices of a triangle. In each, determine the vector from A to B and the vector from A to C. Find the angle between these vectors. Similarly, find the other interior angles of the triangle.

17. $A(2,4,3)$, $B(1,7,1)$, $C(-5,3,-2)$

18. $A(1,-1,-2)$, $B(-2,0,1)$, $C(1,-3,0)$

10.4

THE EQUATION OF A PLANE

We discovered in Section 9.1 that a linear equation in one or two variables represents a plane; and we stated, without proof, that a linear equation in three variables also represents a plane (Theorem 9.3). We now prove that a linear equation in one, two, or three variables represents a plane.

THEOREM 10.8 ● *Any plane in a three-dimensional rectangular-coordinate system can be represented by a linear equation. Conversely, the graph of a linear equation is a plane.*

PROOF. Suppose that a point $P_1(x_1, y_1, z_1)$ is in a given plane and that a nonzero vector

$$\mathbf{N} = A\mathbf{i} + B\mathbf{j} + C\mathbf{k}$$

is perpendicular, or normal, to the plane (Fig. 10.20). A point $P(x, y, z)$ will lie in the given plane if, and only if, the vector

$$\overrightarrow{P_1P} = (x - x_1)\mathbf{i} + (y - y_1)\mathbf{j} + (z - z_1)\mathbf{k}$$

is perpendicular to $\mathbf{N}$. Setting the scalar product of these vectors equal to zero, we obtain the equation

$$\mathbf{N} \cdot \overrightarrow{P_1P} = 0$$

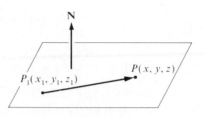

FIGURE 10.20

or

$$A(x - x_1) + B(y - y_1) + C(z - z_1) = 0. \tag{10.12}$$

This is the equation of the plane that passes through $P_1(x_1, y_1, z_1)$ and is perpendicular to the vector $\mathbf{N} = A\mathbf{i} + B\mathbf{j} + C\mathbf{k}$. Substituting D for the constant $-Ax_1 - By_1 - Cz_1$, we write the equation in the form

$$Ax + By + Cz + D = 0. \tag{10.13}$$

Conversely, any linear equation of the form (10.13) represents a plane. Starting with this equation, we can find a point $P_1(x_1, y_1, z_1)$ whose coordinates satisfy it. Then we have

$$Ax_1 + By_1 + Cz_1 + D = 0.$$

This equation and Eq. (10.13) yield, by subtraction,

$$A(x - x_1) + B(y - y_1) + C(z - z_1) = 0,$$

which is of the form (10.12). Hence Eq. (10.13) represents a plane perpendicular to the vector $\mathbf{N} = A\mathbf{i} + B\mathbf{j} + C\mathbf{k}$.

EXAMPLE 1 ● Write the equation of the plane that contains the point $P_1(4, -3, 2)$ and is perpendicular to the vector $\mathbf{N} = 2\mathbf{i} - 3\mathbf{j} + 5\mathbf{k}$.

SOLUTION. We use the coefficients of $\mathbf{i}$, $\mathbf{j}$, and $\mathbf{k}$ as the coefficients of x, y, and z and write the equation

$$2x - 3y + 5z + D = 0.$$

For any value of D this equation represents a plane perpendicular to the given vector. The equation will be satisfied by the coordinates of the given point if

$$8 + 9 + 10 + D = 0 \qquad \text{or} \qquad D = -27.$$

The required equation therefore is

$$2x - 3y + 5z - 27 = 0. \; ●$$

EXAMPLE 2 ● Find the equation of the plane determined by the points $P_1(1, 2, 6)$, $P_2(4, 4, 1)$ and $P_3(2, 3, 5)$.

SOLUTION. A vector that is perpendicular to two sides of the triangle $P_1P_2P_3$ is normal to the plane of the triangle. To find such a vector, we write

$$\overrightarrow{P_1P_2} = 3\mathbf{i} + 2\mathbf{j} - 5\mathbf{k},$$
$$\overrightarrow{P_1P_3} = \mathbf{i} + \mathbf{j} - \mathbf{k},$$
$$\mathbf{N} = A\mathbf{i} + B\mathbf{j} + C\mathbf{k}.$$

The coefficients A, B, and C are to be found so that $\mathbf{N}$ is perpendicular to each of the other vectors. Thus

$$\mathbf{N} \cdot \overrightarrow{P_1P_2} = 3A + 2B - 5C = 0,$$
$$\mathbf{N} \cdot \overrightarrow{P_1P_3} = A + B - C = 0.$$

These equations give $A = 3C$ and $B = -2C$. Choosing $C = 1$, we have $\mathbf{N} = 3\mathbf{i} - 2\mathbf{j} + \mathbf{k}$. The plane $3x - 2y + z + D = 0$ is normal to $\mathbf{N}$, and passes through the given points if $D = -5$. Hence

$$3x - 2y + z - 5 = 0. \bullet$$

EXAMPLE 3 $\bullet$ Find the angle θ between the planes $4x - 8y - z + 5 = 0$ and $x + 2y - 2z + 3 = 0$.

SOLUTION. The angle between two planes is equal to the angle between their normals. The vectors

$$\mathbf{N}_1 = \frac{4\mathbf{i} - 8\mathbf{j} - \mathbf{k}}{9} \quad \text{and} \quad \mathbf{N}_2 = \frac{\mathbf{i} + 2\mathbf{j} - 2\mathbf{k}}{3}$$

are unit vectors normal to the given planes. The dot product yields

$$\cos \theta = \mathbf{N}_1 \cdot \mathbf{N}_2 = -\tfrac{10}{27} \quad \text{and} \quad \theta = 112°.$$

The planes intersect, making a pair of angles equal (approximately) to $112°$, and a second pair equal to $68°$. Choosing the smaller angle, we give the angle between the planes as $68°$. $\bullet$

THEOREM 10.9 $\bullet$ *Let $Ax + By + Cz + D = 0$ be a plane and $P_1(x_1, y_1, z_1)$ a point not on the plane. Then the perpendicular distance d from the plane to P_1 is given by*

$$d = \frac{|Ax_1 + By_1 + Cz_1 + D|}{\sqrt{A^2 + B^2 + C^2}}. \tag{10.14}$$

PROOF. We let $P_2(x_2, y_2, z_2)$ be some point on the given plane and let $\mathbf{N} = \pm(A\mathbf{i} + B\mathbf{j} + C\mathbf{k})$, which is perpendicular to the plane, have its foot at P_2. The sign for $\mathbf{N}$ is to be chosen so that it will be on the same side of the plane as P_1, as pictured in Fig. 10.21. From the figure, we see that the desired distance d is equal to the scalar projection of $\overrightarrow{P_2P_1}$ on $\mathbf{N}$. Observing that

$$\overrightarrow{P_2P_1} = (x_1 - x_2)\mathbf{i} + (y_1 - y_2)\mathbf{j} + (z_1 - z_2)\mathbf{k},$$

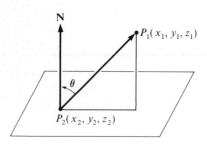

FIGURE 10.21

we have

$$d = |\overrightarrow{P_2P_1}|\cos\theta = \frac{\mathbf{N} \cdot \overrightarrow{P_2P_1}}{|\mathbf{N}|}$$

$$= \frac{\pm(A\mathbf{i} + B\mathbf{j} + C\mathbf{k}) \cdot [(x_1 - x_2)\mathbf{i} + (y_1 - y_2)\mathbf{j} + (z_1 - z_2)\mathbf{k}]}{\sqrt{A^2 + B^2 + C^2}}$$

$$= \frac{\pm[A(x_1 - x_2) + B(y_1 - y_2) + C(z_1 - z_2)]}{\sqrt{A^2 + B^2 + C^2}}$$

$$= \frac{\pm(Ax_1 + By_1 + Cz_1 - Ax_2 - By_2 - Cz_2)}{\sqrt{A^2 + B^2 + C^2}}.$$

Since P_2 is on the plane, $-Ax_2 - By_2 - Cz_2 = D$. Now to remove the ambiguity as to sign, we take the absolute value of the numerator and have

$$d = \frac{|Ax_1 + By_1 + Cz_1 + D|}{\sqrt{A^2 + B^2 + C^2}}.$$

The reader should observe that formula (10.14) for the distance from a point to a plane is a straightforward generalization of formula (2.19) for the distance from a point in a plane to a line.

Likewise, formula (1.1) for the distance between two points in a line, $\sqrt{(x_1 - x_2)^2}$, generalizes to formula (1.2), $\sqrt{(x_1 - x_2)^2 + (y_1 - y_2)^2}$ for the distance between points on a plane, and to formula (9.1),

$$\sqrt{(x_1 - x_2)^2 + (y_1 - y_2)^2 + (z_1 - z_2)^2},$$

for the distance between points in space. In higher mathematics courses, the formula is extended to describe the distance between ordered quadruples

$$(x_1, y_1, z_1, w_1) \qquad \text{and} \qquad (x_2, y_2, z_2, w_2)$$

or between ordered n-tuples, where n is any positive integer.

EXAMPLE 4 ● Find the distance from the plane $2x + 3y - 6z - 2 = 0$ to the point $(4, -6, 1)$.

SOLUTION. Substituting the proper values for A, B, C, D, x_1, y_1, and z_1 in formula (10.14), we get

$$d = \frac{|2(4) + 3(-6) + (-6)(1) - 2|}{\sqrt{2^2 + 3^2 + (-6)^2}}$$

$$= \frac{|8 - 18 - 6 - 2|}{\sqrt{4 + 9 + 36}}$$

$$= \frac{18}{7}. \quad ●$$

Exercises

Write the equation of the plane that satisfies the given conditions in Exercises 1 through 8.

1. Perpendicular to $\mathbf{N} = 3\mathbf{i} - 2\mathbf{j} + 5\mathbf{k}$ and passes through the point $(1, 1, 2)$.

2. Perpendicular to $\mathbf{N} = 4\mathbf{i} - \mathbf{j} - \mathbf{k}$ and passes through the origin.

3. Parallel to the plane $2x - 3y - 4z = 5$ and passes through $(1, 2, -3)$.

4. Perpendicular to the line segment joining $(4, 0, 6)$ and $(0, -8, 2)$ and passing through the midpoint of the segment.

5. Passes through the origin and is perpendicular to the line through $(2, -3, 4)$ and $(5, 6, 0)$.

6. Passes through the points $(0, 1, 2)$, $(2, 0, 3)$, $(4, 3, 0)$.

7. Passes through the points $(2, -2, -1)$, $(-3, 4, 1)$, $(4, 2, 3)$.

8. Passes through $(0, 0, 0)$, $(3, 0, 0)$, $(1, 1, 1)$.

Find the cosine of the acute angle between each pair of planes in Exercises 9 through 12.

9. $2x + y + z + 3 = 0$, $2x - 2y + z - 7 = 0$

10. $2x + y + 2z - 5 = 0$, $2x - 3y + 6z + 5 = 0$

11. $3x - 2y + z - 9 = 0$, $x - 3y - 9z + 4 = 0$

12. $x - 8y + 4z - 3 = 0$, $4x + 2y - 4z + 3 = 0$

13. Show that the planes

$$A_1x + B_1y + C_1z + D_1 = 0 \quad \text{and} \quad A_2x + B_2y + C_2z + D_2 = 0$$

are perpendicular if, and only if,

$$A_1A_2 + B_1B_2 + C_1C_2 = 0.$$

14. Determine the value of C so that the planes $2x - 6y + Cz = 5$ and $x - 3y + 2z = 4$ are perpendicular.

Find the distance from the given plane to the given point in Exercises 15 through 18.

15. $2x - y + 2z + 3 = 0$, $(0, 1, 3)$

16. $6x + 2y - 3z + 2 = 0$, $(2, -4, 3)$

17. $4x - 2y + z - 2 = 0$, $(-1, 1, 2)$

18. $3x - 4y - 5z = 0$, $(5, -1, 3)$

19. Find the equation of the sphere with center at $(1, 1, 7)$ and tangent to the plane $x + 2y - 4z + 3 = 0$.

10.5

VECTOR EQUATION OF A LINE

Let L be a line that passes through a given point $P_1(x_1, y_1, z_1)$ and is parallel to a given nonzero vector

$$\mathbf{V} = A\mathbf{i} + B\mathbf{j} + C\mathbf{k}.$$

If $P(x, y, z)$ is a point on the line, then the vector $\overrightarrow{P_1P}$ is parallel to $\mathbf{V}$ (Fig. 10.22). Conversely, if $\overrightarrow{P_1P}$ is parallel to $\mathbf{V}$, the point P is on the line L. Hence P is on L if, and only if, there is a scalar t such that

$$\overrightarrow{P_1P} = t\mathbf{V}$$

FIGURE 10.22

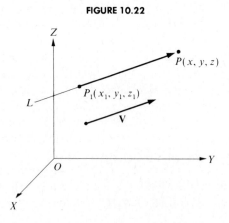

or

$$(x - x_1)\mathbf{i} + (y - y_1)\mathbf{j} + (z - z_1)\mathbf{k} = At\mathbf{i} + Bt\mathbf{j} + Ct\mathbf{k}. \qquad (10.15)$$

Equating corresponding coefficients of $\mathbf{i}$, $\mathbf{j}$, and $\mathbf{k}$, we obtain the equations

$$x - x_1 = At, \qquad y - y_1 = Bt, \qquad z - z_1 = Ct,$$

or, transposing,

$$x = x_1 + At, \qquad y = y_1 + Bt, \qquad z = z_1 + Ct. \qquad (10.16)$$

When t is given any real value, Eqs. (10.16) determine the coordinates (x, y, z) of a point on the line L. Also there is a value of t corresponding to any point of the line. Equations (10.16) are called **parametric equations** of the line.

By solving each of the parametric equations for t and equating the equal values, we get

$$\frac{x - x_1}{A} = \frac{y - y_1}{B} = \frac{z - z_1}{C}, \qquad \text{if } ABC \neq 0. \qquad (10.17)$$

These are called the **symmetric equations** of the line.

The planes that contain a line and are perpendicular to the coordinate planes are called **projecting planes**.

Equations (10.17) represent three projecting planes. This becomes evident when we write the equations as

$$\frac{x - x_1}{A} = \frac{y - y_1}{B}, \qquad \frac{x - x_1}{A} = \frac{z - z_1}{C}, \qquad \frac{y - y_1}{B} = \frac{z - z_1}{C}.$$

These equations, each in two variables, represent planes perpendicular, respectively, to the xy, xz, and yz planes. These equations represent a line, and hence the line is the intersection of the planes. Any two of the equations, of course, determine the line. We note also that any one of the equations can be obtained from the other two.

A line in space may be defined by two planes that pass through the line. Hence there are infinitely many ways of defining a line, since infinitely many planes pass through a line. However, it is usually convenient to deal with the projecting planes.

If a line is parallel to a coordinate plane, one of the quantities A, B, and C in Eqs. (10.17) is zero, and conversely. In this instance, one member of the equation would have zero in the denominator and could not be used. If, for example, $A = 0$ and B and C are not zero, then the line passing through $P_1(x_1, y_1, z_1)$ is parallel to the vector $\mathbf{V} = B\mathbf{j} + C\mathbf{k}$. Hence the line is parallel to the yz plane and, consequently, the plane $x = x_1$ contains the line. If two of A, B, and C are zero, say $A = B = 0$, then the line is parallel to the z axis. Thus the line is the intersection of the planes $x = x_1$ and $y = y_1$. So we see that when a denominator of a member of equations (10.17) is zero, the corresponding numerator equated to zero represents a plane through the line in question.

EXAMPLE 1 ● Find symmetric equations that represent a line passing through the point $(2, -1, 3)$ and parallel to the vector $\mathbf{V} = 2\mathbf{i} - 5\mathbf{j} + 6\mathbf{k}$. Also, find a set of parametric equations that represent the line.

SOLUTION. The coefficients of $\mathbf{i}$, $\mathbf{j}$, and $\mathbf{k}$ of the given vector become the denominators in formula (10.17) and x_1, y_1, and z_1 are to be replaced, respectively, by 2, -1, and 3. Thus we have

$$\frac{x - 2}{2} = \frac{y + 1}{5} = \frac{z - 3}{6.}$$

The parametric equations, obtained by setting each member of these equations equal to a scalar t and solving for x, y, and z, are

$$x = 2 + 2t, \qquad y = -1 - 5t, \qquad z = 3 + 6t.$$

The scalar t is a parameter that may be assigned any real value. Here we passed from the symmetric equations to the parametric equations. Conversely, of course, we can find the symmetric equations from the parametric equations. ●

EXAMPLE 2 ● A line passes through the points $P_1(2, -4, 5)$ and $P_2(-1, 3, 1)$. Find symmetric equations of the line.

SOLUTION. The vector from P_1 to P_2 is parallel to the line. Thus, we have

$$\overrightarrow{P_1P_2} = -3\mathbf{i} + 7\mathbf{j} - 4\mathbf{k},$$

and, consequently, the equations

$$\frac{x - 2}{-3} = \frac{y + 4}{7} = \frac{z - 5}{-4}. ●$$

EXAMPLE 3 ● Write the equations of the line passing through the points $P_1(2, 6, 4)$ and $P_2(3, -2, 4)$.

SOLUTION. The vector from P_1 to P_2 is

$$\overrightarrow{P_1P_2} = \mathbf{i} - 8\mathbf{j}.$$

Hence the required line is parallel to the xy plane. The plane $z = 4$ contains the line. This plane is perpendicular to two of the coordinate planes. We use the first two members of Eqs. (10.17) to get another plane containing the line. Thus we have defining equations

$$z = 4, \qquad \frac{x - 3}{1} = \frac{y + 2}{-8},$$

or

$$z = 4, \qquad 8x + y - 22 = 0.$$

Note that we could not use the third member of the symmetric equations because its denominator would be zero. We did, however, set the numerator of that member equal to zero to obtain one of the planes. ●

EXAMPLE 4 ● Find equations, in symmetric form, of the line of intersection of the planes

$$x + y - z - 7 = 0 \quad \text{and} \quad x + 5y + 5z + 5 = 0.$$

SOLUTION. We multiply the first equation by 5 and add to the second equation to eliminate z. We subtract the first equation from the second to eliminate x. This gives the equations

$$6x + 10y - 30 = 0 \quad \text{and} \quad 4y + 6z + 12 = 0.$$

By solving each of these for y, we find

$$y = \frac{-3x + 15}{5} \quad \text{and} \quad y = \frac{-3z - 6}{2}.$$

Therefore,

$$\frac{-3x + 15}{5} = y = \frac{-3z - 6}{2}.$$

These symmetric equations can be converted to a simpler form if we divide each member by -3. This gives

$$\frac{x - 5}{5} = \frac{y}{-3} = \frac{z + 2}{2}.$$

The symmetric equations can also be written by first finding the coordinates of two points on the line defined by the given equations. When $y = 0$ the equations become $x - z - 7 = 0$ and $x + 5z + 5 = 0$. The solution of these equations is $x = 5$, $z = -2$. Hence the point $P_1(5, 0, -2)$ is on the line of intersection of the given planes. Similarly, we find that $P_2(0, 3, -4)$ is on the line. Then the vector

$$\overrightarrow{P_2 P_1} = 5\mathbf{i} - 3\mathbf{j} + 2\mathbf{k}$$

is parallel to the line whose equations we seek. Consequently, we obtain, as before, the symmetric equations

$$\frac{x - 5}{5} = \frac{y}{-3} = \frac{z + 2}{2}. \quad ●$$

Direction Angles, Direction Cosines, and Direction Numbers

We used the vector $\mathbf{V} = A\mathbf{i} + B\mathbf{j} + C\mathbf{k}$ in deriving the parametric equations of a line. The quantities A, B, and C, as we now point out, have a geometric significance with respect to the line.

DEFINITION 10.7 ● *The angles α, β, and γ that a directed line makes with the positive x, y, and z axes are called* **direction angles**. *The cosines of the direction angles are called* **direction cosines**.

The direction cosines of a line represented by equations of the form (10.16) or (10.17) may be found by the use of vectors. The vector

$$V = A\mathbf{i} + B\mathbf{j} + C\mathbf{k}$$

is parallel to the line. Thus we may choose the positive direction of the line to be V or $-V$. Choosing V as the positive direction, we take the scalar product of V and each of the unit vectors $\mathbf{i}$, $\mathbf{j}$, and $\mathbf{k}$. We note that the angle formed by two vectors that do not intersect is defined to be equal to the angle formed by two vectors that do intersect and are parallel to the given vectors and similarly directed.

Letting d stand for the length of V, we have by definition of the scalar product,

$$(A\mathbf{i} + B\mathbf{j} + C\mathbf{k}) \cdot \mathbf{i} = |A\mathbf{i} + B\mathbf{j} + C\mathbf{k}| |\mathbf{i}| \cos\alpha,$$
$$A = d\cos\alpha.$$

Similarly, $B = d\cos\beta$ and $C = d\cos\gamma$. Hence

$$\cos\alpha = \frac{A}{d}, \qquad \cos\beta = \frac{B}{d}, \qquad \cos\gamma = \frac{C}{d}.$$

The quantities A, B, and C are called **direction numbers**. Also, the products of the direction cosines by any positive number are called direction numbers.

THEOREM 10.10 ● *The sum of the squares of the direction cosines of a line is equal to 1. That is,*

$$\cos^2\alpha + \cos^2\beta + \cos^2\gamma = 1.$$

We leave the proof of this theorem to the student.

EXAMPLE 5 ● The numbers 4, 1, and 8 are direction numbers of a line. Find the direction cosines of the line.

SOLUTION. To get the direction cosines, we divide each direction number by $\sqrt{4^2 + 1^2 + 8^2} = 9$. This gives

$$\cos\alpha = \tfrac{4}{9}, \qquad \cos\beta = \tfrac{1}{9}, \qquad \cos\gamma = \tfrac{8}{9}. \quad ●$$

EXAMPLE 6 ● A line makes an angle of 60° with the positive x axis and 45° with the positive y axis. What angle does the line make with the positive z axis?

SOLUTION. Using Theorem 10.10, we have

$$\cos^2 60° + \cos^2 45° + \cos^2\gamma = 1,$$
$$\tfrac{1}{4} + \tfrac{1}{2} + \cos^2\gamma = 1.$$

Hence $\cos^2\gamma = \tfrac{1}{4}$ and, therefore, $\cos\gamma = \tfrac{1}{2}$ or $\cos\gamma = -\tfrac{1}{2}$. So $\gamma = 60°$ or 120°. ●

The direction angles for a given line depend on the chosen positive direction of the line. We let α_1, β_1, and γ_1 denote the angles for one direction and α_2, β_2, and γ_2 for the opposite direction. Then

$$\alpha_2 = 180° - \alpha_1, \qquad \beta_2 = 180° - \beta_1, \qquad \gamma_2 = 180° - \gamma_1.$$

The equations yield

$$\cos \alpha_2 = -\cos \alpha_1, \qquad \cos \beta_2 = -\cos \beta_1, \qquad \cos \gamma_2 = -\cos \gamma_1.$$

So there are two possible sets of direction cosines for a line, one set being the negative of the other set.

EXAMPLE 7 ● A line passes through $P_1(-4, 9, 5)$ and $P_2(2, 12, 3)$. Find the direction cosines if the line is directed from P_1 to P_2.

SOLUTION. The vector from P_1 to P_2 is

$$\overrightarrow{P_1 P_2} = 6\mathbf{i} + 3\mathbf{j} - 2\mathbf{k}.$$

Hence 6, 3, and -2 is a set of direction numbers. Since $\sqrt{6^2 + 3^2 + (-2)^2} = 7$, we have

$$\cos \alpha = \tfrac{6}{7}, \qquad \cos \beta = \tfrac{3}{7}, \qquad \cos \gamma = -\tfrac{2}{7}. ●$$

EXAMPLE 8 ● Assign a positive direction for the line represented by the equations

$$\frac{x - 1}{4} = \frac{y + 3}{-3} = \frac{z - 5}{-2}$$

and find the direction cosines.

SOLUTION. We can select 4, -3, -2 as a set of direction numbers or the set -4, 3, 2. Selecting the second set and noting that $\sqrt{(-4)^2 + 3^2 + 2^2} = \sqrt{29}$, we have

$$\cos \alpha = \frac{-4}{\sqrt{29}}, \qquad \cos \beta = \frac{3}{\sqrt{29}}, \qquad \cos \gamma = \frac{2}{\sqrt{29}}. ●$$

Exercises

In Exercises 1 through 8, find symmetric equations and parametric equations of the line that passes through P and is parallel to the given vector.

1. $P(4, -3, 5)$; $-2\mathbf{i} + 3\mathbf{j} + 4\mathbf{k}$ 2. $P(0, 1, -2)$; $\mathbf{i} - \mathbf{j} + 2\mathbf{k}$

3. $P(1, 1, 2)$; $2\mathbf{i} + 3\mathbf{j} - \mathbf{k}$ 4. $P(-2, -2, 3)$; $5\mathbf{i} + 4\mathbf{j} + \mathbf{k}$

5. $P(2, -1, 1)$; $2\mathbf{i} + \mathbf{j}$ 6. $P(3, 3, 3)$; $\mathbf{i} + \mathbf{j}$

7. $P(0, 0, 0)$; $\mathbf{j}$ 8. $P(0, 0, 0)$; $\mathbf{k}$

Write symmetric equations for the line through P_1 and P_2 in Exercises 9 through 16.

9. $P_1(1, 2, 3)$, $P_2(-2, 4, 0)$

10. $P_1(0, 0, 0)$, $P_2(3, 4, 5)$

11. $P_1(1, 0, 2)$, $P_2(0, 2, 1)$

12. $P_1(2, 4, 0)$, $P_2(1, 2, 8)$

13. $P_1(2, 5, 4)$, $P_2(2, 4, 3)$

14. $P_1(0, 4, 3)$, $P_2(0, 4, 4)$

15. $P_1(-1, 3, 4)$, $P_2(3, 3, 4)$

16. $P_1(0, 0, 2)$, $P_2(0, 0, 4)$

Find a symmetric form for each pair of equations in Exercises 17 through 20.

17. $x - y - 2z + 1 = 0$,
 $x - 3y - 3z + 7 = 0$

18. $x + y - 2z + 8 = 0$,
 $2x - y - 2z + 4 = 0$

19. $x + y + z - 9 = 0$,
 $2x + y - z + 3 = 0$

20. $x + y - z + 8 = 0$,
 $2x - y + 2z + 6 = 0$

21. A line makes an angle of $45°$ with the positive x axis and $60°$ with the positive z axis. What angle does it make with the positive y axis?

22. A line makes an angle of $135°$ with the positive x axis and $60°$ with the positive y axis. What angle does it make with the positive z axis?

A line passes through P_1 and P_2 in Exercises 23 through 28. Find the direction cosines if the line is directed from P_1 to P_2.

23. $P_1(-3, 4, -2)$, $P_2(1, 5, 6)$

24. $P_1(1, -1, 4)$, $P_2(3, 1, 5)$

25. $P_1(-5, 1, 8)$, $P_2(5, 3, -3)$

26. $P_1(3, 1, -6)$, $P_2(0, 5, 6)$

27. $P_1(3, 2, 1)$, $P_2(4, 5, 6)$

28. $P_1(2, -3, 4)$, $P_2(5, -6, 7)$

29. The numbers 4, 1, 8, and 2, 2, 1 are direction numbers of two lines. Find the cosine of the acute angle θ between the lines.

30. The numbers 3, 4, 12, and 6, 3, 2 are direction numbers of two lines. Find the cosine of the acute angle θ between the lines.

By setting x, y, and z in turn equal to zero, find the coordinates of the points where the line cuts the coordinate planes in Exercises 31 through 34.

31. $\dfrac{x - 6}{2} = \dfrac{y + 2}{1} = \dfrac{z + 3}{3}$

32. $\dfrac{x}{-2} = \dfrac{y - 2}{1} = \dfrac{z - 3}{1}$

33. $\dfrac{x - 3}{3} = \dfrac{y}{-1} = \dfrac{z - 4}{2}$

34. $\dfrac{x - 2}{1} = \dfrac{y + 1}{2} = \dfrac{z - 4}{3}$

10.6

THE VECTOR PRODUCT

We now introduce another kind of product of two vectors. Let **A** and **B** be nonparallel, nonzero vectors forming an angle θ, where $0 < \theta < 180°$ (Fig. 10.23). The **vector product**, or **cross product**, denoted by **A** × **B**, is defined by the following statements:

1. **A** × **B** is a vector perpendicular to the plane determined by **A** and **B**.
2. **A** × **B** points in the direction in which a right-threaded screw would advance when its head is turned from **A** (the first vector) to **B** through the angle θ.
3. $|\mathbf{A} \times \mathbf{B}| = |\mathbf{A}||\mathbf{B}|\sin\theta$.

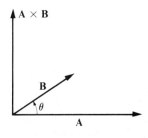

FIGURE 10.23

If **A** and **B** are parallel ($\theta = 0°$ or $\theta = 180°$), the vector product is defined by statement 3. This makes the product equal to zero since $\sin\theta = 0$. The vector product is zero if either **A** or **B**, or both, is equal to zero.

From statement 2, we see that the interchange of the factors in vector multiplication reverses the direction of the product (Fig. 10.24).

FIGURE 10.24

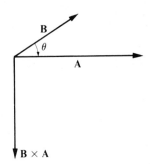

Hence

$$\mathbf{A} \times \mathbf{B} = -\mathbf{B} \times \mathbf{A},$$

and multiplication of this kind is not commutative.

The magnitude of the cross product has a simple geometric interpretation. The area of the parallelogram in Fig. 10.25 is $|\mathbf{A}|h$. But $h = |\mathbf{B}|\sin\theta$ and therefore

$$|\mathbf{A} \times \mathbf{B}| = |\mathbf{A}||\mathbf{B}|\sin\theta = |\mathbf{A}|h.$$

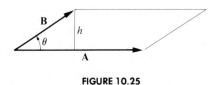

FIGURE 10.25

Hence the area of the parallelogram of which two adjacent sides are vectors is equal to the absolute value of the cross product of the vectors. One-half the absolute value of the cross product is, of course, equal to the area of the triangle determined by the vectors.

It can be shown that vector multiplication is distributive. (See Exercise 34, at the end of this section.) The distributive property is expressed by the equation

$$\mathbf{A} \times (\mathbf{B} + \mathbf{C}) = \mathbf{A} \times \mathbf{B} + \mathbf{A} \times \mathbf{C}.$$

Suppose we apply the definition of vector multiplication to the unit vectors **i**, **j**, and **k** (Fig. 10.26).

FIGURE 10.26

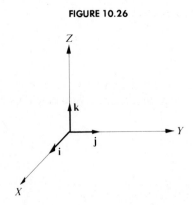

Clearly it follows that

$$\mathbf{i} \times \mathbf{j} = \mathbf{k} \text{ and } \mathbf{j} \times \mathbf{i} = -\mathbf{k},$$
$$\mathbf{j} \times \mathbf{k} = \mathbf{i} \text{ and } \mathbf{k} \times \mathbf{j} = -\mathbf{i},$$
$$\mathbf{k} \times \mathbf{i} = \mathbf{j} \text{ and } \mathbf{i} \times \mathbf{k} = -\mathbf{j},$$
$$\mathbf{i} \times \mathbf{i} = \mathbf{j} \times \mathbf{j} = \mathbf{k} \times \mathbf{k} = 0.$$

These equations and the distributive property of vector multiplication permit us to derive a convenient formula for the vector product when the vectors are expressed in terms of $\mathbf{i}$, $\mathbf{j}$, and $\mathbf{k}$. Thus, if

$$\mathbf{A} = a_1\mathbf{i} + b_1\mathbf{j} + c_1\mathbf{k} \qquad \text{and} \qquad \mathbf{B} = a_2\mathbf{i} + b_2\mathbf{j} + c_2\mathbf{k},$$

then

$$\begin{aligned}
\mathbf{A} \times \mathbf{B} &= (a_1\mathbf{i} + b_1\mathbf{j} + c_1\mathbf{k}) \times (a_2\mathbf{i} + b_2\mathbf{j} + c_2\mathbf{k}) \\
&= a_1a_2\mathbf{i} \times \mathbf{i} + a_1b_2\mathbf{i} \times \mathbf{j} + a_1c_2\mathbf{i} \times \mathbf{k} \\
&\quad + a_2b_1\mathbf{j} \times \mathbf{i} + b_1b_2\mathbf{j} \times \mathbf{j} + b_1c_2\mathbf{j} \times \mathbf{k} \\
&\quad + a_2c_1\mathbf{k} \times \mathbf{i} + b_2c_1\mathbf{k} \times \mathbf{j} + c_1c_2\mathbf{k} \times \mathbf{k} \\
&= 0 + a_1b_2\mathbf{k} - a_1c_2\mathbf{j} - a_2b_1\mathbf{k} + 0 \\
&\quad + b_1c_2\mathbf{i} + a_2c_1\mathbf{j} - b_2c_1\mathbf{i} + 0.
\end{aligned}$$

Hence

$$\mathbf{A} \times \mathbf{B} = (b_1c_2 - b_2c_1)\mathbf{i} + (a_2c_1 - a_1c_2)\mathbf{j} + (a_1b_2 - a_2b_1)\mathbf{k}.$$

This equation provides a formula for the cross product; however, a more convenient form* results when the right member of the equation is expressed as a determinant. Thus, alternatively, we discover that

$$\mathbf{A} \times \mathbf{B} = \begin{vmatrix} \mathbf{i} & \mathbf{j} & \mathbf{k} \\ a_1 & b_1 & c_1 \\ a_2 & b_2 & c_2 \end{vmatrix}.$$

EXAMPLE 1 ● The vectors $\mathbf{A}$ and $\mathbf{B}$ make an angle of $30°$. If the length of $\mathbf{A}$ is 5 and the length of $\mathbf{B}$ is 8, find the area of the parallelogram of which $\mathbf{A}$ and $\mathbf{B}$ are consecutive sides.

---

*See Appendix A for the definition of determinant.

SOLUTION. The area of the parallelogram is equal to the absolute value of $\mathbf{A} \times \mathbf{B}$. From the definition of a cross product,

$$|\mathbf{A} \times \mathbf{B}| = |\mathbf{A}||\mathbf{B}|\sin \theta = (5)(8)\sin 30° = 20.$$

The area of the parallelogram is 20 square units. ●

EXAMPLE 2 ● The points $A(1,0,-1)$, $B(3,-1,-5)$, and $C(4,2,0)$ are vertices of a triangle. Find the area of the triangle.

SOLUTION. The vectors $\overrightarrow{AB}$ and $\overrightarrow{AC}$ form two sides of the triangle. The magnitude of the cross product of the vectors is equal to the area of the parallelogram of which the vectors are adjacent sides. The area of the triangle, however, is one-half the area of the parallelogram. Thus we find

$$\overrightarrow{AB} = 2\mathbf{i} - \mathbf{j} - 4\mathbf{k}, \qquad \overrightarrow{AC} = 3\mathbf{i} + 2\mathbf{j} + \mathbf{k},$$

and

$$\overrightarrow{AB} \times \overrightarrow{AC} = \begin{vmatrix} \mathbf{i} & \mathbf{j} & \mathbf{k} \\ 2 & -1 & -4 \\ 3 & 2 & 1 \end{vmatrix} = 7\mathbf{i} - 14\mathbf{j} + 7\mathbf{k}.$$

The magnitude of this vector is $\sqrt{49 + 196 + 49} = 7\sqrt{6}$. Therefore the area of the triangle is $\frac{7}{2}\sqrt{6}$. ●

EXAMPLE 3 ● The points $A(2,-1,3)$, $B(4,2,5)$, and $C(-1,-1,6)$ determine a plane. Find the distance from the plane to the point $D(5,4,8)$.

SOLUTION. The vectors $\overrightarrow{AB}$ and $\overrightarrow{AC}$ determine a plane as indicated in Fig. 10.27. The cross product $\overrightarrow{AB} \times \overrightarrow{AC}$ is perpendicular to the plane, and $\overrightarrow{AD}$ is the vector from A to the given point D. The scalar product of $\overrightarrow{AD}$ and a unit vector

FIGURE 10.27

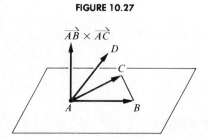

perpendicular to the plane yields the required distance d. Accordingly, we write

$$\vec{AD} = 3\mathbf{i} + 5\mathbf{j} + 5\mathbf{k}, \qquad \vec{AB} = 2\mathbf{i} + 3\mathbf{j} + 2\mathbf{k}, \qquad \vec{AC} = -3\mathbf{i} + 3\mathbf{k},$$

$$\vec{AB} \times \vec{AC} = \begin{vmatrix} \mathbf{i} & \mathbf{j} & \mathbf{k} \\ 2 & 3 & 2 \\ -3 & 0 & 3 \end{vmatrix} = 9\mathbf{i} - 12\mathbf{j} + 9\mathbf{k}.$$

The vector $9\mathbf{i} - 12\mathbf{j} + 9\mathbf{k}$ is perpendicular to $\vec{AB}$ and $\vec{AC}$ and therefore is perpendicular to the plane of A, B, and C. We divide this vector by its length, $3\sqrt{34}$, to obtain a unit vector. Finally, the distance d from the plane to D is given by

$$d = (3\mathbf{i} + 5\mathbf{j} + 5\mathbf{k}) \cdot \frac{3\mathbf{i} - 4\mathbf{j} + 3\mathbf{k}}{\sqrt{34}} = \frac{2\sqrt{34}}{17}. \quad \bullet$$

EXAMPLE 4 ● Find the equations, in symmetric form, of the line of intersection of the planes $2x - y + 4z = 3$ and $3x + y + z = 7$.

SOLUTION. We may write the desired equations immediately if we know the coordinates of any point of the line and any vector parallel to the line. By setting $z = 0$ in the equations of the planes and solving for x and y, we find that $(2, 1, 0)$ is a point of the line. Normals to the planes are given by the vectors

$$\mathbf{N}_1 = 2\mathbf{i} - \mathbf{j} + 4\mathbf{k} \qquad \text{and} \qquad \mathbf{N}_2 = 3\mathbf{i} + \mathbf{j} + \mathbf{k}.$$

The cross product of the vectors is parallel to the line of intersection of the planes. We find

$$\mathbf{N}_1 \times \mathbf{N}_2 = \begin{vmatrix} \mathbf{i} & \mathbf{j} & \mathbf{k} \\ 2 & -1 & 4 \\ 3 & 1 & 1 \end{vmatrix} = -5\mathbf{i} + 10\mathbf{j} + 5\mathbf{k}.$$

We divide this vector by 5 and write the equation of the line as

$$\frac{x - 2}{-1} = \frac{y - 1}{2} = \frac{z}{1}.$$

We note that this is the same kind of problem as that of Example 4, Section 10.5.

●

The Distance from a Line To a Point

Let L (Fig. 10.28) be a line passing through the point $P_0(x_0, y_0, z_0)$ and parallel to a unit vector $\mathbf{u}$. Then let $P_1(x_1, y_1, z_1)$ be any point not on the line and let $\mathbf{V}$ be the vector $\vec{P_0 P_1}$. To find the perpendicular distance from line L to P_1, we take the vector product of $\mathbf{V}$ and $\mathbf{u}$. Thus

$$d = |\mathbf{V}| \sin \theta = |\mathbf{u}| |\mathbf{V}| \sin \theta = |\mathbf{u} \times \mathbf{V}|.$$

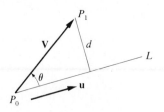

FIGURE 10.28

EXAMPLE 5 ● Find the distance from the line passing through $P_2(3, 0, 6)$ and $P_3(5, -2, 7)$ to the point $P_1(8, 1, -3)$.

SOLUTION. The vector $\overrightarrow{P_3 P_2} = -2\mathbf{i} + 2\mathbf{j} - \mathbf{k}$, and $\mathbf{u} = \frac{1}{3}(-2\mathbf{i} + 2\mathbf{j} - \mathbf{k})$. The vector $\overrightarrow{P_2 P_1} = \mathbf{V} = 5\mathbf{i} + \mathbf{j} - 9\mathbf{k}$. Hence

$$\mathbf{u} \times \mathbf{V} = \frac{1}{3} \begin{vmatrix} \mathbf{i} & \mathbf{j} & \mathbf{k} \\ -2 & 2 & -1 \\ 5 & 1 & -9 \end{vmatrix} = \frac{1}{3}(-17\mathbf{i} - 23\mathbf{j} - 14\mathbf{k})$$

and

$$d = |\mathbf{u} \times \mathbf{V}| = \frac{13}{3}\sqrt{6}. \quad ●$$

In Section 2.4, we derived a formula for the distance from a line to a point when both are in the xy plane. Now we give a much shorter derivation by employing vectors.

The equation $Ax + By + C = 0$, in three-space, represents a plane parallel to the z axis (Theorem 9.2); the vector $A\mathbf{i} + B\mathbf{j}$ is perpendicular to the plane (Section 10.4). The intersection of the plane and the xy plane is a line, and the x and y coordinates of all points of the intersection satisfy the equation $Ax + By + C = 0$. Restricting ourselves to the xy plane, we have a line and a vector perpendicular to the line. Now let $P_1(x_1, y_1)$ be any point in the xy plane (Fig. 10.29). If $B \neq 0$, the line cuts the y axis at $P(0, -C/B)$. The scalar product of the vector $\overrightarrow{PP_1}$ and a unit vector $\mathbf{u}$, perpendicular to the line, gives the distance d.

FIGURE 10.29

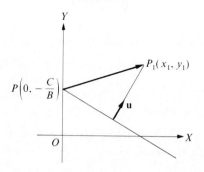

Since

$$\overrightarrow{PP_1} = x\mathbf{i} + \left(y_1 + \frac{C}{B}\right)\mathbf{j} \quad \text{and} \quad \mathbf{u} = \frac{A\mathbf{i} + B\mathbf{j}}{\pm\sqrt{A^2 + B^2}},$$

we have

$$d = \left[x_1\mathbf{i} + \left(y_1 + \frac{C}{B}\right)\mathbf{j}\right] \cdot \frac{A\mathbf{i} + B\mathbf{j}}{\pm\sqrt{A^2 + B^2}} = \frac{Ax_1 + By_1 + C}{\pm\sqrt{A^2 + B^2}}.$$

Thus a scalar product immediately yields the desired formula. The ambiguity as to sign is discussed in Section 2.4.

Exercises

In Exercises 1 through 6, find the cross product, $\mathbf{A} \times \mathbf{B}$, of the vectors and a unit vector perpendicular to the two given vectors.

1. $\mathbf{A} = 3\mathbf{i} - 4\mathbf{j} - 2\mathbf{k}$,
   $\mathbf{B} = \mathbf{i} - 2\mathbf{j} - 2\mathbf{k}$

2. $\mathbf{A} = 6\mathbf{i} - 3\mathbf{j} + 14\mathbf{k}$,
   $\mathbf{B} = 3\mathbf{i} - 2\mathbf{j} + 3\mathbf{k}$

3. $\mathbf{A} = \mathbf{i} + \mathbf{j} + \mathbf{k}$,
   $\mathbf{B} = \mathbf{i} - \mathbf{j} - \mathbf{k}$

4. $\mathbf{A} = 2\mathbf{i} - \mathbf{k}$,
   $\mathbf{B} = \mathbf{j} + 2\mathbf{k}$

5. $\mathbf{A} = 4\mathbf{i} - 3\mathbf{j}$,
   $\mathbf{B} = 3\mathbf{i} + 4\mathbf{j}$

6. $\mathbf{A} = \mathbf{i} - 2\mathbf{j} + 3\mathbf{k}$,
   $\mathbf{B} = 4\mathbf{i} + 5\mathbf{j} - 6\mathbf{k}$

Compute the area of the parallelogram described in Exercises 7 through 10.

7. $\mathbf{A} = 3\mathbf{i} + 2\mathbf{j}$ and $\mathbf{B} = \mathbf{i} - \mathbf{j}$ are adjacent sides.

8. $\mathbf{A} = 4\mathbf{i} - \mathbf{j} + \mathbf{k}$ and $\mathbf{B} = 3\mathbf{i} + \mathbf{j} + \mathbf{k}$ are adjacent sides.

9. The points $A(4, 1, 0)$, $B(1, 2, 1)$, $C(0, 0, 6)$, and $D(3, -1, 5)$ are vertices of the parallelogram $ABCD$.

10. The points $A(1, -1, 1)$, $B(3, 3, 1)$, $C(4, -1, 4)$, and $D(2, -5, 4)$ are vertices of the parallelogram $ABCD$.

Find the area of the triangle described in each of Exercises 11 through 14.

11. The vertices are $A(-1, 3, 4)$, $B(1, 2, 5)$, and $C(2, -3, 1)$.

12. The vertices are $A(1, 0, 4)$, $B(3, -3, 0)$, and $C(0, 1, 2)$.

13. The vectors $\mathbf{A} = 2\mathbf{i} - 3\mathbf{j}$ and $\mathbf{B} = \mathbf{i} - \mathbf{j}$, drawn from the origin, are two sides.

14. The tips of the vectors $\mathbf{A} = 4\mathbf{i} + \mathbf{j}$, $\mathbf{B} = \mathbf{i} + 2\mathbf{j} + \mathbf{k}$, and $\mathbf{C} = 3\mathbf{i} - \mathbf{j} + 5\mathbf{k}$, drawn from the origin, are the vertices.

15. Find the area of the triangle in the xy plane whose vertices are the points (x_1, y_1), (x_2, y_2) and (x_3, y_3). Use a vector product.

16. Write the equation of the plane that passes through the points $(2, -4, 3)$, $(-3, 5, 1)$, and $(4, 0, 6)$.

17. Write the equation of the plane that contains the points $(2, 1, 0)$, $(3, 0, 2)$, and $(0, 4, 3)$.

18. Find the distance from the point $(6, 7, 8)$ to the plane of the points $(-1, 3, 0)$, $(2, 2, 1)$, $(1, 1, 3)$.

19. Find the distance from the point $(-4, -5, -3)$ to the plane passing through the points $(4, 0, 0)$, $(0, 6, 0)$, and $(0, 0, 7)$.

20. Find the equations of the line passing through the point $(3, 1, -2)$ and parallel to each of the planes $x - y + z = 4$ and $3x + y - z = 5$.

21. Find the equations of the line that passes through the origin and is parallel to the line of intersection of the planes $2x - y - z = 2$ and $4x + 2y - 4z = 1$.

22. Find a parametric representation of the line of intersection of the two planes of Exercise 20 and also of Exercise 21.

23. A plane passes through the point $(1, 1, 1)$ and is perpendicular to each of planes $2x + 2y + z = 3$ and $3x - y - 2z = 5$. Find its equation.

24. A plane passes through the point $(0, 0, 0)$ and is perpendicular to the line of intersection of the planes $5x - 4y + 3z = 2$ and $x + 2y - 3z = 4$. Find the equation of the plane.

25. A line passes through the points $(1, 3, 1)$ and $(3, 4, -1)$. Find the distance from the line to the point $(4, 4, 4)$.

26. A line passes through $(3, 2, 1)$ and is parallel to the vector $2\mathbf{i} + \mathbf{j} - 2\mathbf{k}$. Find the distance from the line to $(-3, -1, 3)$.

27. Find the distance from the line $x/2 = y/3 = z/1$ to the point $(3, 4, 1)$.

28. Find the distance from the line $(x - 2)/2 = y/2 = (z - 1)/1$ to the point $(0, 0, 0)$.

29. Let L_1 and L_2 (Fig. 10.30) be lines that are not parallel and do not intersect. Let $\overrightarrow{P_1 P_2}$ be a vector extending from any point on L_1 to any point on L_2, $\mathbf{V}_1$ a vector parallel to L_1, and $\mathbf{V}_2$ parallel to L_2. Then $\mathbf{V}_1 \times \mathbf{V}_2$ is perpendicular to both lines. Show that the perpendicular distance between L_1 and L_2 is

$$d = |\overrightarrow{P_1 P_2} \cdot \mathbf{u}|,$$

where $\mathbf{u}$ is a unit vector parallel to $\mathbf{V}_1 \times \mathbf{V}_2$.

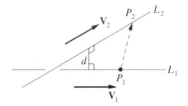

FIGURE 10.30

Find the perpendicular distance between the two lines in Exercises 30 through 33.

30. L_1 passes through $P_1(2, 3, 1)$ in the direction of $V_1 = i + 2j - 3k$, and L_2 passes through $P_2(4, 2, 0)$ in the direction of $V_2 = 3i - j + k$.

31. L_1 passes through $(2, 1, -1)$ and $(-1, 3, 2)$. L_2 passes through $(4, 0, 5)$ and $(3, 4, 0)$.

32. The equations of L_1 are $(x + 2)/3 = (y + 3)/2 = z/2$. The equations of L_2 are $(x - 1)/2 = (y + 4)/3 = (z - 2)/4$.

33. L_1 passes through $(4, 1, 0)$ and $(0, 0, 6)$. L_2 passes through $(1, 2, 1)$ and $(3, -1, 5)$. Do the lines intersect?

34. In Fig. 10.31, the vector **E** is the projection of **B** on a plane perpendicular to **A**. Vector **E** is rotated through 90°, as indicated, and then multiplied by $|A|$ to yield $A \times B$. Tell why these operations lead to $A \times B$.

FIGURE 10.31

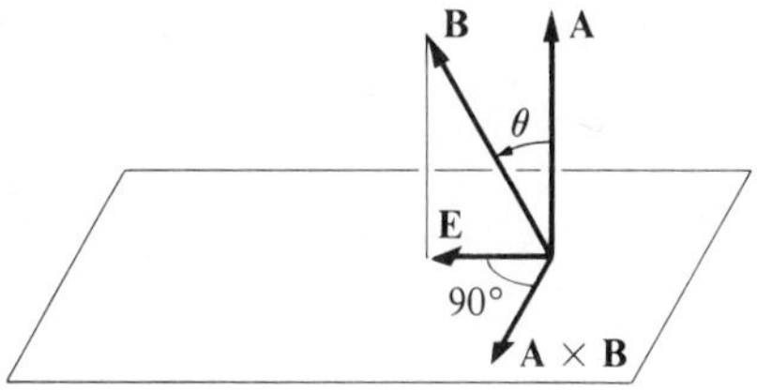

Next, let **B**, **C**, and **B + C** form a triangle. Project this triangle on the plane perpendicular to **A**, thus forming another vector triangle. Rotate the new triangle through 90° and multiply each side by $|A|$. Observe that the sides of the final triangle are $A \times B$, $A \times C$, and $A \times (B + C)$, and that

$$A \times (B + C) = A \times B + A \times C.$$

REVIEW EXERCISES

In each of Exercises 1 through 4, find the lengths of **A** and **B**, **A** · **B**, **A** × **B**, the scalar projection of **B** on **A**, the vector projection of **B** on **A**, the cosine of the angle between the vectors, and a unit vector in the direction of **A**.

1. **A** = 2**i** − 2**j** + **k**,
   B = **i** + **j** + **k**

2. **A** = **i** − 2**j** − **k**,
   B = 2**i** − 3**j** + 6**k**

3. **A** = **i** + 8**j** − 4**k**,
   B = **i** − **j** + 2**k**

4. $\overline{\mathbf{A}}$ = 2**i** + 9**j** − 6**k**,
   $\overline{\mathbf{B}}$ = **i** − **j** − 2**k**

5. Find the equation of the plane perpendicular to the vector 3**i** − 2**j** − 5**k** and passing through the point $(3, -4, -2)$.

6. Find the equation of the plane parallel to the plane $3x - 4y + 7z = 3$ and passing through the point $(1, -1, 3)$.

7. Find the equation of the plane passing through the three points $(1, 2, -1)$, $(-2, 1, 1)$, and $(2, 4, 2)$.

8. Find the distance from the plane $x - 2y + 2z = 5$ to the point $(-1, 3, 2)$.

9. Find symmetric and parametric equations for the line parallel to the vector 3**i** + 2**j** − 4**k** and passing through the point $(1, -2, 3)$.

10. Find symmetric equations for the line passing through the two points $(2, -3, 4)$ and $(-3, 5, 7)$.

11. Find symmetric equations for the line of intersection of the two planes

$$2x + y - z + 3 = 0 \quad \text{and} \quad x - y + 3z + 5 = 0.$$

12. Find the direction cosines of the directed line from $A(-6, 2, 1)$ to $B(3, 5, 4)$.

13. Find the distance from the line passing through $A(1, -2, 3)$ and $B(3, -5, -3)$ to the point $C(1, -3, 0)$.

14. Find the coordinates of the points where the line

$$\frac{x}{3} = \frac{y - 5}{1} = \frac{z - 6}{-4}$$

cuts each of the coordinate planes.

15. If **A** = 2**i** + **j** + **k**, **B** = 2**i** + **j**, **C** = **j** + **k**, find a value for **A** · (**B** × **C**).

Appendix A

ALGEBRA

1. Quadratic Formula

The roots of the equation $ax^2 + bx + c = 0$ $(a \neq 0)$ are $x = \dfrac{-b \pm \sqrt{b^2 - 4ac}}{2a}$.

2. Determinants

The value of a two-by-two determinant is $\begin{vmatrix} a_1 & b_1 \\ a_2 & b_2 \end{vmatrix} = a_1 b_2 - a_2 b_1$.

The value of a three-by-three determinant is

$$\begin{vmatrix} a_1 & b_1 & c_1 \\ a_2 & b_2 & c_2 \\ a_3 & b_3 & c_3 \end{vmatrix} = a_1 \begin{vmatrix} b_2 & c_2 \\ b_3 & c_3 \end{vmatrix} - b_1 \begin{vmatrix} a_2 & c_2 \\ a_3 & c_3 \end{vmatrix} + c_1 \begin{vmatrix} a_2 & b_2 \\ a_3 & b_3 \end{vmatrix}$$

$$= a_1(b_2 c_3 - b_3 c_2) - b_1(a_2 c_3 - a_3 c_2) + c_1(a_2 b_3 - a_3 b_2).$$

TRIGONOMETRY

3. Relation between Degrees and Radians

The entire circumference of a circle subtends a central angle of 2π radians and of 360 degrees.

$$2\pi \text{ radians} = 1 \text{ revolution} = 360°$$

So

$$1 \text{ radian} = \frac{180}{\pi} \text{ degrees,} \qquad \text{and} \qquad 1 \text{ degree} = \frac{\pi}{180} \text{ radians.}$$

4. Definitions and Fundamental Identities

$$\sin\theta = \frac{y}{r} = \frac{1}{\csc\theta}; \qquad \cos\theta = \frac{x}{r} = \frac{1}{\sec\theta}; \qquad \tan\theta = \frac{y}{x} = \frac{1}{\cot\theta} = \frac{\sin\theta}{\cos\theta};$$

$$\sin^2\theta + \cos^2\theta = 1; \qquad 1 + \tan^2\theta = \sec^2\theta; \qquad 1 + \cot^2\theta = \csc^2\theta;$$

$$\sin\theta = \cos(90° - \theta) = \sin(180° - \theta); \qquad \cos\theta = \sin(90° - \theta) = -\cos(180° - \theta);$$
$$\tan\theta = \cot(90° - \theta) = -\tan(180° - \theta); \qquad \cot\theta = \tan(90° - \theta) = -\cot(180° - \theta);$$

$$\csc\theta = \cot\frac{\theta}{2} - \cot\theta;$$

$$\sin(\theta + \phi) = \sin\theta\cos\phi + \cos\theta\sin\phi; \qquad \sin(\theta - \phi) = \sin\theta\cos\phi - \cos\theta\sin\phi;$$
$$\cos(\theta + \phi) = \cos\theta\cos\phi - \sin\theta\sin\phi; \qquad \cos(\theta - \phi) = \cos\theta\cos\phi + \sin\theta\sin\phi;$$

$$\tan(\theta + \phi) = \frac{\tan\theta + \tan\phi}{1 - \tan\theta\tan\phi}; \qquad \tan(\theta - \phi) = \frac{\tan\theta - \tan\phi}{1 + \tan\theta\tan\phi};$$

$$\sin 2\theta = 2\sin\theta\cos\theta; \qquad \cos 2\theta = \cos^2\theta - \sin^2\theta = 2\cos^2\theta - 1 = 1 - 2\sin^2\theta;$$

$$\tan 2\theta = \frac{2\tan\theta}{1 - \tan^2\theta}; \qquad \cot 2\theta = \frac{\cot^2\theta - 1}{2\cot\theta};$$

$$\sin\frac{\theta}{2} = \pm\sqrt{\frac{1 - \cos\theta}{2}}\ ; \qquad \cos\frac{\theta}{2} = \pm\sqrt{\frac{1 + \cos\theta}{2}}\ ;$$

$$\tan\frac{\theta}{2} = \pm\sqrt{\frac{1 - \cos\theta}{1 + \cos\theta}} = \frac{1 - \cos\theta}{\sin\theta} = \frac{\sin\theta}{1 + \cos\theta}\ ;$$

$$\sin\theta + \sin\phi = 2\sin\frac{\theta + \phi}{2}\cos\frac{\theta - \phi}{2}\ ; \qquad \sin\theta - \sin\phi = 2\cos\frac{\theta + \phi}{2}\sin\frac{\theta - \phi}{2}\ ;$$

$$\cos\theta + \cos\phi = 2\cos\frac{\theta + \phi}{2}\cos\frac{\theta - \phi}{2}\ ; \qquad \cos\theta - \cos\phi = -2\sin\frac{\theta + \phi}{2}\sin\frac{\theta - \phi}{2}\ ;$$

$$\sin\theta\cos\phi = \tfrac{1}{2}[\sin(\theta + \phi) + \sin(\theta - \phi)]; \qquad \cos\theta\cos\phi = \tfrac{1}{2}[\cos(\theta + \phi) + \cos(\theta - \phi)];$$

$$\sin\theta\sin\phi = -\tfrac{1}{2}[\cos(\theta + \phi) - \cos(\theta - \phi)].$$

5. Angles and Sides of Triangles

Law of cosines: $\quad a^2 = b^2 + c^2 - 2bc\cos A.$

Law of sines: $\quad \dfrac{\sin A}{a} = \dfrac{\sin B}{b} = \dfrac{\sin C}{c}.$

Area $= \tfrac{1}{2}bc\sin A = \tfrac{1}{2}ac\sin B = \tfrac{1}{2}ab\sin C.$

ANALYTIC GEOMETRY

6. Basic Formulas

Distance between two points: $\quad P_1P_2 = \sqrt{(x_2 - x_1)^2 + (y_2 - y_1)^2}.$

Midpoint: $\quad P\left(\dfrac{x_1 + x_2}{2}, \dfrac{y_1 + y_2}{2}\right).$

Slope of a line: $\quad m = \dfrac{y_2 - y_1}{x_2 - x_1} = \dfrac{y_1 - y_2}{x_1 - x_2}.$

Condition for parallel lines: $\quad m_1 = m_2.$

Condition for perpendicular lines: $\quad m_1 m_2 = -1.$

Angle between two lines: $\quad \tan\theta = \dfrac{m_2 - m_1}{1 + m_2 m_1}.$

Distance between a point and a line: $\quad PQ = \dfrac{|Ax_1 + By_1 + C|}{\sqrt{A^2 + B^2}}.$

7. Equation of Straight Line

General equation: $\quad Ax + By + C = 0.$

Two point form: $\quad y - y_1 = \dfrac{y_2 - y_1}{x_2 - x_1}(x - x_1).$

Point-slope form: $\quad y - y_1 = m(x - x_1).$

Slope-intercept form: $\quad y = mx + b.$

Intercept form: $\quad \dfrac{x}{a} + \dfrac{y}{b} = 1.$

8. Circle

General equation: $\quad x^2 + y^2 + Dx + Ey + F = 0.$

Center at origin, radius r: $\quad x^2 + y^2 = r^2$, or $x = r\cos\theta,\ y = r\sin\theta.$

Equation of line tangent at (x_1, y_1): $\quad x_1 x + y_1 y = r^2.$

Center at (h, k), radius r: $\quad (x - h)^2 + (y - k)^2 = r^2.$

9. Parabola

Vertex at origin, opening in direction of positive x:

$$y^2 = 4ax(a > 0), \quad \text{or} \quad x = at^2, \; y = 2at.$$

Equation of line tangent at (x_1, y_1): $y_1(y - y_1) = 2a(x - x_1)$.
Focus: $(a, 0)$.   Directrix: $x + a = 0$.
Vertex at (h, k), opening in direction of positive x: $(y - k)^2 = 4a(x - h)$.

10. Ellipse

Center at origin: $\dfrac{x^2}{a^2} + \dfrac{y^2}{b^2} = 1$, or $x = a\cos\theta$, $y = b\sin\theta$.

Equation of line tangent at (x_1, y_1): $\dfrac{x_1 x}{a^2} + \dfrac{y_1 y}{b^2} = 1$.

If $a > b$: $c^2 = a^2 - b^2$;   eccentricity $e = c/a$;   foci, $(-c, 0), (c, 0)$.
If $b > a$: $c^2 = b^2 - a^2$;   eccentricity $e = c/b$;   foci, $(0, -c), (0, c)$.

Center at (h, k): $\dfrac{(x - h)^2}{a^2} + \dfrac{(y - k)^2}{b^2} = 1$.

11. Hyperbola

Center at origin: $\dfrac{x^2}{a^2} - \dfrac{y^2}{b^2} = 1$, or $x = a\sec\theta$, $y = b\tan\theta$.

Equation of line tangent at (x_1, y_1): $\dfrac{x_1 x}{a^2} - \dfrac{y_1 y}{b^2} = 1$.

Asymptotes: $y = \dfrac{b}{a}x, \; y = -\dfrac{b}{a}x$.
Foci: $(-c, 0), (c, 0)$, where $c^2 = a^2 + b^2$.
Eccentricity: $e = \dfrac{c}{a}$, where $c^2 = a^2 + b^2$.

Center at (h, k): $\dfrac{(x - h)^2}{a^2} - \dfrac{(y - k)^2}{b^2} = \pm 1$.
With asymptotes $x = a$, $y = b$: $(x - a)(y - b) = k$.

PLANE GEOMETRY

12. Basic Figures and Their Areas

Parallelogram

Area = bh

Rhombus

Area = bh

Trapezoid
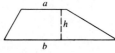
Area = $\frac{1}{2}(a + b)h$

Triangle

Area = $\frac{1}{2}bh$

Tables

TABLE I Powers and roots

No.	Sq.	Sq. Root	Cube	Cube Root	No.	Sq.	Sq. Root	Cube	Cube Root
1	1	1.000	1	1.000	51	2,601	7.141	132,651	3.708
2	4	1.414	8	1.260	52	2,704	7.211	140,608	3.733
3	9	1.732	27	1.442	53	2,809	7.280	148,877	3.756
4	16	2.000	64	1.587	54	2,916	7.348	157,464	3.780
5	25	2.236	125	1.710	55	3,025	7.416	166,375	3.803
6	36	2.449	216	1.817	56	3,136	7.483	175,616	3.826
7	49	2.646	343	1.913	57	3,249	7.550	185,193	3.849
8	64	2.828	512	2.000	58	3,364	7.616	195,112	3.871
9	81	3.000	729	2.080	59	3,481	7.681	205,379	3.893
10	100	3.162	1,000	2.154	60	3,600	7.746	216,000	3.915
11	121	3.317	1,331	2.224	61	3,721	7.810	226,981	3.936
12	144	3.464	1,728	2.289	62	3,844	7.874	238,328	3.958
13	169	3.606	2,197	2.351	63	3,969	7.937	250,047	3.979
14	196	3.742	2,744	2.410	64	4,096	8.000	262,144	4.000
15	225	3.873	3,375	2.466	65	4,225	8.062	274,625	4.021
16	256	4.000	4,096	2.520	66	4,356	8.124	287,496	4.041
17	289	4.123	4,913	2.571	67	4,489	8.185	300,763	4.062
18	324	4.243	5,832	2.621	68	4,624	8.246	314,432	4.082
19	361	4.359	6,859	2.668	69	4,761	8.307	328,509	4.102
20	400	4.472	8,000	2.714	70	4,900	8.367	343,000	4.121
21	441	4.583	9,261	2.759	71	5,041	8.426	357,911	4.141
22	484	4.690	10,648	2.802	72	5,184	8.485	373,248	4.160
23	529	4.796	12,167	2.844	73	5,329	8.544	389,017	4.179
24	576	4.899	13,824	2.884	74	5,476	8.602	405,224	4.198
25	625	5.000	15,625	2.924	75	5,625	8.660	421,875	4.217
26	676	5.099	17,576	2.962	76	5,776	8.718	438,976	4.236
27	729	5.196	19,683	3.000	77	5,929	8.775	456,533	4.254
28	784	5.292	21,952	3.037	78	6,084	8.832	474,552	4.273
29	841	5.385	24,389	3.072	79	6,241	8.888	493,039	4.291
30	900	5.477	27,000	3.107	80	6,400	8.944	512,000	4.309
31	961	5.568	29,791	3.141	81	6,561	9.000	531,441	4.327
32	1,024	5.657	32,768	3.175	82	6,724	9.055	551,368	4.344
33	1,089	5.745	35,937	3.208	83	6,889	9.110	571,787	4.362
34	1,156	5.831	39,304	3.240	84	7,056	9.165	592,704	4.380
35	1,225	5.916	42,875	3.271	85	7,225	9.220	614,125	4.397
36	1,296	6.000	46,656	3.302	86	7,396	9.274	636,056	4.414
37	1,369	6.083	50,653	3.332	87	7,569	9.327	658,503	4.431
38	1,444	6.164	54,872	3.362	88	7,744	9.381	681,472	4.448
39	1,521	6.245	59,319	3.391	89	7,921	9.434	704,969	4.465
40	1,600	6.325	64,000	3.420	90	8,100	9.487	729,000	4.481
41	1,681	6.403	68,921	3.448	91	8,281	9.539	753,571	4.498
42	1,764	6.481	74,088	3.476	92	8,464	9.592	778,688	4.514
43	1,849	6.557	79,507	3.503	93	8,649	9.644	804,357	4.531
44	1,936	6.633	85,184	3.530	94	8,836	9.695	830,584	4.547
45	2,025	6.708	91,125	3.557	95	9,025	9.747	857,375	4.563
46	2,116	6.782	97,336	3.583	96	9,216	9.798	884,736	4.579
47	2,209	6.856	103,823	3.609	97	9,409	9.849	912,673	4.595
48	2,304	6.928	110,592	3.634	98	9,604	9.899	941,192	4.610
49	2,401	7.000	117,649	3.659	99	9,801	9.950	970,299	4.626
50	2,500	7.071	125,000	3.684	100	10,000	10.000	1,000,000	4.642

TABLE II Natural trigonometric functions

Angle					Angle				
De-gree	Ra-dian	Sine	Co-sine	Tan-gent	De-gree	Ra-dian	Sine	Co-sine	Tan-gent
0°	0.000	0.000	1.000	0.000					
1°	0.017	0.017	1.000	0.017	46°	0.803	0.719	0.695	1.036
2°	0.035	0.035	0.999	0.035	47°	0.820	0.731	0.682	1.072
3°	0.052	0.052	0.999	0.052	48°	0.838	0.743	0.669	1.111
4°	0.070	0.070	0.998	0.070	49°	0.855	0.755	0.656	1.150
5°	0.087	0.087	0.996	0.087	50°	0.873	0.766	0.643	1.192
6°	0.105	0.105	0.995	0.105	51°	0.890	0.777	0.629	1.235
7°	0.122	0.122	0.993	0.123	52°	0.908	0.788	0.616	1.280
8°	0.140	0.139	0.990	0.141	53°	0.925	0.799	0.602	1.327
9°	0.157	0.156	0.988	0.158	54°	0.942	0.809	0.588	1.376
10°	0.175	0.174	0.985	0.176	55°	0.960	0.819	0.574	1.428
11°	0.192	0.191	0.982	0.194	56°	0.977	0.829	0.559	1.483
12°	0.209	0.208	0.978	0.213	57°	0.995	0.839	0.545	1.540
13°	0.227	0.225	0.974	0.231	58°	1.012	0.848	0.530	1.600
14°	0.244	0.242	0.970	0.249	59°	1.030	0.857	0.515	1.664
15°	0.262	0.259	0.966	0.268	60°	1.047	0.866	0.500	1.732
16°	0.279	0.276	0.961	0.287	61°	1.065	0.875	0.485	1.804
17°	0.297	0.292	0.956	0.306	62°	1.082	0.883	0.469	1.881
18°	0.314	0.309	0.951	0.325	63°	1.100	0.891	0.454	1.963
19°	0.332	0.326	0.946	0.344	64°	1.117	0.899	0.438	2.050
20°	0.349	0.342	0.940	0.364	65°	1.134	0.906	0.423	2.145
21°	0.367	0.358	0.934	0.384	66°	1.152	0.914	0.407	2.246
22°	0.384	0.375	0.927	0.404	67°	1.169	0.921	0.391	2.356
23°	0.401	0.391	0.921	0.424	68°	1.187	0.927	0.375	2.475
24°	0.419	0.407	0.914	0.445	69°	1.204	0.934	0.358	2.605
25°	0.436	0.423	0.906	0.466	70°	1.222	0.940	0.342	2.748
26°	0.454	0.438	0.899	0.488	71°	1.239	0.946	0.326	2.904
27°	0.471	0.454	0.891	0.510	72°	1.257	0.951	0.309	3.078
28°	0.489	0.469	0.883	0.532	73°	1.274	0.956	0.292	3.271
29°	0.506	0.485	0.875	0.554	74°	1.292	0.961	0.276	3.487
30°	0.524	0.500	0.866	0.577	75°	1.309	0.966	0.259	3.732
31°	0.541	0.515	0.857	0.601	76°	1.326	0.970	0.242	4.011
32°	0.559	0.530	0.848	0.625	77°	1.344	0.974	0.225	4.332
33°	0.576	0.545	0.839	0.649	78°	1.361	0.978	0.208	4.705
34°	0.593	0.559	0.829	0.675	79°	1.379	0.982	0.191	5.145
35°	0.611	0.574	0.819	0.700	80°	1.396	0.985	0.174	5.671
36°	0.628	0.588	0.809	0.727	81°	1.414	0.988	0.156	6.314
37°	0.646	0.602	0.799	0.754	82°	1.431	0.990	0.139	7.115
38°	0.663	0.616	0.788	0.781	83°	1.449	0.993	0.122	8.144
39°	0.681	0.629	0.777	0.810	84°	1.466	0.995	0.105	9.514
40°	0.698	0.643	0.766	0.839	85°	1.484	0.996	0.087	11.43
41°	0.716	0.656	0.755	0.869	86°	1.501	0.998	0.070	14.30
42°	0.733	0.669	0.743	0.900	87°	1.518	0.999	0.052	19.08
43°	0.750	0.682	0.731	0.933	88°	1.536	0.999	0.035	28.64
44°	0.768	0.695	0.719	0.966	89°	1.553	1.000	0.017	57.29
45°	0.785	0.707	0.707	1.000	90°	1.571	1.000	0.000	

TABLE III Exponential functions

x	e^x	e^{-x}	x	e^x	e^{-x}
0.00	1.0000	1.0000	2.5	12.182	0.0821
0.05	1.0513	0.9512	2.6	13.464	0.0743
0.10	1.1052	0.9048	2.7	14.880	0.0672
0.15	1.1618	0.8607	2.8	16.445	0.0608
0.20	1.2214	0.8187	2.9	18.174	0.0550
0.25	1.2840	0.7788	3.0	20.086	0.0498
0.30	1.3499	0.7408	3.1	22.198	0.0450
0.35	1.4191	0.7047	3.2	24.533	0.0408
0.40	1.4918	0.6703	3.3	27.113	0.0369
0.45	1.5683	0.6376	3.4	29.964	0.0334
0.50	1.6487	0.6065	3.5	33.115	0.0302
0.55	1.7333	0.5769	3.6	36.598	0.0273
0.60	1.8221	0.5488	3.7	40.447	0.0247
0.65	1.9155	0.5220	3.8	44.701	0.0224
0.70	2.0138	0.4966	3.9	49.402	0.0202
0.75	2.1170	0.4724	4.0	54.598	0.0183
0.80	2.2255	0.4493	4.1	60.340	0.0166
0.85	2.3396	0.4274	4.2	66.686	0.0150
0.90	2.4596	0.4066	4.3	73.700	0.0136
0.95	2.5857	0.3867	4.4	81.451	0.0123
1.0	2.7183	0.3679	4.5	90.017	0.0111
1.1	3.0042	0.3329	4.6	99.484	0.0101
1.2	3.3201	0.3012	4.7	109.95	0.0091
1.3	3.6693	0.2725	4.8	121.51	0.0082
1.4	4.0552	0.2466	4.9	134.29	0.0074
1.5	4.4817	0.2231	5	148.41	0.0067
1.6	4.9530	0.2019	6	403.43	0.0025
1.7	5.4739	0.1827	7	1096.6	0.0009
1.8	6.0496	0.1653	8	2981.0	0.0003
1.9	6.6859	0.1496	9	8103.1	0.0001
2.0	7.3891	0.1353	10	22026	0.00005
2.1	8.1662	0.1225			
2.2	9.0250	0.1108			
2.3	9.9742	0.1003			
2.4	11.023	0.0907			

TABLE IV Common logarithms of numbers

N	0	1	2	3	4	5	6	7	8	9
0	....	0000	3010	4771	6021	6990	7782	8451	9031	9542
1	0000	0414	0792	1139	1461	1761	2041	2304	2553	2788
2	3010	3222	3424	3617	3802	3979	4150	4314	4624	4624
3	4771	4914	5051	5185	5315	5441	5563	5682	5798	5911
4	6021	6128	6232	6335	6435	6532	6628	6721	6812	6902
5	6990	7076	7160	7243	7324	7404	7482	7559	7634	7709
6	7782	7853	7924	7993	8062	8129	8195	8261	8325	8388
7	8451	8513	8573	8633	8692	8751	8808	8865	8921	8976
8	9031	9085	9138	9191	9243	9294	9345	9395	9445	9494
9	9542	9590	9638	9685	9731	9777	9823	9868	9912	9956
10	0000	0043	0086	0128	0170	0212	0253	0294	0334	0374
11	0414	0453	0492	0531	0569	0607	0645	0682	0719	0755
12	0792	0828	0864	0899	0934	0969	1004	1038	1072	1106
13	1139	1173	1206	1239	1271	1303	1335	1367	1399	1430
14	1461	1492	1523	1553	1584	1614	1644	1673	1703	1732
15	1761	1790	1818	1847	1875	1903	1931	1959	1987	2014
16	2041	2068	2095	2122	2148	2175	2201	2227	2253	2279
17	2304	2330	2355	2380	2405	2430	2455	2480	2504	2529
18	2553	2577	2601	2625	2648	2672	2695	2718	2742	2765
19	2788	2810	2833	2856	2878	2900	2923	2945	2967	2989
20	3010	3032	3054	3075	3096	3118	3139	3160	3181	3201
21	3222	3243	3263	3284	3304	3324	3345	3365	3385	3404
22	3424	3444	3464	3483	3502	3522	3541	3560	3579	3598
23	3617	3636	3655	3674	3692	3711	3729	3747	3766	3784
24	3802	3820	3838	3856	3874	3892	3909	3927	3945	3962
25	3979	3997	4014	4031	4048	4065	4082	4099	4116	4133
26	4150	4166	4183	4200	4216	4232	4249	4265	4281	4298
27	4314	4330	4346	4362	4378	4393	4409	4425	4440	4456
28	4472	4487	4502	4518	4533	4548	4564	4579	4594	4609
29	4624	4639	4654	4669	4683	4698	4713	4728	4742	4757
30	4771	4786	4800	4814	4829	4843	4857	4871	4886	4900
31	4914	4928	4942	4955	4969	4983	4997	5011	5024	5038
32	5051	5065	5079	5092	5105	5119	5132	5145	5159	5172
33	5185	5198	5211	5224	5237	5250	5263	5276	5289	5302
34	5315	5328	5340	5353	5366	5378	5391	5403	5416	5428
35	5441	5453	5465	5478	5490	5502	5514	5527	5539	5551
36	5563	5575	5587	5599	5611	5623	5635	5647	5658	5670
37	5682	5694	5705	5717	5729	5740	5752	5763	5775	5786
38	5798	5809	5821	5832	5843	5855	5866	5877	5888	5899
39	5911	5922	5933	5944	5955	5966	5977	5988	5999	6010
40	6021	6031	6042	6053	6064	6075	6085	6096	6107	6117
41	6128	6138	6149	6160	6170	6180	6191	6201	6212	6222
42	6232	6243	6253	6263	6274	6284	6294	6304	6314	6325
43	6335	6345	6355	6365	6375	6385	6395	6405	6415	6425
44	6435	6444	6454	6464	6474	6484	6493	6503	6513	6522
45	6532	6542	6551	6561	6571	6580	6590	6599	6609	6618
46	6628	6637	6646	6656	6665	6675	6684	6693	6702	6712
47	6721	6730	6739	6749	6758	6767	6776	6785	6794	6803
48	6812	6821	6830	6839	6848	6857	6866	6875	6884	6893
49	6902	6911	6920	6928	6937	6946	6955	6964	6972	6981
50	6990	6998	7007	7016	7024	7033	7042	7050	7059	7067
N	0	1	2	3	4	5	6	7	8	9

TABLE IV Common logarithms of numbers (cont'd.)

N	0	1	2	3	4	5	6	7	8	9
50	6990	6998	7007	7016	7024	7033	7042	7050	7059	7067
51	7076	7084	7093	7101	7110	7118	7126	7135	7143	7152
52	7160	7168	7177	7185	7193	7202	7210	7218	7226	7235
53	7243	7251	7259	7267	7275	7284	7292	7300	7308	7316
54	7324	7332	7340	7348	7356	7364	7372	7380	7388	7396
55	7404	7412	7419	7427	7435	7443	7451	7459	7466	7474
56	7482	7490	7497	7505	7513	7520	7528	7536	7543	7551
57	7559	7566	7574	7582	7589	7597	7604	7612	7619	7627
58	7634	7642	7649	7657	7664	7672	7679	7686	7694	7701
59	7709	7716	7723	7731	7738	7745	7752	7760	7767	7774
60	7782	7789	7796	7803	7810	7818	7825	7832	7839	7846
61	7853	7860	7868	7875	7882	7889	7896	7903	7910	7917
62	7924	7931	7938	7945	7952	7959	7966	7973	7980	7987
63	7993	8000	8007	8014	8021	8028	8035	8041	8048	8055
64	8062	8069	8075	8082	8089	8096	8102	8109	8116	8122
65	8129	8136	8142	8149	8156	8162	8169	8176	8182	8189
66	8195	8202	8209	8215	8222	8228	8235	8241	8248	8254
67	8261	8267	8274	8280	8287	8293	8299	8306	8312	8319
68	8325	8331	8338	8344	8351	8357	8363	8370	8376	8382
69	8388	8395	8401	8407	8414	8420	8426	8432	8439	8445
70	8451	8457	8463	8470	8476	8482	8488	8494	8500	8506
71	8513	8519	8525	8531	8537	8543	8549	8555	8561	8567
72	8573	8579	8585	8591	8597	8603	8609	8615	8621	8627
73	8633	8639	8645	8651	8657	8663	8669	8675	8681	8686
74	8692	8698	8704	8710	8716	8722	8727	8733	8739	8745
75	8751	8756	8762	8768	8774	8779	8785	8791	8797	8802
76	8808	8814	8820	8825	8831	8837	8842	8848	8854	8859
77	8865	8871	8876	8882	8887	8893	8899	8904	8910	8915
78	8921	8927	8932	8938	8943	8949	8954	8960	8965	8971
79	8976	8982	8987	8993	8998	9004	9009	9015	9020	9025
80	9031	9036	9042	9047	9053	9058	9063	9069	9074	9079
81	9085	9090	9096	9101	9106	9112	9117	9122	9128	9133
82	9138	9143	9149	9154	9159	9165	9170	9175	9180	9186
83	9191	9196	9201	9206	9212	9217	9222	9227	9232	9238
84	9243	9248	9253	9258	9263	9269	9274	9279	9284	9289
85	9294	9299	9304	9309	9315	9320	9325	9330	9335	9340
86	9345	9350	9355	9360	9365	9370	9375	9380	9385	9390
87	9395	9400	9405	9410	9415	9420	9425	9430	9435	9440
88	9445	9450	9455	9460	9465	9469	9474	9479	9484	9489
89	9494	9499	9504	9509	9513	9518	9523	9528	9533	9538
90	9542	9547	9552	9557	9562	9566	9571	9576	9581	9586
91	9590	9595	9600	9605	9609	9614	9619	9624	9628	9633
92	9638	9643	9647	9652	9657	9661	9666	9671	9675	9680
93	9685	9689	9694	9699	9703	9708	9713	9717	9722	9727
94	9731	9736	9741	9745	9750	9754	9759	9763	9768	9773
95	9777	9782	9786	9791	9795	9800	9805	9809	9814	9818
96	9823	9827	9832	9836	9841	9845	9850	9854	9859	9863
97	9868	9872	9877	9881	9886	9890	9894	9899	9903	9908
98	9912	9917	9921	9926	9930	9934	9939	9943	9948	9952
99	9956	9961	9965	9969	9974	9978	9983	9987	9991	9996
100	0000	0004	0009	0013	0017	0022	0026	0030	0035	0039
N	0	1	2	3	4	5	6	7	8	9

TABLE V Natural logarithms of numbers

x	$\ln x$	x	$\ln x$	x	$\ln x$
0.0	*	4.5	1.5041	9.0	2.1972
0.1	7.6974	4.6	1.5261	9.1	2.2083
0.2	8.3906	4.7	1.5476	9.2	2.2192
0.3	8.7960	4.8	1.5686	9.3	2.2300
0.4	9.0837	4.9	1.5892	9.4	2.2407
0.5	9.3069	5.0	1.6094	9.5	2.2513
0.6	9.4892	5.1	1.6292	9.6	2.2618
0.7	9.6433	5.2	1.6487	9.7	2.2721
0.8	9.7769	5.3	1.6677	9.8	2.2824
0.9	9.8946	5.4	1.6864	9.9	2.2925
1.0	0.0000	5.5	1.7047	10	2.3026
1.1	0.0953	5.6	1.7228	11	2.3979
1.2	0.1823	5.7	1.7405	12	2.4849
1.3	0.2624	5.8	1.7579	13	2.5649
1.4	0.3365	5.9	1.7750	14	2.6391
1.5	0.4055	6.0	1.7918	15	2.7081
1.6	0.4700	6.1	1.8083	16	2.7726
1.7	0.5306	6.2	1.8245	17	2.8332
1.8	0.5878	6.3	1.8405	18	2.8904
1.9	0.6419	6.4	1.8563	19	2.9444
2.0	0.6931	6.5	1.8718	20	2.9957
2.1	0.7419	6.6	1.8871	25	3.2189
2.2	0.7885	6.7	1.9021	30	3.4012
2.3	0.8329	6.8	1.9169	35	3.5553
2.4	0.8755	6.9	1.9315	40	3.6889
2.5	0.9163	7.0	1.9459	45	3.8067
2.6	0.9555	7.1	1.9601	50	3.9120
2.7	0.9933	7.2	1.9741	55	4.0073
2.8	1.0296	7.3	1.9879	60	4.0943
2.9	1.0647	7.4	2.0015	65	4.1744
3.0	1.0986	7.5	2.0149	70	4.2485
3.1	1.1314	7.6	2.0281	75	4.3175
3.2	1.1632	7.7	2.0412	80	4.3820
3.3	1.1939	7.8	2.0541	85	4.4427
3.4	1.2238	7.9	2.0669	90	4.4998
3.5	1.2528	8.0	2.0794	95	4.5539
3.6	1.2809	8.1	2.0919	100	4.6052
3.7	1.3083	8.2	2.1041		
3.8	1.3350	8.3	2.1163		
3.9	1.3610	8.4	2.1282		
4.0	1.3863	8.5	2.1401		
4.1	1.4110	8.6	2.1518		
4.2	1.4351	8.7	2.1633		
4.3	1.4586	8.8	2.1748		
4.4	1.4816	8.9	2.1861		

Answers to Selected Exercises

Chapter 1

1. $\overrightarrow{AB} = 3$, $\overrightarrow{AC} = 6$, $\overrightarrow{BC} = 3$, $\overrightarrow{CB} = -3$, $\overrightarrow{CA} = -6$, $\overrightarrow{BA} = -3$
2. $\overrightarrow{AB} = 3$, $\overrightarrow{BA} = -3$, $\overrightarrow{AC} = 7$, $\overrightarrow{CA} = -7$, $\overrightarrow{BC} = 4$, $\overrightarrow{CB} = -4$
4. 5
    5. 13.604
7. 13
    8. 5
10. $\sqrt{65}$
    11. $2\sqrt{29}$

13.

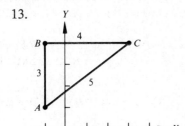

14. $\sqrt{13}$, $\sqrt{10}$, $\sqrt{5}$

16. $3\sqrt{2}$, 5, 1
    31. On a line

32. Not on a line
    34. $x = 13$

35. $y = -\frac{43}{16}$
    37. $(-1, 0)$

13. 3
    14. 4/3, Yes, No

16. 4, 76°

17. $-\frac{6}{17}$, 161°

19. $\frac{5}{3}$, 59°

20. 0.43912, 24°

37. On a line

38. Not on a line

40. $\tan A = \frac{7}{6}$, $A = 49°$; $\tan B = \frac{14}{5}$, $B = 70°$; $\tan C = \frac{7}{4}$, $C = 60°$

41. $\tan A = 5$, $A = 79°$; $\tan B = \frac{20}{9}$, $B = 66°$; $\tan C = \frac{5}{7}$, $C = 36°$

43. $\tan A = \frac{16}{13}$, $A = 51°$; $\tan B = \frac{24}{23}$, $B = 46°$; $\tan C = 8$, $C = 83°$

44. 27°, 153°

46. 34°

47. Parallel

49. Oblique

50. Perpendicular

52. Oblique

53. 8.9 ft

55. 114°

EXERCISES, SECTION 3

1. $(0, 0)$

2. $(2, 4)$

4. $(-1, 1)$

5. $(-1, 2)$

7. $\left(\frac{3}{2}, \frac{7}{2}\right), (4, 4), \left(\frac{7}{2}, \frac{5}{2}\right)$

8. $\left(3, \frac{7}{2}\right), \left(\frac{7}{2}, 2\right), \left(\frac{9}{2}, \frac{5}{2}\right)$

10. $\left(-2, -\frac{11}{2}\right), \left(-\frac{5}{2}, -\frac{7}{2}\right), \left(-\frac{3}{25}, 4\right)$

11. $x_1 = 5$, $y_2 = 0$

13. $\left(\frac{13}{3}, \frac{7}{3}\right)$

14. $\left(-\frac{4}{3}, \frac{2}{3}\right)$

16. $\left(3, \frac{8}{5}\right)$

17. $(-8, 16)$

19. $(18, 6)$

20. $(2.4174, 0.8315)$

22. $\left(5, \frac{17}{5}\right)$

23. $(4, 13)$

25. $\left(\frac{2}{7}, 1\right), \left(\frac{5}{7}, 0\right)$

26. $\left(\frac{26}{7}, -\frac{11}{7}\right)$

28. (a) $(8, 11)$, (b) $(-1, -1)$

29. (a) $\left(\frac{11}{2}, \frac{13}{2}\right)$, (b) $\left(-\frac{9}{2}, -\frac{7}{2}\right)$

31. $(-5, -5), (4, 7)$

32. $\left(6, \frac{3}{2}\right)$

34. $\left(5, -\frac{3}{2}\right)$

35. $\left(1, \frac{1}{2}\right)$

37. $\left(\frac{27}{2}, \frac{13}{2}\right)$

38. $10', 3', 4\frac{1}{2}'$

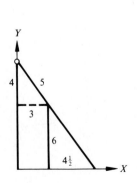

EXERCISES, SECTION 5

1. Yes

2. No, since $(1, 2)$ and $(1, 1)$ are both in the relation.

4. A function

5. A function

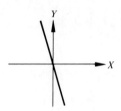

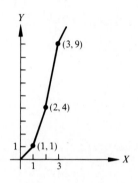

7. $y = |x|$ is a function.

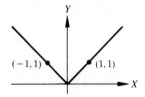

8. $y = x^2$ is a function.

10. $x = y^2$ is *not* a function.

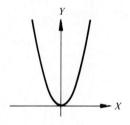

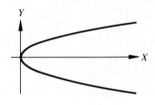

11. This is a function.

13. $y = \dfrac{1}{x}$ is a function.

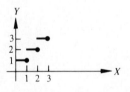

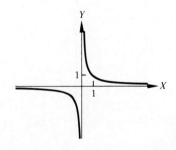

14. $x^2 + y^2 = 1$ is *not* a function.

16. This is a function.

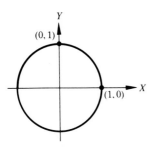

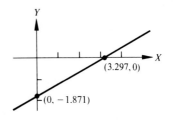

17. (a) No, (b) Yes, (c) Yes, (d) No, (e) No

19.

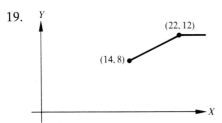

EXERCISES, SECTION 6

1. $x - y = 2$

2. $2x + y = 0$

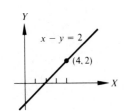

4. $3x + 2y = 29$

5. $3.1786x + y = -3.5759$

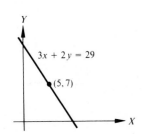

7. $x = 3$

8. $y = 2$

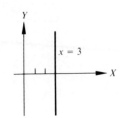

10. $x = -4$

11. $x - 3y - 1 = 0$

13. $y^2 - 8x + 16 = 0$

14. $y^2 - 16x = 0$

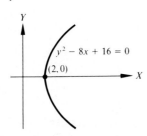

16. $x^2 + y^2 - 32 = 0$

17. $x^2 + y^2 = 9$

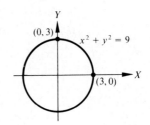

19. $5x^2 + 9y^2 = 180$

20. $25x^2 + 16y^2 = 400$

22. $4x^2 - 140y^2 + 35 = 0$

23. $x^2 + y^2 - 6x - 8y = 0$

25. $|xy| = 5$

26. $x^2y^2 + y^4 - 8y^3 + 16y^2 = 16$

REVIEW EXERCISES

2. $\sqrt{10}, \sqrt{13}, \sqrt{29}$

4. $\tan A = 5$, $A = 79°$; $\tan B = 8$, $B = 83°$; $\tan C = \frac{1}{3}$, $C = 18°$

5. (a) $(-1, 4)$; (b) $(-2, 3)$

7. $12x - 10y + 17 = 0$

Chapter 2

EXERCISES, SECTION 1

1. $m = \frac{1}{4}$, $a = 8$, $b = -2$, $y = \frac{1}{4}x - 2$

2. $m = -1$, $a = 6$, $b = 6$, $y = -x + 6$

4. $m = -\frac{9}{4}$, $a = 4$, $b = 9$, $y = -\frac{9}{4}x + 9$

5. $m = \frac{4}{3}$, $a = 3$, $b = -4$, $y = \frac{4}{3}x - 4$

7. $m = -7$, $a = \frac{11}{7}$, $b = 11$, $y = -7x + 11$

8. $m = -\frac{3}{2}$, $a = \frac{14}{3}$, $b = 7$, $y = -\frac{3}{2}x + 7$

10. $m = 4$, $a = \frac{5}{8}$, $b = -\frac{5}{2}$, $y = 4x - \frac{5}{2}$

11. $m = \frac{8}{3}$, $a = \frac{1}{2}$, $b = -\frac{4}{3}$, $y = \frac{8}{3}x - \frac{4}{3}$

13. $m = -\frac{6}{7}$, $a = 2$, $b = \frac{12}{7}$, $y = -\frac{6}{7}x + \frac{12}{7}$

14. $m = 0.4589$, $a = 2.8183$, $b = -1.2933$, $y = 0.4589x - 1.2933$

16. $y = 2x - 3$ 17. $y = -4x + 7$ 19. $y = -\frac{2}{3}x + 2$

20. $y = \frac{3}{2}x - 4$ 22. $y = \frac{4}{3}x + 3$ 23. $y = 7x$

25. $y = 9x + 11$ 26. $y = -8x + 7$ 28. $y = \frac{5}{3}x - 4$

29. $y = 7.0132x - 4.9875$ 31. $x - y + 8 = 0$

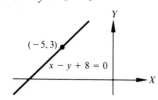

32. $y = -2$ 34. $x - y = -3$ 35. $x + 2y = 8$

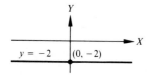

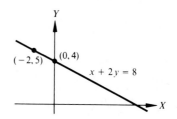

37. $4x - 3y = 0$ 38. $4.90032x + y = 6.29651$ 40. $7x + 6y = 11$

41. $3x + 4y = 19$ 43. $x - 9y = -29$ 44. $3x + 4y = 0$

46. $x = -1$ 47. $42x - 3y = 2$ 49. $x - y = 3$

50. $1.93759x + y = 18.023216$ 52.

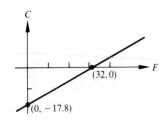

53. (a) $20°C$, (b) $158°F$ 55. $R = 0.0170 + 0.000075T$

56. $T = 20 + 15x, 0 \le x \le 8;$ $T = 140 - 15(8 - x)$ if $0 \le x \le 8$.

58. $\$1.40, \$3900$

EXERCISES, SECTION 2

1. $3x - 2y = 10$

2. $4x + 5y = -13$

4. $2x - 5y = -6$

5. $2x + y = 4$

7. $0.7642x - y = 4.6836$

8. $3x + 4y = -6$

10. $2x + 3y = 6$

11. $4x + y = 4$

13. $3x - 2y = 4$

14. $10x + 9y = -6$

16. $\dfrac{x}{2} + \dfrac{y}{-8} = 1$

17. $\dfrac{x}{2} + \dfrac{y}{3} = 1$

19. $\dfrac{x}{(5/4)} + \dfrac{y}{(5/3)} = 1$

20. $\dfrac{x}{(20/9)} + \dfrac{y}{(-5/9)} = 1$

22. $3x - 2y + 5 = 0,\ 2x + 3y - 14 = 0$

23. $x - 2y - 4 = 0,\ 2x + y - 3 = 0$

25. $x - y - 6 = 0$, $x + y - 6 = 0$

26. $1492x + 1776y - 3552 = 0$, $1776x - 1492y + 2984 = 0$

28. $x + 3y - 8 = 0$

29. $x + 2y - 3 = 0$

31. $1.954x + y + 0.085 = 0$

34. If y = cost to rent and x = miles driven, then $y = 20 + 0.11x$.

EXERCISES, SECTION 3

1. $(5, -2)$ 2. $(3, 0)$ 4. $(1, -\frac{1}{2})$ 5. $(\frac{16}{3}, 7)$

7. $(3, \frac{2}{3})$ 8. $(\frac{3}{2}, -\frac{2}{3})$ 10. $(10, 0)$

11. $(\frac{27}{5}, \frac{16}{5})$, $(5, 3)$, $(\frac{26}{5}, \frac{13}{5})$

13. The mixture should have 150 gallons of unleaded and 350 gallons of super unleaded gasoline.

14. Market equilibrium is to make 12 units at \$44 per unit.

16(a)

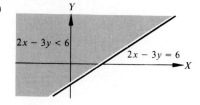

(b)

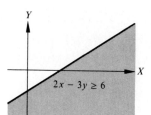

(c)

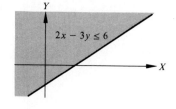

(d)

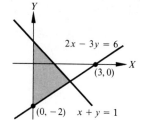

EXERCISES, SECTION 4

1. $\frac{19}{13}$ 2. $-\frac{12}{5}$ 4. $\frac{14}{5}$ 5. $\dfrac{13\sqrt{2}}{2}$ 7. $\dfrac{8\sqrt{13}}{13}$

8. 0

10. $\frac{7}{5}$ 11. -3 13. $\frac{12}{5}$ 14. $\frac{3}{13}$ 16. $2\sqrt{2}$

17. $\frac{19}{13}$ 19. $\dfrac{4\sqrt{17}}{51}$ 20. $\dfrac{9\sqrt{5}}{5}$ 22. $\dfrac{2\sqrt{13}}{13}$ 23. 0.112620

25. $\dfrac{14\sqrt{13}}{13}$ 26. $x = y$ 28. $56x - 32y - 77 = 0$

29. $2x + 11y - 13 = 0$, $11x - 2y - 34 = 0$

31. $(15 + 3\sqrt{10})x - (5 - 4\sqrt{10})y - 25 - 12\sqrt{10} = 0$
    $(15 - 3\sqrt{10})x - (5 + 4\sqrt{10})y - 25 + 12\sqrt{10} = 0$

32. 9.5

34. 7.603 mi, $x - 2y - 8 = 0$

35. $3x + 4y - 72 = 0$, 30 ft

EXERCISES, SECTION 5

1. y intercept $= -4$ 2. Slope $= 3$ 4. Pass through $(-3, 4)$

5. x intercept $= 1$ 7. y intercept $= -3$ 8. x intercept $= 4$

10. $4x - 7y = k$ 11. $y + 4 = m(x - 3)$

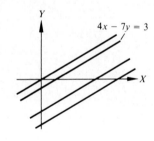

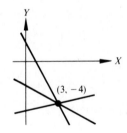

13. $4x + 3y = k$ 14. $x + ky = 3$

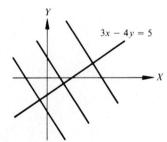

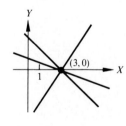

16. $8x + a^2y = 8a$

17. $28x + a^2y = 28a$

19. $\begin{cases} 4x - 3y + k = 0 \\ 4x - 3y + 8 = 0, \quad 4x - 3y - 42 = 0 \end{cases}$

20. $3x + 4y + k = 0$, $\quad 3x + 4y + 4 = 0$, $\quad 3x + 4y - 26 = 0$

22. $2x - 3y + 2 \pm 3\sqrt{13} = 0$ $\qquad$ 23. $3x + 2y + 4 \pm 2\sqrt{13} = 0$

25. $25x - 8y - 10 = 0$ $\qquad$ 26. $3x + y + 14 = 0$ $\qquad$ 28. $x + 3y = 0$

29. $24x + 24y - 13 = 0$ and $9x - 4y = 0$

31. $8x - 11 = 0$

32. $20x - 25y + 62 = 0$, $\quad 27x - 18y + 46 = 0$, $\quad 7x + 7y - 16 = 0$

34. $34x + a^2y = 34a$

EXERCISES, SECTION 6

1. $(x - 3)^2 + (y + 4)^2 = 36$ $\qquad$ 2. $x^2 + (y - 8)^2 = 25$

4. $x^2 + (y - 3)^2 = 49$ $\qquad$ 5. $(x - 5)^2 + (y + 3)^2 = 6$

7. $(x - 3)^2 + (y + \frac{1}{2})^2 = 17$ $\qquad$ 8. $(x - \frac{3}{5})^2 + (y - \frac{1}{3})^2 = 10$

10. $(x + 3)^2 + (y + 1)^2 = 40$ $\qquad$ 11. $(x - \frac{1}{2})^2 + (y + \frac{1}{2})^2 = \frac{49}{2}$

13. $(x - 5)^2 + (y - 3)^2 = 25$ $\qquad$ 14. $(x + 3)^2 + (y + 4)^2 = 16$

16. $(x - 5)^2 + (y + 5)^2 = \frac{3721}{169}$

17. $(x + 3)^2 + (y - 2)^2 = 25$ $\qquad$ 19. $(x - 4)^2 + (y - 1)^2 = 16$

20. $(x - 5)^2 + (y + 2)^2 = 36$ 22. $(x - 5)^2 + (y - 12)^2 = 144$

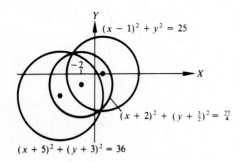

23. $(x + 2)^2 + (y - 6)^2 = 37$ 25. $(x + 3)^2 + (y - \frac{1}{2})^2 = \frac{43}{4}$

26. $(x - 1)^2 + (y + \frac{5}{6})^2 = \frac{61}{36}$ 28. Point

29. Circle   31. Circle   32. Circle

34. Circle   35. Point   37. $(x - 6)^2 + y^2 = 8$

38. $x^2 + (y - 4)^2 = 5$ 40. $(x - 7)^2 + (y + 6)^2 = 26$

41. $(x - 7)^2 + (y - 3)^2 = 25$ 43. $7x^2 + 7y^2 - 25x - 27y + 18 = 0$

44. $x^2 + y^2 - x - 5y = 0$ 46. $x^2 + y^2 + 3x - y - 10 = 0$

47. $8x^2 + 8y^2 - 53x - 54y + 120 = 0$

49. $(x - \frac{5}{6})^2 + (y - \frac{4}{3})^2 = \frac{529}{360}$

50. $x^2 + y^2 = \frac{121}{85}$

EXERCISES, SECTION 7

1. $x^2 + y^2 + 4x + 3y - 13 = 0$

2. $3x^2 + 3y^2 + 14x + 8y + 8 = 0$ *or* $3x^2 + 3y^2 + 10x + 10y - 2 = 0$

4. $4x - y + 2 = 0$ 5. $x + 3 = 0$ 7. $7x - 22 = 0$

8. $10x + 13y + 40 = 0$

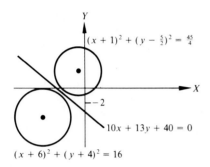

10. $5x^2 + 5y^2 + 32x - 22y = 0$

EXERCISES, SECTION 8

1. $(-1, 1), (-1, -3), (-7, -3), (-7, 1)$

2. $(5, -2), (2, -1), (2, -6), (0, 0)$

4. $(11, 5), (7, -1), (12, 6), (1, 0)$

5. $(7, 1), (4, -5), (9, 2), (-2, -4)$

7. $2x' + y' - 8 = 0$ 8. $x' - 2y' - 5 = 0$

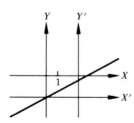

10. $4x' - 3y' + 3 = 0$ 11. $x'^2 + y'^2 + 2x' + 8y' + 12 = 0$

13. $x'^2 + y'^2 - 2x' + 10y' + 16 = 0$

14. $x'^2 + y'^2 + 8x' - 16y' + 60 = 0$

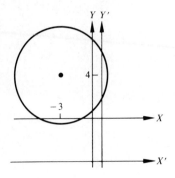

16. $x'^2 + y'^2 - 2x' - 4y' - 54 = 0$

17. $(x' + 11.015)^2 + (y' - 2.807)^2 = 54$

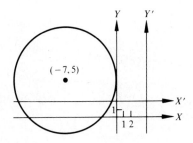

19. $(2, -1)$, $x'^2 + y'^2 = 10$ 20. $(-5, 6)$, $x'^2 + y'^2 = 58$

22. $(4, 1)$, $x'^2 + 4y'^2 = 15$ 23. $(-5, 7)$, $x'^2 + y'^2 = 81$

REVIEW EXERCISES

1. $5x - 4y - 22 = 0$ 2. $-\dfrac{4}{\sqrt{29}}, \dfrac{50}{\sqrt{29}}, -\dfrac{17}{\sqrt{29}}$

4. $5x + 12y + k = 0$, $5x + 12y + 17 = 0$

5. $y + 2 = m(x - 5)$, $3x + 4y = 7$ 7. $(x + 7)^2 + y^2 = 25$

8. $(x - 5)^2 + (y - \frac{7}{2})^2 = \frac{177}{4}$ 10. $(x + 1)^2 + y^2 = 3$

11. $x'^2 + y'^2 + 2x' - 5 = 0$

Chapter 3

EXERCISES, SECTION 1

1. $(1, 0)$, 4, $(1, 2)$, $(1, -2)$, $x = -1$

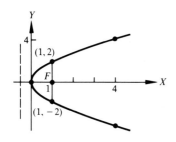

2. $(-4, 0)$, 16, $(-4, 8)$, $(-4, -8)$, $x = 4$

4. $(0, -\frac{5}{2})$, 10, $(-5, -\frac{5}{2})$, $(5, -\frac{5}{2})$, $y = \frac{5}{2}$

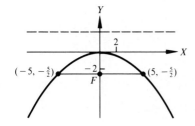

5. $\left(-\frac{3}{4}, 0\right)$, 3, $\left(-\frac{3}{4}, \frac{3}{2}\right)$, $\left(-\frac{3}{4}, -\frac{3}{2}\right)$, $x = \frac{3}{4}$

7. $\left(\frac{7}{8}, 0\right)$, $\frac{7}{2}$, $\left(\frac{7}{8}, \frac{7}{4}\right)$, $\left(\frac{7}{8}, -\frac{7}{4}\right)$, $x = -\frac{7}{8}$

8. $\left(-\frac{3}{8}, 0\right)$, $\frac{3}{2}$, $\left(-\frac{3}{8}, \frac{3}{4}\right)$, $\left(-\frac{3}{8}, -\frac{3}{4}\right)$, $x = \frac{3}{8}$

10. $y^2 = 12x$ 11. $x^2 = 12y$ 13. $x^2 = -12y$

14. $y^2 = 16x$ 16. $y^2 = 10x$ 17. $x^2 = 8y$

19. $y^2 = -\frac{16}{3}x$ 22. $300\sqrt{2}$ ft

EXERCISES, SECTION 2

1. $(x - 3)^2 = 8(y - 2)$ 2. $(x - 3)^2 = -24(y + 2)$

4. $(y - 1)^2 = 8(x - 4)$ 5. $(x - 4)^2 = 16(y - 1)$

7. $(x - 1)^2 = -8(y - 2)$

8. $(x - 3)^2 = 10(y + 2)$

10. $(y + 3)^2 = -8(x - 4)$

11. $y^2 = 12(x - 2)$

13. $(x - 2)^2 = 6(y + \frac{3}{2})$

14. $y^2 = -8(x + 1)$, $V(-1, 0)$, $F(-3, 0)$, $(-3, 4)$, $(-3, -4)$

16. $y^2 = 12(x + 4)$, $V(-4, 0)$, $F(-1, 0)$, $(-1, 6)$, $(-1, -6)$

17. $x^2 = -16(y - 2)$, $V(0, 2)$, $F(0, -2)$, $(8, -2)$, $(-8, -2)$

19. $(y - 3)^2 = 4x$, $V(0, 3)$, $F(1, 3)$, $(1, 1)$, $(1, 5)$

20. $(y + 4)^2 = -6x$, $V(0, -4)$, $F\left(-\frac{3}{2}, -4\right)$, $\left(-\frac{3}{2}, -1\right)$, $\left(-\frac{3}{2}, -7\right)$

22. $(y - 2)^2 = -8(x - 4)$, $V(4, 2)$, $F(2, 2)$, $(2, 6)$, $(2, -2)$

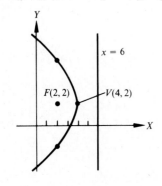

23. $(x + 1)^2 = -12(y + 3)$, $V(-1, -3)$, $F(-1, -6)$, $(-7, -6)$, $(5, -6)$

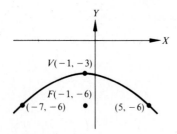

25. $(y + 3)^2 = -10(x - 1)$, $V(1, -3)$, $F\left(-\frac{3}{2}, -3\right)$, $\left(-\frac{3}{2}, 2\right)$, $\left(-\frac{3}{2}, -8\right)$

26. $(y - 6.32)^2 = +21.49(x - 3.73)$, $V(3.73, 6.32)$, $F(9.1, 6.32)$, $(9.1, 17.06)$,
$(9.1, -4.42)$

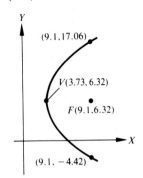

28. $(y + 4)^2 = -(x - 3)$

29. $(x + 1)^2 = 2(y + 2)$

31. $y^2 + 2y - x - 2 = 0$

32. $x^2 - 4x + y - 5 = 0$

34. $7y^2 - 11y - 18x - 14 = 0$

35. $5x^2 + 3x - 6y - 20 = 0$

41. 166 ft

EXERCISES, SECTION 3

1. $F(\pm 4, 0)$; $V(\pm 5, 0)$, $B(0, \pm 3)$, $\left(4, \pm \frac{9}{5}\right)$, $\left(-4, \pm \frac{9}{5}\right)$

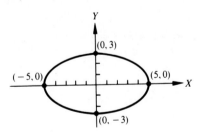

2. $F(0, \pm 4)$; $V(0, \pm 5)$; $B(\pm 3, 0)$; $\left(\pm \frac{9}{5}, 4\right)$, $\left(\pm \frac{9}{5}, -4\right)$

4. $F(0, \pm \sqrt{5})$; $V(0, \pm 3)$; $B(\pm 2, 0)$; $\left(\pm \frac{4}{3}, \sqrt{5}\right)$, $\left(\pm \frac{4}{3}, -\sqrt{5}\right)$

5. $F(\pm \sqrt{24}, 0)$; $V(\pm 7, 0)$; $B(0, \pm 5)$; $\left(\sqrt{24}, \pm \frac{25}{7}\right)$, $\left(-\sqrt{24}, \pm \frac{25}{7}\right)$

7. $F(0, \pm 5)$; $V(0, \pm 13)$; $B(\pm 12, 0)$; $\left(\pm \frac{144}{13}, 5\right), \left(\pm \frac{144}{13}, -5\right)$

8. $F(0, \pm \sqrt{21})$; $V(0, \pm 5)$; $B(\pm 2, 0)$; $\left(\pm \frac{4}{5}, \sqrt{21}\right), \left(\pm \frac{4}{5}, -\sqrt{21}\right)$

10. $F(\pm \sqrt{3}, 0)$; $V(\pm 2, 0)$; $B(0, \pm 1)$; $\left(\sqrt{3}, \pm \frac{1}{2}\right), \left(-\sqrt{3}, \pm \frac{1}{2}\right)$

11. $F(\pm \sqrt{3}, 0)$; $V(\pm 3, 0)$; $B(0, \pm \sqrt{6})$; $(\sqrt{3}, \pm 2), (-\sqrt{3}, \pm 2)$

13. $F(3 \pm \sqrt{7}, 2)$; $V(3 \pm 4, 2)$; $B(3, 2 \pm 3)$; $\left(3 + \sqrt{7}, 2 \pm \frac{9}{4}\right), \left(3 - \sqrt{7}, 2 \pm \frac{9}{4}\right)$

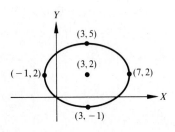

14. $F(3 \pm \sqrt{27}, 2)$; $V(3 \pm 6, 2)$; $B(3, 2 \pm 3)$; $\left(3 + \sqrt{27}, 2 \pm \frac{3}{2}\right)$, $(3 - \sqrt{27}, 2 \pm \frac{3}{2})$

16. $F(6, -3 \pm \sqrt{20})$; $V(6, -3 \pm 6)$; $B(6 \pm 4, -3)$; $\left(6 + \frac{8}{3}, -3 \pm \sqrt{20}\right), (6 - \frac{8}{3}, -3 \pm \sqrt{20})$

17. $\dfrac{(x + 5)^2}{25} + \dfrac{(y + 4)^2}{16} = 1$, Center $(-5, -4)$; $F(-5 \pm 3, -4)$; $V(-5 \pm 5, -4)$; $B(-5, -4 \pm 4)$; $(-5 + 3, -4 \pm \frac{16}{5}), \left(-5 - 3, -4 \pm \frac{16}{5}\right)$

19. $\dfrac{(x + 1)^2}{4} + \dfrac{(y - 2)^2}{16} = 1$, Center $(-1, 2)$; $F(-1, 2 \pm \sqrt{12})$; $V(-1, 2 \pm 4)$; $B(-1 \pm 2, 2)$; $(-1 \pm 1, 2 \pm \sqrt{12}), (-1 \pm 1, 2 - \sqrt{12})$

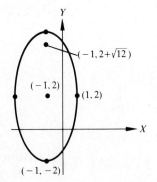

20. Center $(-1.324, 0.417)$
$F(-1.324, 0.417 \pm 1.143)$
$V(-1.324, 0.417 \pm 1.988)$
$B(-1.324 \pm 1.627, 0.417)$

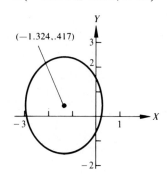

22. $\dfrac{(x+1)^2}{169} + \dfrac{(y-2)^2}{25} = 1$,   Center $(-1, 2)$; $F(-1 \pm 12, 2)$;
$V(-1 \pm 13, 2)$; $B(-1, 2 \pm 5)$;
$(-1 + 12, 2 \pm \frac{25}{13})$, $(-1 - 12, 2 \pm \frac{25}{13})$

23. $\dfrac{(x-2)^2}{289} + \dfrac{(y-3)^2}{225} = 1$,   Center $(2, 3)$; $F(2 \pm 8, 3)$, $V(2 \pm 17, 3)$;
$B(2, 3 \pm 15)$; $(2 + 8, 3 \pm \frac{225}{17})$,
$(2 - 8, 3 \pm \frac{225}{17})$

25. $\dfrac{(x-5)^2}{4} + \dfrac{(y-1)^2}{9} = 1$

26. $\dfrac{x^2}{36} + \dfrac{(y-3)^2}{20} = 1$

28. $\dfrac{(x-2)^2}{9} + \dfrac{(y-3)^2}{4} = 1$

29. $\dfrac{y^2}{36} + \dfrac{x^2}{16} = 1$

31. $\dfrac{(x+2)^2}{4} + \dfrac{(y-2)^2}{16} = 1$

32. $\dfrac{x^2}{25} + \dfrac{(y-2)^2}{9} = 1$

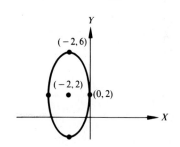

34. $\dfrac{y^2}{289} + \dfrac{x^2}{225} = 1$

35. $\dfrac{x^2}{41} + \dfrac{y^2}{16} = 1$

37. $\left(\dfrac{6}{\sqrt{13}}, \dfrac{6}{\sqrt{13}} \right), \left(\dfrac{6}{\sqrt{13}}, \dfrac{-6}{\sqrt{13}} \right),$

$\left(\dfrac{-6}{\sqrt{13}}, \dfrac{6}{\sqrt{13}} \right), \left(\dfrac{-6}{\sqrt{13}}, \dfrac{-6}{\sqrt{13}} \right)$

38. $\dfrac{y^2}{100} + \dfrac{x^2}{75} = 1$

41. $\dfrac{x^2}{64} + \dfrac{y^2}{16} = 1$

43. $\begin{cases} 91.4 \text{ million miles} \\ 94.6 \text{ million miles} \end{cases}$

44. $\frac{20}{3}\sqrt{5}$ ft

46. $\dfrac{x^2}{100} + \dfrac{y^2}{75} = 1$

47. $\dfrac{x^2}{36} + \dfrac{y^2}{20} = 1$

50. $(\pm 4, 0), (0, \pm 1);$ $(\pm 5, 0), (0, \pm\sqrt{10});$ $(\pm 6, 0), (0, \pm\sqrt{21});$ $(\pm 7, 0), (0, \pm\sqrt{34})$

52. (a) $\frac{9}{5}\sqrt{51}, \frac{18}{5}\sqrt{21}, \frac{27}{5}\sqrt{11}, \frac{36}{5}\sqrt{6}, 9\sqrt{3}, \frac{54}{5}$ feet
(b) $\frac{1}{27}(7500 - 900\pi) = 173$ cu yds

EXERCISES, SECTION 4

1. $V(\pm 4, 0),\ F(\pm\sqrt{20}, 0),\ 2,\ x \pm 2y = 0$

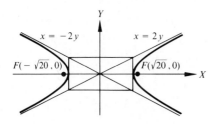

2. $V(0, \pm 4),\ F(0, \pm\sqrt{20}),\ 2,\ y \pm 2x = 0$

4. $V(0, \pm 6),\ F(0, \pm\sqrt{45}),\ 3,\ y \pm 2x = 0$

5. $V(\pm 5, 0),\ F(\pm\sqrt{34}, 0),\ \frac{18}{5},\ 3x \pm 5y = 0$

7. $V(\pm 7, 0),\ F(\pm 7\sqrt{2}, 0),\ 14,\ x \pm y = 0$

8. $V(0, \pm 8),\ F(0, \pm 8\sqrt{2}),\ 16,\ y \pm x = 0$

10. $V(-2, 3 \pm 4)$, $F(-2, 3 \pm 5)$, $\frac{9}{2}$, $3(y-3) \pm 4(x+2) = 0$

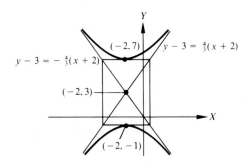

11. $V(4 \pm 5, 5)$, $F(4 \pm 5\sqrt{2}, 5)$, 10, $(y-5) \pm (x-4) = 0$

13. $\dfrac{x^2}{8} - \dfrac{(y-1)^2}{12} = 1$, $C(0,1)$, $V(\pm\sqrt{8}, 1)$, $F(\pm\sqrt{20}, 1)$

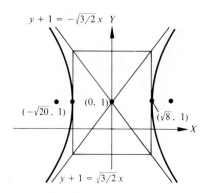

14. $\dfrac{(x+5)^2}{4} - \dfrac{y^2}{9} = 1$, $C(-5, 0)$, $V(-5 \pm 2, 0)$, $F(-5 \pm \sqrt{13}, 0)$

16. $\dfrac{(x+3)^2}{2} - \dfrac{(y-1)^2}{1} = 1$, $C(-3, 1)$, $V(-3 \pm \sqrt{2}, 1)$, $F(-3 \pm \sqrt{3}, 1)$

17. $\dfrac{(y+1)^2}{4} - \dfrac{(x+6)^2}{49} = 1$, $C(-6, -1)$, $V(-6, -1 \pm 2)$, $F(-6, -1 \pm \sqrt{53})$

19. $\dfrac{(x-2)^2}{16} - \dfrac{y^2}{48} = 1$

20. $\dfrac{(y-2)^2}{16} - \dfrac{x^2}{48} = 1$

22. $\dfrac{y^2}{36} - \dfrac{x^2}{18} = 1$

23. $\dfrac{(x-2)^2}{9} - \dfrac{(y-2)^2}{55} = 1$

25. $\dfrac{(y+2.1374)^2}{14.8363} - \dfrac{(x-4.1075)^2}{5.9316} = 1$

26. $\dfrac{(x-2)^2}{9} - \dfrac{(y+2)^2}{25} = 1$

28. $\dfrac{y^2}{25} - \dfrac{(x-6)^2}{36} = 1$

31. $\dfrac{x^2}{4} - \dfrac{y^2}{12} = 1$

32. $\dfrac{x^2}{16} - \dfrac{y^2}{9} = 1$

REVIEW EXERCISES

2. $y^2 = 4ax$

4. $(y-3)^2 = 8(x+2);\ V(-2,3),\ F(0,3),\ (0,7),\ (0,-1)$

5. $F\left(\dfrac{\pm 3\sqrt{10}}{10}, 0\right),\ V\left(\pm\dfrac{\sqrt{10}}{2}, 0\right),\ B\left(0, \pm\dfrac{2\sqrt{10}}{5}\right),$

$\left(\dfrac{3\sqrt{10}}{10}, \pm\dfrac{8\sqrt{10}}{25}\right),\ \left(\dfrac{-3\sqrt{10}}{10}, \pm\dfrac{8\sqrt{10}}{25}\right)$

7. $V(\pm 6, 0),\ F(\pm 10, 0),\ B(0, \pm 8),\ (10, \pm\frac{32}{8}),\ 4x \pm 3y = 0$

8. Center $(-3,2),\ V(-3 \pm 2, 2),\ F(-3 \pm 4, 2),\ (1, 2 \pm 6),\ (-7, 2 \pm 6)$

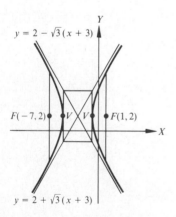

Chapter 4

EXERCISES, SECTION 1

1. $3x' - 2y' + 12 = 0$

2. $5x' + 4y' + 16 = 0$

4. $x'^2 + 7y' + 38 = 0$

5. $3x'^2 + 4y'^2 - 16y' + 8 = 0$

7. $4y'^2 - 5x'^2 + 16y' - 4 = 0$

8. $4x'^2 - y'^2 + 12y' - 52 = 0$

10. $3x'y' - 12y' - 89 = 0$

11. $3.015x'^2 + 3.208x' + 0.005y' = 0$

13. $(-4, 2)$, $x'y' + 4 = 0$

14. $(-3, -3)$, $x'y' = 12$

16. $(-3, 1)$, $x'^2 + 2y'^2 = 9$

17. $(-4, 2)$, $3x'^2 - 2y'^2 = 74$

19. $(8, 7)$, $x'^2 - x'y' + y'^2 = 84$

20. $(-1, -1)$, $2x'^2 - 3x'y' - y'^2 + 2 = 0$

22. $(3.0019, -2.8761)$ $'y' - 1 = 0$

23. $(1, -3)$, $y'^2 + 4x' = 0$

25. $(-2, 2)$, $y'^2 + 10x' = 0$

26. $(-3, -2)$, $y'^2 - x' = 0$

28. $(2, 1)$, $3y'^2 + 11x' = 0$

EXERCISES, SECTION 2

1. $x' = 3\sqrt{2}$

2. $y' + 2 = 0$

4. $x'^2 - y'^2 = 2$

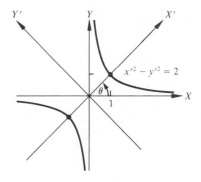

5. $x'^2 + 5y'^2 = 4$ 7. $5x'^2 + y'^2 = 6$ 8. $5y'^2 - 16x' + 8y' = 0$

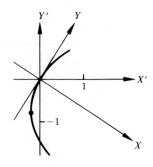

10. $45°$ 11. $\frac{1}{2}$ arctan 3

EXERCISES, SECTION 3

1. $\theta = $ Arctan 2
   $x'^2 = y'$

2. $45°$, $y''^2 - 4x'' = 0$

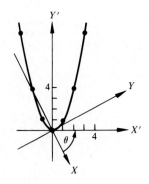

4. $\theta = 45°$,
   $y'' = y' - \sqrt{2}$,
   $x'' = x'$;
   $\dfrac{y''^2}{2} - \dfrac{x''^2}{2} = 1$

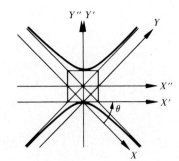

5. $\theta = \text{Arctan } \frac{1}{2}$,
   $x'' = x' + 1$,
   $y'' = y' - \frac{1}{2}$;
   $$\frac{x''^2}{3} - \frac{y''^2}{2} = 1$$

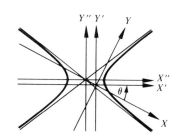

8. $\theta = \text{Arctan } \frac{4}{3}$,
   $x'' = x' + 2$,
   $y'' = y' + 2$;
   $x''^2 + 4y''^2 = 4$

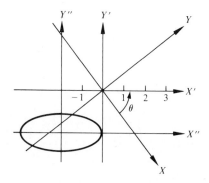

10. $\theta = \text{Arctan } \frac{2}{5}$,
    $$x'' = x' - \frac{2}{\sqrt{29}},$$
    $$y'' = y' - \frac{5}{\sqrt{29}}$$
    $$x''^2 + \frac{y''^2}{4} = 1$$

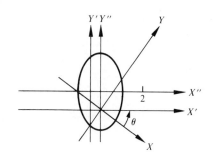

11. $\theta = \text{Arctan } \frac{1}{3}$,
    $x'' = x' + 3.091$,
    $y'' = y' - 0.047$;
    $y''^2 = -0.0316x''$

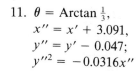

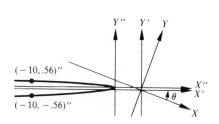

EXERCISES, SECTION 4

1. Hyperbola   2. Ellipse   4. Hyperbola

5. Parabola   7. Ellipse   8. Ellipse

10. Lines $3x - 2y + 1 = 0$ and $2x + y = 0$

11. Lines $x + y - 2 = 0$ and $x + y = 0$

13. No graph   14. Line $2x - y + 1 = 0$

16. Point $(-1, -\frac{3}{2})$ 17. Lines $3x - y + 1 = 0$; $3x - y = 0$

19. $3x - y = 0$ 20. Parabola   22. Hyperbola

23. Hyperbola

REVIEW EXERCISES

1. $5x' - 3y' + 9 = 0$ 2. $x'^2 + 11x' + 2y' + 24 = 0$

4. $x'^2 + 3y'^2 = 2$ 5. $7x'^2 + 3y'^2 = 4$

7. $12x'^2 + 18y'^2 = 5$ 8. $x'^2 + 5y'^2 = 4$

10. $75x''^2 + 45y''^2 = 2072$ 11. $x'^2 - y'^2 = 2$

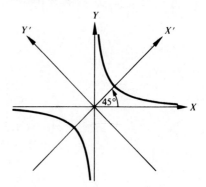

Chapter 5

EXERCISES, SECTION 1

2.

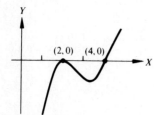

7.

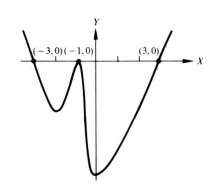

14.

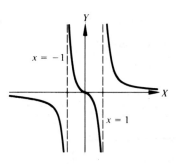

22. Asymptotes: $x = -1$, $x = 1$, $x = 4$, $y = 0$

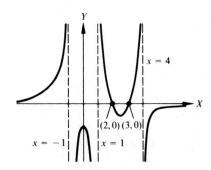

25. None   26. None   28. $y = 0$

29. $y = 0$ 31. None   32. None

EXERCISES, SECTION 2

1. $y = \sqrt{9 - x^2}$

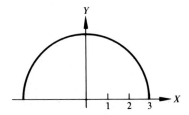

4. Asymptote: $y = 0$

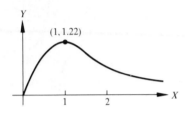

5. Asymptotes: $x = -3, y = 0$

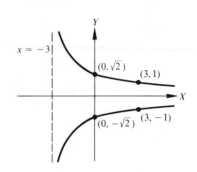

EXERCISES, SECTION 3

1. Asymptotes: $y = 0, y = -x$

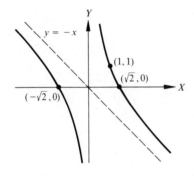

11. Asymptotes: $y = x + 4, x = 2$

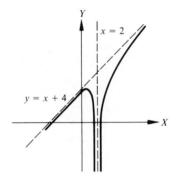

23.

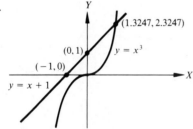

25. $(2, 2\sqrt{3}), (2, -2\sqrt{3})$

26. $(3, 2), (-3, -2)$

28. $(3, 4), (-4, -3)$

29. Each week, 10,000 calculators, to sell for $10 each, should be made.

REVIEW EXERCISES

7. $(4, 3)$, $(-3, -4)$

8.

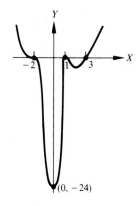

Chapter 6

EXERCISES, SECTION 1

1. $\dfrac{\pi}{2}$, 1   2. 4π, 3

4. $\dfrac{\pi}{3}$, 2   5. 2, $\frac{1}{2}$

7. 1, 1   8. $\frac{1}{4}$, 2

10. 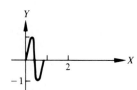   11.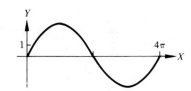

13. 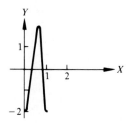   14.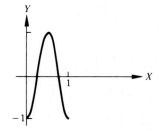

16.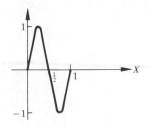

17.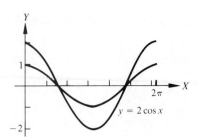

$y = 2\cos x$

19.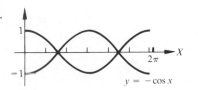

$y = -\cos x$

20.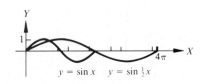

$y = \sin x \quad y = \sin \tfrac{1}{2}x$

22.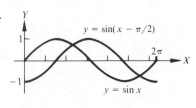

$y = \sin(x - \pi/2)$

$y = \sin x$

23.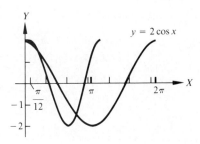

$y = 2\cos x$

25.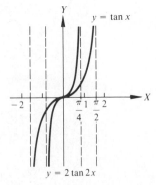

$y = \tan x$

$y = 2\tan 2x$

26.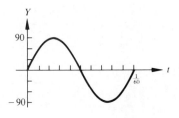

EXERCISES, SECTION 2

1.

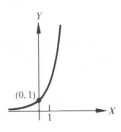

2.

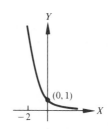

4.

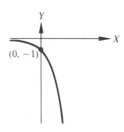

5.

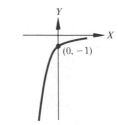

8.

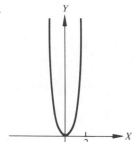

10. The account has $1221.38 after two years, and $1490.40 after four years.

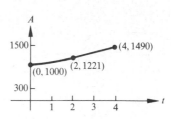

11.

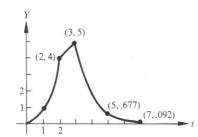

EXERCISES, SECTION 3

1. $\log_3 27 = 3$ 2. $\log_2 16 = 4$ 4. $\log_8 8 = 1$

5. $\log_{15} 1 = 0$ 7. $\log_4 \frac{1}{8} = -\frac{3}{2}$ 8. $\log_{32} 4 = \frac{2}{5}$

10. $\log_{2/3} \frac{9}{4} = -2$ 11. $\log_{3/2} \frac{27}{8} = 3$ 13. $3^0 = 1$

14. $6^1 = 6$ 16. $9^0 = 1$ 17. $2^{-4} = \frac{1}{16}$

19. $2^{-3} = \frac{1}{8}$ 20. $4^{-2} = \frac{1}{16}$ 22. $\left(\frac{1}{4}\right)^{-2} = 16$

23. $16^{3/2} = 64$ 25. 25   26. 16

28. $\frac{1}{27}$ 29. 8   31. 8

32. 2   34. 256   35. $\frac{1}{81}$

37. $\frac{1}{16}$ 38. 4   40. -2

41. 3

43.    44.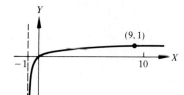

46. 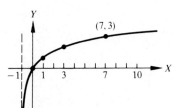   47.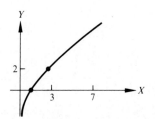

49. An earthquake of magnitude 5 will release approximately 10^{19} ergs of energy.

50. Conversation-level sound is about 65 decibels. A sound of 30 dB has 10^{-13} watts/square cm of power.

EXERCISES, SECTION 4

1. 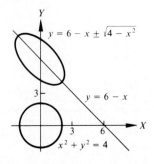   2.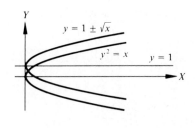

4.

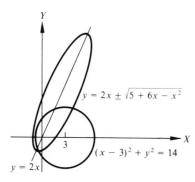

$y = 2x \pm \sqrt{5 + 6x - x^2}$

$(x - 3)^2 + y^2 = 14$

$y = 2x$

3

7.

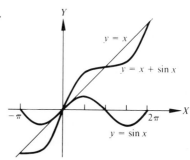

$y = x$

$y = x + \sin x$

$-\pi$

2π

$y = \sin x$

8.

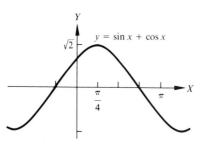

$y = \sin x + \cos x$

$\sqrt{2}$

$\dfrac{\pi}{4}$

π

10.

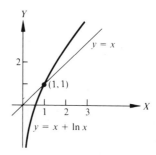

$y = x$

2

$(1, 1)$

1   2   3

$y = x + \ln x$

11.

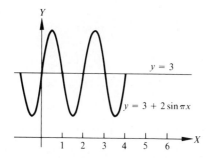

$y = 3$

$y = 3 + 2 \sin \pi x$

1   2   3   4   5   6

REVIEW EXERCISES

2. Amplitude is 2, period is 2.

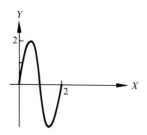

2

2

3.

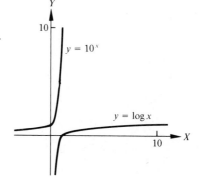

10

$y = 10^x$

$y = \log x$

10

4. $x = \frac{1}{125}$

5. $a = 512$

8.

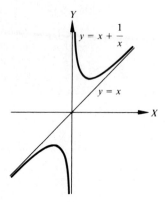

Chapter 7

EXERCISES, SECTION 1

1. $\dfrac{\pi}{6}, \dfrac{\pi}{4}, \dfrac{\pi}{2}$

2. $\dfrac{\pi}{3}, \dfrac{2}{3}\pi, \dfrac{5}{4}\pi$

4. $18°, 12°, 270°$

5. $135°, 20°, 12°$

7. $(25\frac{5}{7})°, (16\frac{4}{11})°, (13\frac{11}{13})°$

34. $(-3, -120°), (3, -300°), (-3, 240°)$

35. $(-6, -210°), (-6, 150°), (6, 330°)$

37. $(-4, -165°), (-4, 195°), (4, -345°)$

38. $\left(-1, -\dfrac{7}{6}\pi\right), \left(1, -\dfrac{\pi}{6}\right), \left(1, \dfrac{11}{6}\pi\right)$

40. $(-4, -\pi), (4, 0), (4, 2\pi)$

41. $\left(-4, -\dfrac{\pi}{2}\right), \left(-4, \dfrac{3}{2}\pi\right), \left(4, \dfrac{\pi}{2}\right)$

EXERCISES, SECTION 2

2. $(0, 3)$

4. $(5, 0)$

5. $(0, 0)$

7. $(-1, \sqrt{3})$

8. $(-2\sqrt{3}, 2)$

10. $(-6, 0)$

11. $\left(\frac{3}{2}\sqrt{3}, \frac{3}{2}\right)$

13. $\left(-\frac{5}{2}\sqrt{2}, -\frac{5}{2}\sqrt{2}\right)$

14. $(4, 90°)$

16. $(0, 0°)$

17. $(2, 180°)$

19. $(2, 45°)$

20. $(4, 330°)$

22. $(2, 225°)$

23. $(4\sqrt{2}, 225°)$

25. $(\sqrt{3}, 325°)$

26. $(5, 143°)$

28. $(13, 247°)$

29. $r\cos\theta = -3$

31. $r = \dfrac{3}{2\cos\theta + \sin\theta}$

32. $r(3\cos\theta - \sin\theta) = 0$

34. $r\cos^2\theta = 4\sin\theta$

35. $r = 3$

37. $r\cos^2\theta - 9\sin\theta = 0$

38. $r^2(1 - 3\sin^2\theta) = 4$

40. $r = 2\sin\theta$

41. $x^2 + y^2 = 9$

43. $x = \sqrt{3}\,y$

44. $x = 4$

46. $x^2 + y^2 = 6x$

47. $x^2 + y^2 = 8y$

49. $x^2 - y^2 = a^2$

50. $y^2 = 4(x + 1)$

52. $y^2 - 3x^2 = 12x + 9$

53. $4x + 3y = 3$

55. $x + 3y = 1$

56. $x^2 - 3y^2 + 8y = 4$

EXERCISES, SECTION 3

2.

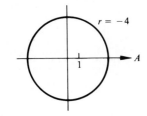

4.

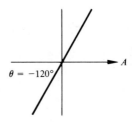

8.

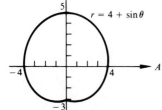

14.

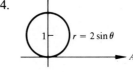

17.

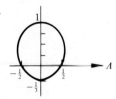

EXERCISES, SECTION 4

2.

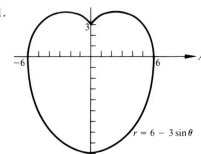

$r = 3(1 - \sin\theta)$

5.

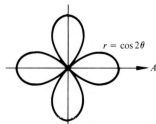

$r = \cos 2\theta$

11.

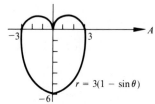

$r = 6 - 3\sin\theta$

13.

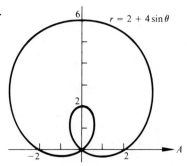

$r = 2 + 4\sin\theta$

19.

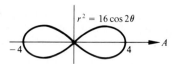

$r^2 = 16\cos 2\theta$

EXERCISES, SECTION 5

10. $r\sin\theta = -3$

11. $r\cos\theta = 4$

13. $\theta = 90°$

14. $r\cos\left(\theta - \dfrac{\pi}{3}\right) = 3$

16. $(2, 0°), 2$

17. $(3, 90°), 3$

19. $(-1, 0°), 1$

20. $(4\frac{1}{2}, 0°), 4\frac{1}{2}$

22. $r = 8\cos\theta$

23. $r = 10\cos\theta$

25. $r + 8\cos\theta = 0$

26. $r^2 - 10r\cos\theta + 9 = 0$

28. $r^2 - 10r\cos(\theta - 45°) + 9 = 0$

29. $r^2 - 6r\cos(\theta - 120°) + 5 = 0$

31. $r^2 - 14r\cos(\theta - 120°) + 24 = 0$

32. $r^2 - 18r\cos(\theta - 225°) + 17 = 0$

EXERCISES, SECTION 6

1.
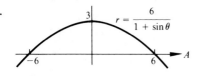

5.
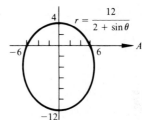

10.
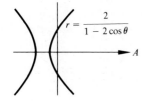

EXERCISES, SECTION 7

1. $\left\{\dfrac{\pi}{12}, \dfrac{5}{12}\pi, \dfrac{13}{12}\pi, \dfrac{17}{12}\pi\right\}$

2. $\left\{\dfrac{\pi}{6}, \dfrac{5}{6}\pi, \dfrac{7}{6}\pi, \dfrac{11}{6}\pi\right\}$

4. $\left\{\dfrac{\pi}{4}, \dfrac{5}{12}\pi, \dfrac{11}{12}\pi, \dfrac{13}{12}\pi, \dfrac{19}{12}\pi, \dfrac{7}{4}\pi\right\}$

5. $\left\{\dfrac{2}{9}\pi, \dfrac{4}{9}\pi, \dfrac{8}{9}\pi, \dfrac{10}{9}\pi, \dfrac{14}{9}\pi, \dfrac{16}{9}\pi\right\}$

7. $\left\{\dfrac{\pi}{6}, \dfrac{5}{6}\pi\right\}$

8. $\{36°, 108°, 252°, 324°\}$

10. $\left\{\dfrac{\pi}{12}, \dfrac{5}{12}\pi, \dfrac{3}{4}\pi, \dfrac{13}{12}\pi, \dfrac{17}{12}\pi, \dfrac{7}{4}\pi\right\}$

11. $\{34°, 65°, 115°, 146°, 214°, 245°, 295°, 326°\}$

13. $\left\{\dfrac{2}{3}\pi\right\}$

14. No solution in the interval $0 \le \theta < 2\pi$.

EXERCISES, SECTION 8

1. $(1, 30°), (1, 150°)$

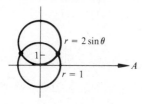

2. $(1, 0°), (1, 180°)$. Also intersect at $(1, 90°)$ and $(1, 270°)$ but no set of coordinates for each will satisfy both equations.

4. $(-3\sqrt{2}, 45°)$ and the pole.

5. $(3\sqrt{2}, 45°), (3\sqrt{2}, 135°)$

7. $(2, 30°), (2, 150°)$

8. $(2, 45°), (2, 225°)$

10. $\left(\dfrac{4}{3}, 30°\right), \left(\dfrac{4}{3}, 150°\right)$

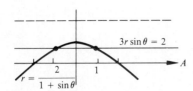

11. $(3, 90°)$ and $(1, 19°), (1, 161°)$ with the angles to the nearest degree.

13. $\left(\dfrac{a}{\sqrt{2}}, 30°\right), \left(\dfrac{a}{\sqrt{2}}, 150°\right)$, and the pole.

14. $(0, 90°)$, $(2\sqrt{3}, 30°)$, $(-2\sqrt{3}, 150°)$, and the pole.

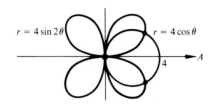

16. $(1, 60°)$, $(1, 300°)$. Also intersect at $(-1, 420°)$ and $(-1, 660°)$ but no set of coordinates for each will satisfy both equations.

17. $(a, 0°)$, $(a, 180°)$, and the pole.

19. $(3, 0°)$

20. $(2, 30°)$, $(2, 150°)$, $(-1, 270°)$

22. $(2, 60°)$, $(2, 300°)$

23. $\left(\dfrac{3}{\sqrt{2}}, 45°\right)$ and the pole.

25. The pole.

26. $(2, 0°)$ and the pole.

28. $\left(-\dfrac{12}{5}, 37°\right)$ and the pole.

29. $\left(\dfrac{3}{2}, 60°\right)$, $\left(\dfrac{3}{2}, 300°\right)$, and the pole.

31. $(2, 90°)$ and the pole.

REVIEW EXERCISES

1. $(-5, 195°)$, $(-5, -165°)$, $(5, -345°)$

2. $\left(-3, -\dfrac{\pi}{3}\right), \left(-3, \dfrac{5}{3}\pi\right), \left(3, \dfrac{2}{3}\pi\right)$

4. $(2, 315°)$; $(\sqrt{41}, 231°)$ angle to nearest degree.

5. $r = \dfrac{4}{3\cos\theta - 2\sin\theta}$

7. $r \sin \theta = \dfrac{3\sqrt{3}}{2}$

8. $r^2 - 12r\cos(\theta - 30°) + 20 = 0$

10. $\{30°, 150°, 270°\}$

11. $(3, 0°)$, $(3, 180°)$, and the pole.

Chapter 8

1.

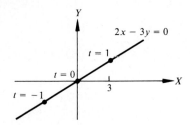

8.

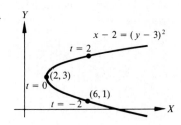

10. $(x - 2)^2 + (y - 2)^2 = 1$

11. $\dfrac{x^2}{9} + \dfrac{y^2}{16} = 1$

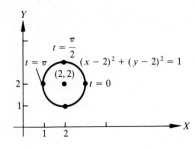

13. $x^2 + 2y = 2$

14. $y^2 - x^2 = 1$

16. $(x + 2)^2 + (y - 3)^2 = 4$

17. $x^2 + y^2 = 2x$

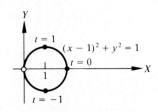

19. $2x + 3y = 6$, part between $(0, 2)$ and $(3, 0)$

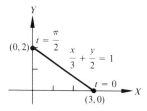

20. $x - y = 1$, part where $y \geq 0$

22. $x^2 = -4(y - 1)$, part for which $0 \leq y \leq 1$

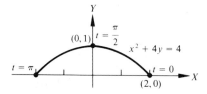

23. $x^2 = -8(y - 2)$, part for which $0 \leq y \leq 2$

25. $x = \dfrac{1}{t + 4}$, $y = t + 2$

26. $x = \dfrac{-3t}{1 + t^3}$, $y = \dfrac{-3t^2}{1 + t^3}$

28. $x = \dfrac{-3t^3}{t^2 - 1}$, $y = \dfrac{-3t^2}{t^2 - 1}$

29. $x = \tan\theta$, $y = \sin\theta$

EXERCISES, SECTION 2

1. $\left\{ \begin{array}{l} x = 48\sqrt{2}\, t \\ y = 48\sqrt{2}\, t - 16t^2 \end{array} \right.$ $\left. \begin{array}{l} \text{Distance 288 ft} \\ \text{Height 72 ft} \end{array} \right\}$

2. (a) $(96\sqrt{3}, 80)$ (b) $(288\sqrt{3}, 144)$ (c) $(480\sqrt{3}, 80)$, $3 \pm \sqrt{3}$ sec.

4. 0.98 ft; 2.64 ft   7. $x = 4\pi t - \frac{5}{2}t \sin \pi t$, $y = 4 - \frac{5}{2}t \cos \pi t$, where t is the elapsed time in seconds.

2. $\dfrac{x^2}{16} + \dfrac{y^2}{9} = 1$

4. It hits the ground 2 sec. after being dropped, 32 ft south of the drop spot.

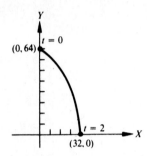

5. $\begin{cases} x = 40\sqrt{2}\,t \\ y = 40\sqrt{2}\,t - 16t^2 \end{cases}$ 　Distance 200 ft 　Height 50 ft

Chapter 9

EXERCISES, SECTION 1

10. 11

11. $5\sqrt{5}$

13. $(1,1,1)$

14. $(4, -4, 10)$

16. $(6, \frac{1}{2}, \frac{7}{2})$

17. $(-5, 7, 5)$

19. $\left(\frac{8}{5}, \frac{1}{5}, \frac{11}{10}\right)$

20. $(17, 2, -7)$

22. $(5,0,0), (0,6,0), (0,0,8), (5,0,8), (5,6,0), (0,6,8)$

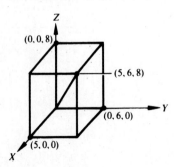

32.

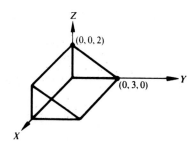

37.

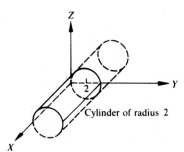

40.

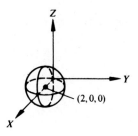

EXERCISES, SECTION 2

1. Center $(2, 0, 0)$
   radius 2

2. Center $(0, 1, 3)$
   radius $\sqrt{10}$

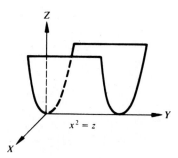

4. Center $(-2, -3, -4)$
   radius $\sqrt{29}$

5. $\dfrac{x^2}{a^2} + \dfrac{y^2}{b^2} + \dfrac{z^2}{a^2} = 1$

7.

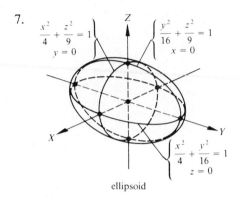

$$\dfrac{x^2}{4} + \dfrac{z^2}{9} = 1 \\ y = 0$$

$$\dfrac{y^2}{16} + \dfrac{z^2}{9} = 1 \\ x = 0$$

$$\dfrac{x^2}{4} + \dfrac{y^2}{16} = 1 \\ z = 0$$

ellipsoid

REVIEW EXERCISES

1. $2\sqrt{51}$, $(6, 7, 4)$, $(3, 4, \frac{17}{5})$, $(51, 52, 13)$

2. $5\sqrt{6}$, $\left(\frac{5}{2}, \frac{1}{2}, -2\right)$, $(1, -1, 1)$, $(25, 23, -47)$

4. $5\sqrt{14}$, $\left(\frac{1}{2}, 2, \frac{7}{2}\right)$, $(-1, -1, -1)$, $(23, 47, 71)$

10. $Ax + By + Cz = 0$

11. (a) $z = 0$ (b) $x = 0$ (c) $y = 0$

13. x intercept: $-\dfrac{D}{A}$

   y intercept: $-\dfrac{D}{B}$

   z intercept: $-\dfrac{D}{C}$

   Trace in xy plane: $Ax + By + D = 0$
   Trace in xz plane: $Ax + Cz + D = 0$
   Trace in yz plane: $By + Cz + D = 0$

Chapter 10

EXERCISES, SECTION 1

1. $\mathbf{P} = 3\mathbf{i} + 2\mathbf{j}$, $\mathbf{Q} = 5\mathbf{i} - 4\mathbf{j}$, $\overrightarrow{PQ} = 2\mathbf{i} - 6\mathbf{j}$, $\mathbf{P} + \mathbf{Q} = 8\mathbf{i} - 2\mathbf{j}$,
   $\mathbf{P} - \mathbf{Q} = -2\mathbf{i} + 6\mathbf{j}$

2. $\mathbf{P} = 4\mathbf{i}$, $\mathbf{Q} = 5\mathbf{j}$, $\overrightarrow{PQ} = -4\mathbf{i} + 5\mathbf{j}$, $\mathbf{P} + \mathbf{Q} = 4\mathbf{i} + 5\mathbf{j}$, $\mathbf{P} - \mathbf{Q} = 4\mathbf{i} - 5\mathbf{j}$

4. $\mathbf{P} = 5\mathbf{i} + 6\mathbf{j}$, $\mathbf{Q} = -3\mathbf{i} - 3\mathbf{j}$, $\overrightarrow{PQ} = -8\mathbf{i} - 9\mathbf{j}$, $\mathbf{P} + \mathbf{Q} = 2\mathbf{i} + 3\mathbf{j}$,
   $\mathbf{P} - \mathbf{Q} = 8\mathbf{i} + 9\mathbf{j}$

5. $P = 3j$, $Q = -4i - 2j$, $\overrightarrow{PQ} = -4i - 5j$, $P + Q = -4i + j$,
    $P - Q = 4i + 5j$

7. $P = -i - 3j$, $Q = 3i + 4j$, $\overrightarrow{PQ} = 4i + 7j$, $P + Q = 2i + j$,
    $P - Q = -4i - 7j$

8. $P = 3i - 12j$, $Q = 4i - j$, $\overrightarrow{PQ} = i + 11j$, $P + Q = 7i - 13j$,
    $P - Q = -i - 11j$

10. $\dfrac{5}{13}i + \dfrac{12}{13}j$ 11. $\dfrac{i}{\sqrt{26}} + \dfrac{5}{\sqrt{26}}j$ 13. $\dfrac{24}{5}i + \dfrac{32}{5}j$

14. $\dfrac{75}{13}i + \dfrac{180}{13}j$ 16. $\dfrac{20}{\sqrt{13}}i - \dfrac{30}{\sqrt{13}}j$ 17. $(4, -1)$

19. $(2, 1)$ 20. $\left(\dfrac{a_1 + a_2}{2}, \dfrac{b_1 + b_2}{2}\right)$ 22. $(-5, 1), (-2, -3)$

23. $(0, 1), (-3, -1)$ 25. $V = (9t - 4)i + (6t - 4)j$

26. $V = (10t + 13)i + (8 - 6t)j$

28. $V = bti + (1 - t)aj$

29. $R = 5i + 12j$, $14°$

EXERCISES, SECTION 2

1. $P = i + 2j + 2k$, $Q = i - 4j - 2k$, $\overrightarrow{PQ} = -6j - 4k$,
    $P + Q = 2i - 2j$, $P - Q = 6j + 4k$, $|P| = 3$

2. $P = 3i + 4k$, $Q = 7j + 8k$, $\overrightarrow{PQ} = -3i + 7j + 4k$,
    $P + Q = 3i + 7j + 12k$, $P - Q = 3i - 7j - 4k$, $|P| = 5$

4. $P = 7i + 8j + 5k$, $Q = 3i - 3j + 2k$, $\overrightarrow{PQ} = -4i - 11j - 3k$,
    $P + Q = 10i + 5j + 7k$, $P - Q = 4i + 11j + 3k$, $|P| = \sqrt{138}$

5. $P = -4i - 4j - 4k$, $Q = i + 3j + 7k$, $\overrightarrow{PQ} = 5i + 7j + 11k$,
    $P + Q = -3i - j + 3k$, $P - Q = -5i - 7j - 11k$, $|P| = 4\sqrt{3}$

7. $P = 8j - 6k$, $Q = 4i - 3j + 6k$, $\overrightarrow{PQ} = 4i - 11j + 12k$,
    $P + Q = 4i + 5j$, $P - Q = -4i + 11j - 12k$, $|P| = 10$

8. $P = 4i - 2j - 4k$, $Q = i + 9j + 12k$, $\overrightarrow{PQ} = -3i + 11j + 16k$,
    $P + Q = 5i + 7j + 8k$, $P - Q = 3i - 11j - 16k$, $|P| = 6$

10. $\dfrac{3}{\sqrt{41}}i - \dfrac{4}{\sqrt{41}}j + \dfrac{4}{\sqrt{41}}k$ 11. $\dfrac{3}{5}j - \dfrac{4}{5}k$

13. $4\mathbf{i} + 8\mathbf{j} - 8\mathbf{k}$

14. $\dfrac{3}{\sqrt{26}}\mathbf{i} + \dfrac{9}{\sqrt{26}}\mathbf{j} - \dfrac{12}{\sqrt{26}}\mathbf{k}$

16. $\dfrac{20}{\sqrt{29}}\mathbf{i} - \dfrac{30}{\sqrt{29}}\mathbf{j} - \dfrac{40}{\sqrt{29}}\mathbf{k}$

17. $4\mathbf{i} - 5\mathbf{j} + 4\mathbf{k}$, $(4, -5, 4)$

19. $\frac{11}{2}\mathbf{i} + \frac{11}{2}\mathbf{j} - \frac{1}{2}\mathbf{k}$, $\left(\frac{11}{2}, \frac{11}{2}, -\frac{1}{2}\right)$

20. $\frac{7}{2}\mathbf{i} + \frac{3}{2}\mathbf{j} + \frac{11}{2}\mathbf{k}$, $\left(\frac{7}{2}, \frac{3}{2}, \frac{11}{2}\right)$

22. $-4\mathbf{i} + 7\mathbf{j} + 2\mathbf{k}$, $-6\mathbf{i} + 11\mathbf{j} + 8\mathbf{k}$, $(-4, 7, 2)$, $(-6, 11, 8)$

23. $\mathbf{V} = (4t + 3)\mathbf{i} + (5 - 4t)\mathbf{j} + (6 - 2t)\mathbf{k}$

25. $(4, 1, -9)$, $(-5, 4, 12)$

EXERCISES, SECTION 3

1. $4, \dfrac{\sqrt{14}}{42}$

2. $1, \dfrac{\sqrt{3}}{9}$

4. $11, \dfrac{11\sqrt{3}}{27}$

5. $14, \frac{7}{27}$

7. $\frac{3}{3}, \frac{4}{9}\mathbf{i} - \frac{4}{9}\mathbf{j} + \frac{2}{9}\mathbf{k}$

8. $\frac{1}{3}, \frac{8}{27}\mathbf{i} + \frac{4}{27}\mathbf{j} - \frac{1}{27}\mathbf{k}$

10. $\frac{1}{3}, \frac{2}{9}\mathbf{i} - \frac{2}{9}\mathbf{j} - \frac{1}{9}\mathbf{k}$

11. Perpendicular

13. Parallel

14. Perpendicular

16. $35°$ (nearest degree)

17. $64°, 26°$ (nearest degrees), and $90°$

EXERCISES, SECTION 4

1. $3x - 2y + 5z = 11$

2. $4x - y - z = 0$

4. $x + 2y + z = -2$

5. $3x + 9y - 4z = 0$

7. $2x + 3y - 4z = 2$

8. $y - z = 0$

10. $\frac{13}{21}$

11. 0

14. $C = -10$

16. $\frac{3}{7}$

17. $\dfrac{2\sqrt{21}}{7}$

19. $(x - 1)^2 + (y - 1)^2 + (z - 7)^2 = 21$

EXERCISES, SECTION 5

1. $\dfrac{x - 4}{-2} = \dfrac{y + 3}{3} = \dfrac{z - 5}{4}$; $x = 4 - 2t$, $y = -3 + 3t$, $z = 5 + 4t$

2. $\dfrac{x}{1} = \dfrac{y - 1}{-1} = \dfrac{z + 2}{2}$; $x = t$, $y = 1 - t$, $z = -2 + 2t$

4. $\dfrac{x + 2}{5} = \dfrac{y + 2}{4} = \dfrac{z - 3}{1}$; $x = -2 + 5t$, $y = -2 + 4t$, $z = 3 + t$

5. $\dfrac{x-2}{2} = \dfrac{y+1}{1}$, $z - 1 = 0$, $x = 2 + 2t$, $y = -1 + t$, $z = 1$

7. $x = 0$, $z = 0$; $x = 0$, $y = t$, $z = 0$

8. $x = 0$, $y = 0$; $x = 0$, $y = 0$, $z = t$

10. $\dfrac{x}{3} = \dfrac{y}{4} = \dfrac{z}{5}$

11. $\dfrac{x-1}{-1} = \dfrac{y}{2} = \dfrac{z-2}{-1}$

13. $x - 2 = 0$, $\dfrac{y-5}{1} = \dfrac{z-4}{1}$

14. $x = 0$, $y - 4 = 0$

16. $x = 0$, $y = 0$

17. $\dfrac{x-2}{3} = \dfrac{y-3}{-1} = \dfrac{z}{2}$

19. $\dfrac{x}{2} = \dfrac{y-3}{-3} = \dfrac{z-6}{1}$

20. $\dfrac{x}{1} = \dfrac{y+22}{-4} = \dfrac{z+14}{-3}$

22. $60°$ or $120°$ 23. $\frac{4}{9}, \frac{1}{9}, \frac{8}{9}$ 25. $\frac{2}{5}, \frac{2}{15}, -\frac{11}{15}$

26. $-\frac{3}{13}, \frac{4}{13}, \frac{12}{13}$ 28. $\dfrac{1}{\sqrt{3}}, -\dfrac{1}{\sqrt{3}}, \dfrac{1}{\sqrt{3}}$ 29. $\frac{2}{3}$

31. xy plane at $(8, -1, 0)$, yz plane at $(0, -5, -12)$, xz plane at $(10, 0, 3)$

32. xy plane at $(6, -1, 0)$, yz plane at $(0, 2, 3)$, xz plane at $(4, 0, 1)$

34. xy plane at $\left(\frac{2}{3}, -\frac{11}{3}, 0\right)$, yz plane at $(0, -5, -2)$, xz plane at $\left(\frac{5}{2}, 0, \frac{11}{2}\right)$

EXERCISES, SECTION 6

1. $4\mathbf{i} + 4\mathbf{j} - 2\mathbf{k}$, $\frac{2}{3}\mathbf{i} + \frac{2}{3}\mathbf{j} - \frac{1}{3}\mathbf{k}$

2. $19\mathbf{i} + 24\mathbf{j} - 3\mathbf{k}$, $\dfrac{19}{\sqrt{946}}\mathbf{i} + \dfrac{24}{\sqrt{946}}\mathbf{j} - \dfrac{3}{\sqrt{946}}\mathbf{k}$

4. $\mathbf{i} - 4\mathbf{j} + 2\mathbf{k}$, $\dfrac{1}{\sqrt{21}}\mathbf{i} - \dfrac{4}{\sqrt{21}}\mathbf{j} + \dfrac{2}{\sqrt{21}}\mathbf{k}$

5. $25\mathbf{k}$, $\mathbf{k}$ 7. 5 8. $3\sqrt{6}$

10. 18 11. $\dfrac{9\sqrt{3}}{2}$ 13. $\frac{1}{2}$ 14. $\dfrac{7\sqrt{6}}{2}$

16. $35x + 11y - 38z + 88 = 0$

17. $9x + 7y - z = 25$ 19. $\dfrac{274}{\sqrt{781}}$

20. $x - 3 = 0$, $y - z = 3$

22. $x = \frac{9}{4}$, $y = -\frac{7}{4} + t$, $z = t$; $x = \frac{5}{8} + 3t$, $y = -\frac{3}{4} + 2t$, $z = 4t$

23. $3x - 7y + 8z = 4$

25. $\dfrac{\sqrt{170}}{3}$

26. $\dfrac{4\sqrt{5}}{3}$

28. $\dfrac{2\sqrt{5}}{3}$

31. $\dfrac{43\sqrt{227}}{227}$

32. $\dfrac{8\sqrt{93}}{31}$

REVIEW EXERCISES

1. 3, $\sqrt{3}$, $\mathbf{A} \cdot \mathbf{B} = 1$, $\mathbf{A} \times \mathbf{B} = -3\mathbf{i} - \mathbf{j} + 4\mathbf{k}$, $\frac{1}{3}$,
   $\mathbf{u} = \dfrac{2\mathbf{i} - 2\mathbf{j} + \mathbf{k}}{3}$, $\dfrac{1}{9}(2\mathbf{i} - 2\mathbf{j} + \mathbf{k})$, $\cos\theta = \dfrac{\sqrt{3}}{9}$

2. $\sqrt{6}$, 7, $\mathbf{A} \cdot \mathbf{B} = 2$, $\mathbf{A} \times \mathbf{B} = -15\mathbf{i} - 8\mathbf{j} + \mathbf{k}$, $\dfrac{\sqrt{6}}{3}$,
   $\mathbf{u} = \dfrac{\mathbf{i} - 2\mathbf{j} - \mathbf{k}}{\sqrt{6}}$, $\dfrac{1}{3}(\mathbf{i} - 2\mathbf{j} - \mathbf{k})$, $\cos\theta = \dfrac{\sqrt{6}}{21}$

4. 11, $\sqrt{6}$, $\mathbf{A} \cdot \mathbf{B} = 5$, $\mathbf{A} \times \mathbf{B} = -24\mathbf{i} - 2\mathbf{j} - 11\mathbf{k}$, $\frac{5}{11}$,
   $\mathbf{u} = \dfrac{2\mathbf{i} + 9\mathbf{j} - 6\mathbf{k}}{11}$, $\dfrac{5}{11} \dfrac{(2\mathbf{i} + 9\mathbf{j} - 6\mathbf{k})}{11}$, $\cos\theta = \dfrac{5\sqrt{6}}{66}$

5. $3x - 2y - 5z - 27 = 0$

7. $7x - 11y + 5z + 20 = 0$

8. $\frac{8}{3}$

10. $\dfrac{x - 2}{5} = \dfrac{y + 3}{-8} = \dfrac{z - 4}{-3}$

11. $\dfrac{x}{-2} = \dfrac{y + 7}{7} = \dfrac{z + 4}{3}$

13. 1

14. yz plane at $(0, 5, 6)$, xz plane at $(-15, 0, 26)$, xy plane at $\left(\frac{9}{2}, \frac{13}{2}, 0\right)$

Index

Abscissa, 5
Absolute value, 2
Addition of ordinates, 190
Addition of vectors, 285
Aids, graphing, 210
Analytic proofs, 29–33
Angle, direction, 315
  between two lines, 16, 17
  between two planes, 309
  between two vectors, 301
  vectorial, 196
Applications of conics, 108ff
Asymptotes, 128
Axes, coordinate, 4
  of ellipse, 113
  of hyperbola, 126
  rotation of, 140
  translation of, 90, 136
  $x, y,$ and z in space, 255
Axis, horizontal, 4
  of parabola, 96
  polar, 196
  radical, 88
  vertical, 4

Cardioid, 211, 216
Center-radius equation of a
  circle, 77
Circle, center-radius equation of
  a, 77
  definition of a, 77
  determined by geometric
    conditions, 80
  general equation of a, 78
  involute of a, 252
  parametric equations of a,
    241
  point, 79
  polar equation of a, 222
Circles, family of, 85
Cissoid, 253
Cone, elliptic, 280
  right circular, 95
Conic, 95
Conics, 95–135
  applications of, 108ff

degenerate, 96
identification of, 149
in polar coordinates, 224
Coordinate axes, 4
Coordinates, 4
  polar, 196
  polar to rectangular, 202
  rectangular, 4, 5
  rectangular to polar, 203
  space, 255
  transformation of, 90
Cosines, direction, 315
Cycloid, parametric equations
  of a, 249

Degenerate conic, 96
Directed distance, 2, 4
  from a line to a point, 63
Directed line, 1
Directed line segment, 1
Direction angles, 315
Direction numbers, 316
Directrix, 96, 116
Distance, between two points,
  7, 8, 9, 257
  from a line to a point, 63, 323
  from a plane to a point, 309
Division of a line segment, 23
Dot product, 301

Eccentricity, 129
  of ellipse, 117
Ellipse, 112
  axes of an, 113
  definition of an, 112
  equation of an, 114, 120
  foci of an, 112
  focus-directrix property of
    an, 116
  parametric equations of an,
    242
  polar equations of an, 225
Ellipsoid, 271
Elliptic, paraboloid, 277
Elliptic cone, 280
Equation of a graph, 42

Equation of a line, 51
  intercept form, 58
  point-slope form, 52
  polar form, 220
  slope-intercept form, 50, 51
Equations of a line, parametric
  form, 313
  symmetric form, 313
Exponential and logarithmic
  equations, 185
Exponential function, 181

Family, of circles, 85
  of lines, 71
Foci, of ellipse, 112
  of hyperbola, 125
Focus-directrix property, of an
  ellipse, 116
  of a hyperbola, 133, problem
    33
Focus of a parabola, 96
Function, definition of a, 35
  exponential, 181
  graph of a, 37
  inverse, 186
  logarithmic, 185, 186
  transcendental, 172

General equation of a circle, 78
General first-degree equation,
  51, 262
General second-degree equation,
  95
Graph, equation of a, 42
  of an equation, 38, 39
  of a function, 37
  of parametric equations, 239
  of a relation, 37
  of polar coordinate equations,
    206
Graphing aids, 210
Graphs, intersections of, 60,
  232

Horizontal axis, 4
Horizontal line, 7

Hyperbola, 125
  asymptotes of a, 128
  axes of a, 126
  branches of a, 126
  definition of a, 125
  equation of a, 126, 128
  foci of a, 125
  focus-directrix property of a,
    133, problem 33
  parametric equations of a,
    243
  standard form equations of a,
    128
Hyperbolic paraboloid, 278
Hyperboloid, of one sheet, 273
  of two sheets, 275

Identification of conics, 149
Inclination of a line, 12
Intercept form of equations, 58
Intercepts, 40
Intersection of polar coordinate
  graphs, 232
Intersections of graphs, 60, 232
Intersections of lines, 60
Inverse of a function, 186
Involute of a circle, 252

Latus rectum, 97, 113, 127
Line, directed, 1
  horizontal, 7
  inclination of a, 12
  slant, 7
  slope of a, 13
  vertical, 7
Line segment, 1
  directed, 1
  division of a, 23
  undirected, 1, 2
Linear equation, 38, 262
Lines, family of, 71
Logarithmic function, 185

Numbers, direction, 316

Octant, 255
Ordinate, 5
Ordinates, addition of, 190
Origin, 3, 196

Pair, ordered, 4
Parabola, 96
  axis of a, 96
  definition of a, 96
  directrix of a, 96
  equations of a, 100, 101, 104
  focus of a, 96
  polar equations of a, 225
  standard form equations of a,
    104
Parameter, 71, 85, 239
Parametric equations, 239
  of circles, 241
  of a cycloid, 249
  of an ellipse, 242
  graphs of, 243
  of a hyperbola, 243
  of a line, 313
Period function, 173
Plane, coordinate, 4
Plane, equation of a, 307
Point circle, 79
Point-slope form equation, 52
Polar axis, 196
Polar coordinates, 196
  to rectangular coordinates,
    202
Polar equations, of circles, 219
  of conics, 224
  of lines, 219
Pole, 196
Product, scalar, 301
Projection, vector, 301
Proofs, analytic, 29–33

Quadrant, 6
Quadric surfaces, 265, 268

Radical axis, 88
Radius vector, 196
Real number line, 3
Rectangular coordinate system,
  4
Rectangular coordinates
  to polar coordinates, 203
Relation, 35
  definition of a, 35
  graph of a, 37
Rose curve, 214, 215
Rotation of axes, 140
Rotation formulas, 140

Scalar product, 301
Second-degree equation,
  general, 95
Simplification of equations, 136
  by rotation of axes, 140
  by rotations and translations,
    143
  by translations of axes, 136
Slant line, 7
Slope-intercept form of
  equation, 50
Slope of a line, 13
Space coordinates, 255
Standard equations, of ellipse,
  120
  of hyperbola, 128
  of parabola, 100, 101, 104
Surface, 262
  cylindrical, 262
  quadric, 265, 268
  of revolution, 265
Symmetric equations of a line,
  313
Symmetry, definition of, 107
  tests for, 108, 212

Tangent line of the origin, 211
Traces, 262
Transcendental functions, 172
Transformation of coordinates,
  90
Translation of axes, 90, 136

Undirected line segment, 1, 2
Unit vector, 289

Value, absolute, 2
Vector addition, 285
Vector, definition of a, 284
  projection, 301
  unit, 289
  zero, 289
Vectorial angle, 196
Vertex, 96
Vertical axis, 4
Vertical line, 7

Witch of Agnesi, 253

Zero vector, 289